International Library of Ethics, Law, and the New Medicine

Volume 62

Series editor

David N. Weisstub
University of Montreal Fac. Medicine, Montreal, Québec, Canada

The book series International Library of Ethics, Law and the New Medicine comprises volumes with an international and interdisciplinary focus. The aim of the Series is to publish books on foundational issues in (bio) ethics, law, international health care and medicine. The 28 volumes that have already appeared in this series address aspects of aging, mental health, AIDS, preventive medicine, bioethics and many other current topics. This Series was conceived against the background of increasing globalization and interdependency of the world's cultures and governments, with mutual influencing occurring throughout the world in all fields, most surely in health care and its delivery. By means of this Series we aim to contribute and cooperate to meet the challenge of our time: how to aim human technology to good human ends, how to deal with changed values in the areas of religion, society, culture and the self-definition of human persons, and how to formulate a new way of thinking, a new ethic. We welcome book proposals representing the broad interest of the interdisciplinary and international focus of the series. We especially welcome proposals that address aspects of 'new medicine', meaning advances in research and clinical health care, with an emphasis on those interventions and alterations that force us to re-examine foundational issues.

More information about this series at http://www.springer.com/series/6224

Dennis R. Cooley

Death's Values and Obligations: A Pragmatic Framework

Springer

Dennis R. Cooley
Department of History, Philosophy,
and Religious Studies
North Dakota State University
Fargo, ND
USA

ISSN 1567-8008 ISSN 2351-955X (electronic)
International Library of Ethics, Law, and the New Medicine
ISBN 978-94-017-7263-1 ISBN 978-94-017-7264-8 (eBook)
DOI 10.1007/978-94-017-7264-8

Library of Congress Control Number: 2015942753

Springer Dordrecht Heidelberg New York London

Printed on acid-free paper

Springer Science+Business Media B.V. Dordrecht is part of Springer Science+Business Media (www.springer.com)

To all my family and friends—whether living or dead.

Preface

Every living thing has a narrative that ends with its death. These narratives, which are merely the living thing's life story from its beginning as a living being to its end as that being, range from relatively simple to extremely complex. Found among the simpler stories are simpler living things, such as the unicellular organisms of bacteria, protozoa, and some algae. For a generally short period of time, these beings interact with their environment and then cease to exist as a living being. Their stories tend not to hold much interest for people unless they somehow affect those beings that have much more complicated narratives.

Among those with complicated narratives are, of course, human beings or persons. Their narratives are created by causes sometimes within their power and sometimes outside their control. Involuntary external forces act upon people, such as their environment, as well as internal powers, such as their genetics. In addition and perhaps most interesting is the artificial power each person has to help craft her own stories. What she chooses to be and how she lives her overall life are at least a small bit within her control, unlike the unicellular organisms which are limited to the influences of their nurture and nature. These more complicated stories hold our interests when it comes to death and dying and what it means because these are the individuals we value the most.

In this work, I will construct a pragmatic moral framework to think about intrinsically valuable individuals and their narratives. I do not claim that the framework can answer all questions that might be raised in death and dying issues, but it is useful to justify positions so that reasonable people can see that the framework's outcomes are reasonable, although they might not be universally adopted as something each person is permitted or obligated to do. In order to build the framework, I start with the same approach found in John Locke's *An Essay Concerning Human Understanding*: The ground covering must first be cleared so that a strong foundation can be laid. To remove the obscuring brush, I will address the value of using intuitions in reasoning. Although this might seem far afield in a book devoted to examining death's values, the literature is rife with the use of intuitions and "common sense" views on which many erroneous conclusions have been based. Before getting to the heart of death's values, we need to see what should be used and what should be put aside at the most basic level of how to think about these issues.

Secondly, if we are going to have a pragmatic discussion of death's moral values, then it is vital to delve into how morality is possible in the first place. This investigation allows us to see some important limitations and requirements morality has so that a proper understanding of death's values in bioethics does not venture into the realm of the conceivable but causally impossible. These will be the building blocks of the superstructure that comes later. After the materials are identified and sorted, the pragmatic moral framework foundation is laid using them. The Practical Principle created is based upon two central truths about how we think about morality: Things with intrinsic value should be respected according to their worth, and we should make the world a better place. In order to make sense of these two sub-principles, there has to be a value theory. After all, if we cannot identify intrinsic value, then it will be impossible for us to respect things with such worth as they should be or to make the world a better place unless that happens by sheer accident. However, for morality to exist in the first place, our narratives as people are not merely accidents or fully outside of our control. I will establish a plausible, pragmatic case as to why certain entities have intrinsic value and how much they have, as well as showing how such value can be reasonably determined. The remaining chapters are devoted to explaining what is lost when someone's life narrative comes to an end, and what if anything survives the event of death.

I offer a bit of a warning to the reader: In what follows, I spend a great deal of time developing those ideas in death and dying which I think require much more depth than they are often given. The reason for this approach is that ethics and bioethics deal with the most important issues. Therefore, we need a more nuanced way of thinking about things that reflects each situations' reality and gravity. In other words, we require Fox thinking over Hedgehog thinking. According to Philip Tetlock, the Hedgehog knows one big thing and then tries to force all decisions to fit that idea or principle, regardless of whether they can be adequately accommodated by it. For example, a Hedgehog might expect standard act utilitarianism to fully answer every bioethics question without residue, even though standard act utilitarianism ignores morally relevant information, such as the facts that people deserve to be respected for their own sakes and that relationships matter in ways unconducive to number crunching cost-benefit analysis. Although appealing because Hedgehog thinkers exhibit so much confidence in their cutting of bioethics' Gordian knots, they can be rather dangerous if we are truly interested in doing and being what we should (Tetlock 2006, 219).

Fox thinkers do not have this dynamic, harmful characteristic because of how they do what they do. First, Foxes have more nuanced views of situations, unlike Hedgehogs who try to make everything fit their too few pieces of information and rules. In addition, Hedgehogs dismiss anyone who does not fully agree with them and their beliefs, positions, and conclusions because to disagree is to challenge the one thing that they know. Any challenge to her blind certainty then becomes an attack on the Hedgehog's identity—the Hedgehog cannot make compromises since she knows one thing, and to challenge that is to attack who she is. On the other hand, Foxes are open to re-examining and altering their views as new information becomes available. To challenge them does not attack their identity given they are willing to be wrong and to have altered one or more of the many things they know (Tetlock 2006). Finally, Foxes are unlikely to attack others with different viewpoints. For the former, it is not a dichotomy between the totally compliant and everyone else as found with the Hedgehog, who must demonize any challenge to her views lest they be also a challenge to who she is. Foxes know that the situation is complex with many different interconnected factors and relationships involved; therefore, many different stakeholders must be consulted for relevant bits of information. From these strands of moral factors, the Fox weaves a complex solution that works for many, if not most, of the stakeholders. In other words, a Fox crafts a solution to fit reality instead of forcing reality to fit the solution.

Even though a great deal more time and resources are available now for Fox thinking, and most people are in agreement that such thinking is vital in bioethics since the situations are so complicated and the moral stakes are so high, there has not been sufficient environmental pressure to abandon the Hedgehog way. The problem is that being a Hedgehog continues to work well enough in many cases.[1] In fact, Hedgehog thinking outperforms Fox cognition if being successful is measured by influence in one's sphere and certainty in one's position. Comfort comes from the beliefs that there is no chaos to threaten us. Any problem can be solved easily by those who are strong and powerful; we should not worry or be involved.

Of course, bioethics and ethics cannot afford to form beliefs, think about, or make decisions primarily from a desire for comfort and need for security. Its subject matter is far too important for that. Hence, in what follows it is vital to go where bioethics' gravity and need for high standards and depth understanding takes us.

References

Kahneman, D. 2011. *Thinking, fast and slow.* New York: Farrar, Straus and Giroux.
Tetlock, P.E. 2006. *Expert political judgment: How good is it? How can we know?* Princeton: Princeton University Press.

[1]Daniel Kahneman's System 1 explains how a Hedgehog would think generally. System 1 "operates automatically and quickly with little or no effort and no sense of voluntary control." (2011, 20) System 1 appears to be the result of innate and deeply learned responses and patterns.

Acknowledgments

Adam Taylor, Catherine Cater, Tony Flood, and Brad Morris were wonderful help in discussing and challenging some of my ideas and arguments.

Cheryl Barrett did a masterful job copyediting, and two of my students, Katherine Thoreson and Tyler Waltz, provided assistance editing and creating the references.

Contents

Chapter 1
A Pragmatic Method

Abstract Many arguments in ethics and bioethics, especially those in death and dying, appear to be driven by intuitions or what some authors claim is common sense. Very rarely questioned, if ever, are the tools from which many arguments and positions are built, unless by a person who has a different set of intuitions or notions of common sense. In this chapter, I evaluate intuitions and common sense to determine if they can have a role in the morality of death and dying. I contend that those who most closely reflect the moral platforms developed in our brains or minds by our genetic natures and environmental experiences generally serve as the best evidence that the intuition makes pragmatic sense, and therefore may be included in our reasoning.

1.1 Introduction

Ethics' purpose was always to be a practical enterprise. It was intended to provide concepts, ideas, justifications, and other information and skills that would enable people to make better decisions in order to live good lives. Generally, the initial attempt was to find essential truths about the most significant matters in life that, because they were necessary, would also be true in the sense of Truth with a capital "T". In addition to ethics, epistemology, metaphysics, logic, philosophy of religion, social and political philosophy and their sub-fields examined the big questions we confront in our lives, such as what is real, how do I know it, and what type of person should I be? Each of these areas is crucial to understanding what we need to understand, including the world, ourselves, and others, and then making the best choices, especially when it comes to death and dying. If one does not have knowledge or justified belief, for example, then it becomes impossible to form and

D.R. Cooley, *Death's Values and Obligations: A Pragmatic Framework*,
International Library of Ethics, Law, and the New Medicine 62,
DOI 10.1007/978-94-017-7264-8_1

implement solutions to questions about what type of society we should live in, how we should live or end our lives or, for that matter, what is possible at all.[1]

For most of ethics' existence, this is precisely what philosophers did. The ancient writers not only addressed the standard studies in our contemporary fields, but also delved into many other areas, such as the natural and social sciences and mathematics, because each area was needed to inform the other areas. That is, the interrelatedness and interconnectedness of the subject matter made ethics an interdisciplinary activity. More recently, Peirce's Critical Common-Sensism highlighted the philosophy and natural sciences connection: "It is above all the normative sciences, esthetics, ethics, and logic, that men are in dire need of having severely criticized, in their relation to the new world created by science." (Peirce 1955, 297). In addition to math and natural science, sociology, psychology, and economics all were part of ethics' realm. Hence, to do ethics well, one had to understand other areas as well.

Over time, disciplines that the ethicists initiated and developed, such as bioethics, departed from being ethics sub-fields as they became mature disciplines in their own rights. This was a good consequence because each field created is worthy of its own systematic study by those who are experts in it. To do an adequate job of such work requires depth, but depth in the subject matter can only be acquired through specialization by those engaged in studying it.

Ethics' generation process brought negative consequences as well. The area's increasing narrowness has too often caused it to be less sufficiently informed by the other disciplines relevant to the subject matter being studied. Peirce, for example, was concerned that philosophers were building elaborate theoretical fantasies of what the world is instead of making their ideas, principles, theories, and other work conform to scientific reality. As a result, ideologies based more on theory and abstraction rather than how things actually work became accepted as the benchmarks by which to evaluate statements and arguments. If one was a utilitarian, for example, then the only concern comes from cost-benefit analysis, and all other values and principles could be safely swept under the rug if they could not be accommodated by the theory itself. J.S. Mill, for example, unsuccessfully struggled to make sense of intellectual value, mental states, and virtues and vices by turning them into qualitative pleasures and pains rather than quantitative pleasures and pains.[2] The result was that the qualitative pleasures' true worth and why they have it were mangled because Mill could not account for them in their own right. In general, the values that cannot be adequately accommodated by particular ideology are corrupted to grim shadows of their former selves just so they can fit the theory rather than the theory being built to fit them.

[1]The work here sets up the pragmatic framework for the examination of death's values. The reason I spend such a great deal of time with this foundational groundwork is because of its impact on how we should think about death and dying.

[2]See Mill's *Utilitarianism*.

Value theory was not the only area negatively affected by the attempt to reduce morality to far simpler components and then fully explicate it. Moral principles about right and wrong, and good and bad received the same treatment. R.M. Hare's and Mill's theories of ethics are paradigm cases of impractical ethical principles with thinking about morality taking the place of the pragmatic.[3] Hare, for example, argues that there are two levels of moral thinking: the Archangel and the Prole. The Archangel has superhuman powers of thought and knowledge without any human weakness.

> When presented with a novel situation, [this entity] will be able at once to scan all its properties, including the consequences of alternative actions, and frame a universal principle...which he can accept for the action in that situation, no matter what role he himself were to occupy in it. (Hare 1981, 44)

The Prole, on the other hand, is incapable of critical thought. Among other things, he relies upon intuitions and *prima facie* principles taught to him by others, but his mental deficit makes it impossible for him to figure out morality when his *prima facie* principles come into conflict (Ibid, 45).

For Mill, most people do not use the utilitarian principle he espouses in their daily lives because it is too complicated or there is too little time for them to make the rather extraordinarily detailed cost-benefit calculations required to know which one of an infinite number of alternatives with infinite consequences they should take to do the best they can. Saving them from continuously doing the wrong thing through no fault of their own are the normative guidelines or rules of thumb that human beings over the course of their history have developed (Mill 1972, 24). Luckily for us, these rules generally capture the same action or set of actions as that of the utilitarian principle so that we can act rightly even if we cannot know with any form of certainty that we are doing what we should.

When there is a conflict between the Prole's *prima facie* principles or the rules of thumb, then both Hare and Mill argue that their first order principles and critical thinking can solve the dilemma. In fact, first order principles ***must*** be appealed to in order to solve the conflict (Mill 1972, 27; Hare 1981, 45). Therefore, the Archangel's principles and utilitarianism are necessary and practical to morality; without them, we would not have the second order principles nor would we be able to find the right thing to do when secondary principles give us contradictory results.

A puzzle develops. If we value Occam's razor, then there seems to be no need for primary principles. In fact, we could eliminate first order principles in favor of the general rules of thumb and an understanding of what will work well in the situation to achieve some desired goal, which when combined together will become the moral code. That is, all we really need is a mechanism to sort out the pluralistic rules we clearly have in a way that allows us to be ethical. We do not need to abandon pluralism for a primary rule that governs all situations, such as the

---

[3]I contend that they are paradigmatic cases on the grounds that the two philosophers could see that there was a serious problem between ideal morality and practical morality, but chose ideal morality although it was difficult if not impossible for most, if not all, people to use.

requirement to maximize utility whenever acting, regardless of whether one can ever know or have adequate reason to believe that what one is doing is actually fulfilling the consequentialist requirement. The end result might be messier than we desire because it could state that several different, contradictory actions can each be morally right in the situation. However, this might be merely an indication that morality is not logic with the clear, indisputable answers coming after a proof.[4]

Perhaps the greatest cause of the movement away from the actual to idealized theory was the Enlightenment's influence. The high ideals of being able to reduce all issues to their component parts, and then the parts being fully explicable to pure reason or scientific demonstration, were nothing more than blind faith (not fully blind faith because that approach did work well in many natural science fields, such as physics). The universe and its workings can be reduced to their rational components and forces for many situations. Science did a marvelous job in its reductionist model of inquiry, and other disciplines, such as ethics, followed suit.

Despite this, even physicists find their paradigm limited. That is, not everything is capable of certainty or fully explicable, such as knowing both the velocity and position of electrons. In addition, we will be unable to have a Unified Field Theory because the universe is not capable of being explained in that way. These are just a few of the limitations to physics and attempting to reduce the universe into only explicable bits.

At times, natural scientists have to use other disciplines to advance their thinking about physics. Much of string theory is based on philosophy since we will be unlikely to establish the basic tenets of string theory given its very nature and our limitations in confirming the hypotheses.[5] In other words, to understand how the universe works, we have to understand different areas of study. Therefore, since morality is part of the universe and how we perceive and interact with it, we have to craft a moral code and set of beliefs and decision procedures that are useful in living our lives.

If even the most fundamental level of science requires other disciplines, then we should return to what ethics actually was to move forward with issues facing us today. In addition, we must try to combine all areas of knowledge when those areas are relevant to developing a subject. Doing so avoids the danger of creating complete and consistent systems that work only in ideal worlds that can never exist. Another benefit is that we rid ourselves of those who insist that their ideological framework is the primary or sole paradigm. Instead of having individuals competing with each other for whose paradigm wins in intuition battles, we can instead have various narratives that work well enough for those using them so that they and others, including societies and communities, can flourish.

---

[4]In fact, deontic logic was a short-lived attempt to create a logic for morality. The ability to constantly generate contradictions using the axioms appears to be one main reason it was abandoned; another seems to be that the loss of content through abstraction lost the essence of what morality is.

[5]This is why Steven Hawking's dismissal of philosophy's value is contradictory. He is only able to argue for his position by using a philosophical argument.

We should also use this pragmatic interdisciplinary approach to avoid a second pitfall into which ethicists sometimes stumble when it comes to death issues-intuition pumps and the odd certainty they bring to the minds of those who use them that they have to be correct. An intuition pump substitutes the intuitions of an individual over actual fact finding because intuitions are assumed to have some sort of evidentiary force that make them unquestionable and superior to doing empirical research.

> The argument of [intuition pump] writers will not survive scrutiny, unless the scrutiny is conducted by sympathetic people – that is, by people who already share the writer's convictions. Against those who do not, no arguments are provided. So we are bound to conclude that they are placing reliance on their ability to discern the moral truth without argument, but exercising their power of moral cognition, and then think that all who similarly exercise it will come to the same conclusions. (Hare 1989, 101)

That is, intuitions provide internal evidence that eliminates the need for external, empirical data. The results of intuition pumps in philosophy are definitions, principles, and arguments, as well as everything developed from them; generally, they merely reflect the ideological bent of the ethicist, bioethicist, or writer without guaranteeing accuracy or usability.[6,7]

If we assume that ethics, especially bioethics, is a practical exercise, then this consequence needs to be avoided. We should be learning and expanding our understanding of how morality works rather than remaining continually engaged in confirmation bias affirming conflicts that are all sound and fury and signify nothing.

1.2 The Usefulness of Intuitions

Intuitions are odd sorts of things to use for evidence in part because people do not seem to know what they are or how they work, but place great reliance on them for providing adequate evidence on the most important ethical situations, such as death. With this in mind, it is worth considering what intuitions are and whether they can play a legitimate role in moral discourse or if they merely lead to a form of moral relativism.

First, intuitions cannot be self-evident truths. A self-evident truth appears to be an analytic a priori proposition, such as "All red things are red"; "Each unicorn has one horn"; and "1 + 1 = 2 in a decimal system." If one understands the sentence,

---

[6]What is odd is that many folks never seem to notice that an intuition fails to bring new information to the person with the intuition; it is already what the person thinks, so why not call these biases rather than intuitions that provide some form of usable evidence? There appears to be far better evidence that they are merely what the person already thinks more than an "eureka" moment of revelation of some truth.

[7]I am subject to the same charge, although I am willing to change positions provided that such a change is justified on pragmatic grounds.

then one has to know that the proposition is necessarily true because of the relation of ideas.

If intuitions were a self-evident truth, as is sometimes the case, then they would be called "self-evident truths" or "a priori analytic propositions." Intuitions cannot be self-evident truths because of a lack of universal agreement on intuitions considered to be basic moral thinking at work.[8] If universal agreement is possible, then the Trolley Problem, created by Judith Jarvis Thomson and Philippa Foote, and all the variations of the problem described since, would show that each moral agent has the same intuition.

In Trolley Problem cases, reasonable people are asked to decide a moral dilemma in which the only two alternatives lead to the death of one or more people. In one version, survey participants are asked to imagine that they are in charge of a run-away trolley, and are fast approaching a branch in the track. If the trolley stays on its current course, then five men will die, while changing the trolley to the side track will kill the man standing there. Most people select to change the trolley's path to the side track. For a slightly different version of the problem, each person is asked to imagine that she is standing to the side of the track of a run-away trolley next to a very fat man. She sees that if she does nothing, then the trolley will kill five men on the track. However, if she pushes the fat man in front of the trolley, then he will be killed while the five survive as the trolley grinds to a halt.

The problem, if described in the first way, appeals to many people's consequentialist beliefs in that they are willing to sacrifice a small number of people—usually one—in order to save a significantly greater number of individuals. If the problem is depicted in the second manner, then it becomes one that is solved primarily through deontological reasoning. For instance, the general refusal to push a fat man off a bridge to save five people who would otherwise be run over by the out-of-control trolley results in one person being saved at the cost of the lives of many more people. Regardless of the description, there is rarely, if ever, 100 % agreement on the right course of action to take. If there were self-evident responses or intuitions, then this outcome would be impossible.

Moreover, in Trolley Problems, if a person selects an alternative that runs counter to what the vast majority of individuals choose, then the person has not chosen the wrong solution.[9] Although it might not be something that we would do, we would not find the person choosing differently from us as blameworthy for making what appears to us as the wrong decision. This lack of blame shows why intuitions are not self-evident truths. If someone gets the number of sides to a triangle wrong, then we know that he has made a mistake, but intuitions do not prove that a mistake has been made—at least, not in the same obvious way.

[8]More complicated intuitions, of course, will have fewer adherents because there are more conditions that have to be held in common.

[9]If we already know which answer is the right or wrong one, then we would not need intuitions as evidence. We would merely use the principle or decision procedure that gave us the correct information.

Therefore, a variation of Locke's argument against innate ideas can be used to show that intuitions are not self-evident truths (Locke 1992).

Although intuitions are not self-evident truths, they are given a great deal of evidentiary weight by some very smart people. Perhaps it is a process we undergo that legitimizes them: "[W]hen an ethical philosopher claims to know by intuition he is claiming that there is some unique process in the world to which he refers, which entitles him to knowledge claims." (Sterling 1994, 77). That is, intuitions provide adequate or reliable evidence because they are validated through careful cognitive introspection or are innate ideas discovered when needed.[10] Jonathan Haidt argues that there are moral intuitions that are "the sudden appearance in consciousness of a moral judgment, including an affective valence…without any conscious awareness of having gone through steps of searching, weighing evidence, or inferring a conclusion." (Haidt 2001, 818). Although intuitions do not provide certainty, they might still be reliable information because their source produces only justified beliefs. Therefore, an intuition would be able to provide evidence for a position based on that intuition.

In order to give intuitions' reliability as much plausibility as possible, let us assume that the moral cognitive process for intuitions is much like logical cognitive processes. If rational agents know a form of logic, then each and every one of them can come to a valid conclusion by using the rules of the stipulated logic, although each might arrive at the conclusion along different pathways. Those who do not come to that conclusion may be correctly stated to have erred in their use of the logic's rules. If the moral cognitive process did not have logic's level of legitimacy to convey onto the intuitions' credibility, then they would always be susceptible to the attack that the intuitions are merely biased beliefs of the person thinking about them.

Geoffrey Scarre seems to use the careful cognitive introspection approach when he incorporates intuitions into definitions, evaluations of claims, and arguments. First, he states that "[t]hinking about death should help us identify the conditions for authentic living." (Scarre 2007, 3).[11] This is a strong claim to make; it requires framing authentic living in terms of death rather than thinking about authentic living in a way that might not require looking at death. The assumption is that an individual thinking about the issue will somehow bring her adequate, incontrovertible evidence to assist her in her search for the conditions of authentic living.

The justification for using individual intuitions found through introspection for defining, evaluating, and discoursing on death and other issues appear to be other intuitions. That is, it is intuitive that intuitions tell us the truth or that the processes

---

[10]Sterling (1994) argues that we use introspection, which cannot be checked against objective evidence, although we do check these against other people's introspections. According to Sterling, intuitions should be treated similarly.

I think that Sterling's position is too broad. The only introspections that should count as adequate evidence are the types examined by Panayot Butchvarov, viz, those in which it is impossible for the person introspecting to be mistaken about, such as being in pain or being appeared to redly.

[11]Scarre would be unlikely to agree with how I characterize his position.

using and validating them tell us about reality. Although we do not want to go down the path to Pyrrhonian skepticism, we need to question why "exercising [our] power of moral cognition" (Hare 1989, 101). or discovering what appears to be an innate idea is sufficient to lead us to the truth, especially since other people doing the very same thing come to different conclusions.

The other oddity from the use of intuitions is making the reality fit the intuition rather than the intuition fit the reality. For example, "[i]t is very doubtful whether it is possible to arrive at a single and univocal definition of death that will accord with our intuitions in all circumstances." (Scarre 2007, 6). Rather than taking this difficulty as at least *prima facie* evidence that using individual intuitions might be illegitimate beyond general conclusions about how humans make ethical decisions, Scarre tries to accommodate these intuitions in a way that accounts for as many as reasonably can be incorporated. This approach is interesting because he offers no evidence for why the coherence approach incorporating the most intuitions is *the* method to use when trying to figure out issues about death, or any other issue for that matter.[12]

More interesting is the evidentiary weight Scarre gives to intuitions. We could delete some of the intuitions to achieve the single and universal definition, which we would have to do anyway because we cannot accommodate all of them. Deletion, however, does not seem to be a live option here, possibly because there is no method to tell which intuitions could be legitimately ignored and why that is the case. If we allow for a range from weak to powerful intuitions, then we need to identify some property or set of characteristics that would enable us to classify or grade the intuitions' evidentiary power. We would need to know what privileges one intuition over another, and therefore justifies its inclusion over those excluded.

Why should the individual think that intuitions offer this evidence? What makes them reliable enough to justify using them to begin, support, or decide an argument? At best, intuitions are merely not irrational beliefs; they are not contradictions, deeply implausible, or defeated by available empirical evidence. Moreover, if they were rational, then the evidence for their rationality would be given and the term "intuition" would fall to the wayside in favor of the far stronger "justified belief". Therefore, they must be beliefs that a rational person can have because they seem reasonable, but ultimately they are non-rational. When someone states that something is intuitively reasonable or plausible, what they are legitimately saying is that a reasonable person can hold this belief and still be thought reasonable by other reasonable people.

Of course, this is an extraordinary foundation on which to build critical arguments, especially if they are about life and death matters in ethics and elsewhere. Intuitions here seem to be, basically, what the speaker believes to be true without adequate evidence for the belief being justified, (it might merely be that it "feels" right to the individual). The "feel" of its rightness stems from the fact that the person already believes it, or that the belief coheres to his other beliefs, even if

[12]To be fair, the intuition evidentiary approach is so common that it has become habit; hence, there would be little reason to see a need to justify it.

those other beliefs are unexamined and unsupported. For these individuals, much like those favoring the Intelligent Design argument seeing design in local or global aspects of the universe, the intuition is one of their existing central beliefs. Rarely, if ever, does an intuition provide any new evidence that will make the person realize that she has made a mistake, and should therefore rectify her position accordingly.

If intuitions are merely existing beliefs, then using biased intuitions will result in a form of moral relativism in which whatever the person thinks is true is true and whatever the person thinks is false is false merely because the person thinks or feels that way. Intuitions would be unreliable on these grounds, and therefore, should be abandoned.

Perhaps we are making a category mistake when thinking about intuitions. Instead of being justified beliefs based on their origin, self-evident truths, or Cartesian innate ideas, they might be some form of basic instinct—created by nature and environment and influenced by each individual—that help make morality possible.

1.3 Natural-Evolutionary Intuitions

We still have the issue of why there appears to be a universal morality in which "all people possess conceptions of morality that include the idea that certain forms of conduct are right and wrong and that certain character traits are good and bad." (Krebs 2011, 19).[13,14] Moreover, moral communities are not only universal for *Homo sapiens* with a moral code comprised of normative principles and values, but also communities have the ability to remove deviants, provide incentives for people to obey the code, and engage in information exchanges to evaluate individual adherence to the social code (Boehm 2004, 83). Universal morality and its abilities are not possible unless there is something each person shares in common to create the common system.

One hypothesis is to claim that morality is intimately linked to the human brain and its natural workings: "the human capacity for morality is fundamentally evolutionary." (Zamulinksi 2007, 4). That is, how *Homo sapiens* and sufficiently similar social animals have evolved over time through group selection has created a brain that will have the morality that we do (Richerson and Boyd 2004, 62). Some neurophysiologists, for example, appear to accept the view that "the brain…can be subdivided into a large number of portions…with different functions, which are

[13]Paul Ekman's faculty psychology on the universality of emotions and ability of others to recognize such emotions from an evolutionary perspective might help explain the universality of morality. See, for instance, Ekman's "A methodological discussion of nonverbal behavior." The same sort of faculty of the mind idea is seen in Pinker's Swiss Army Knife analogy.

[14]Lisa Barrett's psychological constructivism is closer to my final position. "Psychological construction relies on a similar kind of population thinking. Emotions are not physical (morphological) types, but are cognitive categories that contain a variety of unique instances." (Barrett 2013, 381).

independent of each other and can be isolated." (Legrenzi and Umilta 2011, 4).[15] If the area of the brain that functions in a certain way is sufficiently damaged, then the person might be unable to think or act in certain ways, including being moral.

Additionally, genes do appear to have an enormous influence on our intelligence, personality, psychological interests, and social attitude as borne out through a myriad of scientific studies, including those performed on twins raised in different environments (Bouchard 2007, 74–83). Paul Bloom argues that studies on babies show that people are born with a moral sense that is "the capacity to make certain types of judgments – to distinguish between good and bad, kindness and cruelty." (Bloom 2013, 31).[16] In order to make these judgments, babies—and adults for that matter—have specific feeling and motivations, such as compassion, empathy, fairness, status, punishment, and a natural tendency to favor one's own group and those for whom one cares over others (Ibid.). Thus, it is not an implausible leap of faith to think that morality is fundamentally dependent on mental faculties, innate ideas, or some other brain activity or structure that is the result of evolutionary forces.

Brian Zamulinski develops his ethical intuitionism based on how he thinks evolution affected the hard wiring of the human brain. Ethical intuitionism claims all of the following:

1. Moral realism is true;
2. Moral objectivism is true;
3. Moral facts are knowable: cognitivism is true;
4. Since there are knowable moral facts, our moral judgments can be true;
5. The Is/Ought gap is unbridgeable;
6. Morality exists for reasons beyond pure instrumentality;
7. Moral agents can be motivated to act in a morally acceptable manner without an extrinsic pay-off.
8. We have intuitive access to moral truths, and consequently, we can know what is right intuitively under certain conditions (Zamulinski 2007, 24).

Although all eight claims are important, the latter especially shows the thinking behind morality being genetic and therefore internal. First, there have to be moral truths that are innate; otherwise, we would not have innate access to them. Moreover, such truths are found if our innate process of introspection is working in the correct way, which means that we must have innate processing abilities. Third, it cannot only be rational to believe that something is right or wrong or good or bad,

[15] Alva Noe argues that our minds cannot be identical or reduce to our brains because consciousness is something that we do rather than something internal to us (Noe 2009).

[16] Although Bloom calls these capacities and rejects the notion that we cannot improve on them, there is still a concern that too strong a position is being advocated. Just as we should never anthropomorphize animals, we could agree that these capacities exist in some nascent way, but we should not believe that they are more developed than they are. The judgments are relatively complex, and make sense if we are talking about an adult moral agent making them. Where the problem arises is saying that babies are making sufficiently similar distinctions and judgments.

but also we can know it. Knowledge requires a higher evidentiary standard than rational belief, which is generally that the belief is plausible or more likely than not given the circumstances. Knowledge's standard is beyond a reasonable doubt, which means that the evidence is so valuable that it can defeat competing evidence, if any such exists.[17] Hence, the innate moral truth need not be self-evident, but it must carry a lot of evidentiary weight, which could mean that we virtually have to think that the moral truth is, in fact, true.

Unlike adaptive evolutionists who argue that morality is only possible because of adaptive features of the mind that helped make organisms better fit to survive and reproduce, Zamulinski claims that morality is a by-product of evolutionary forces on the pre and *Homo sapiens* populations (2007, 3 and 21). If correct, then the altruism problem faced by adaptive evolutionists—how could sacrificing one's best interests for the sake of another be evolutionarily beneficial?—drops away. Altruism is innate in the human brain as a genetic trait that came along as a "free-rider" on an evolutionary beneficial gene or set of genes which were selected for.

Regardless of the particular cause of the moral ideas or faculties that we share in common as human moral agents—adaptive or by-product evolution—the idea is that we have them as a result of evolution, and such ideas and faculties are innate in some way (Zamulinski 2007, 3; Haidt 2001, 818; Hauser 2006, 43–55; Pinker 1994, 45–58; De Waal 2006, 24; Broom 2003, Chap. 2; Krebs 2011, 10). For the moment, we will consider a naturalist or nativist theory of morality, as it is sometimes called, to understand what intuitions are as either a faculty or faculties, or innate ideas.

First, the concept of innate ideas and how they function needs some examination. Gabriel Segal defines an innate idea as one not acquired by a psychological process—it is not learned—but canalized where "a trait is canalized to the extent that its development is causally insensitive to environmental and genetic variation." (Segal 2007, 91–2). Language, which might be able to teach us something about morality, is an innate faculty. Segal argues language can be shown to be a genetic faculty not only by the lack of data for the competing hypothesis of a general-purpose learning mechanism, but through seeing examples in which learning cannot explain the phenomenon but genetic faculties can, such as acquired aphasia in which subjects lose a learned language and cannot relearn another (Ibid., 94 and 96–7). Others have suggested a faculty of morality similar to language—such as Marc Hauser and Steven Pinker—or something less developed, such as a "sense of morality" (Krebs 2011, 204).

A sense of morality[18] is comprised from a sense of moral obligation, a conscience, a sense of rights, moral sentiments about others, a sense of justice, and abstract ideas about morality (Krebs 2011, Chap. 16). There might not be a faculty,

[17]This standard is controversial but pragmatically plausible.

[18]Krebs argues that there are four moral senses, each of which is a dichotomy: evaluative feelings and thoughts; positive and negative aspects; pertaining to self or others; and thoughts and feeling agents have before and after they make a moral decision (Krebs 2011, 204).

but there is some innate brain structure that allows us to have the common morality that we, in fact, have. According to Kreb's adaptive evolutionary account, the sense of morality evolved to solve the problem of individual behavior that violated group social order and welfare of the group and other members (Krebs 2011, 205–6). The art in Krebs's view is determining which traits are more likely to be canalized. It seems plausible to believe that the more primitive the trait is, then the more likely canalization has occurred. After all, more complex traits would require greater genetic platforms to be able to exist and function well, whereas simpler traits need fewer resources.

Then again, innate moral ideas or faculties might be a much weaker sort of thing, such as an innate instinct or reaction that can be overcome with additional information, as has been shown in a preliminary study of why babies are wary of unknown plants (Wertz and Wynn 2014, 49). It is hypothesized that a brain that is reluctant to come into contact with unknown plants was evolutionarily favored because many plants are toxic. However, the avoidance mechanism or desire can be overcome if a learning process is employed in which the infant is familiarized with the plant. The same sort of thing might happen with innate ideas that are merely moral reactions. They provide some incentive and motivation, but are not authoritative as to what morality demands.

Even though the genetics of morality are rather ambiguous and can go in a variety of directions, we can start making some headway in thinking about how naturalism could plausibly work in ethics. Instead of individual moral intuitions being thought of as adequate evidence, let us suppose that they are instincts of some sort that result from evolutionary influences on our ancestors' brains. Instincts have been understood in a variety of ways, including being a complex activity such as foraging or having a unique neural module, which can make the use of "instincts" rather confusing. Among other things, it has been proposed that instincts are:

1. Present at birth (or at a particular stage of development);
2. Not learned;
3. Developed before they can be used;
4. Unchanged once developed;
5. Shared by all members of the species (or the same sex and age);
6. Organized into a distinct behavioral system (such as foraging);
7. Served by a distinct neural module;
8. Adapted during evolution; or
9. Differentiated between individuals due to genetics (Bateson 2000, 164).

For my purposes, I will try to incorporate as many of these conditions as I can into the definition to be as inclusive as practical.[19] Instincts are innate fixed patterns of emotional responses and behavior in response to particular stimuli. Because they

[19] I earlier criticized Scarre for taking this approach, but will try to justify it here on pragmatic grounds. The more inclusive the definition without sacrificing intelligibility, then the more likely it is that the definition will be useful to more people.

are the result of evolution, they are shared by a sufficient number of the species to make it a general trait, which the absence of is very unusual. By instinct, I do not mean the fully developed faculty that Chomsky and others talk about for language, but more of a biological platform in the brain that allows us to develop the complicated moral system that we as moral agents have. In order to account for the commonality of ethical systems and beliefs about morality, there has to be some commonality, and for morality to exist for us in the first place, there has to be some shared mental platform based on genetics that enables us to think about ethics in the manner that we do. This is not to say that the mental platform is the full story of how ethics is possible, but just as hardware is necessary for a computer to function, the biological platform is necessary for morality to function.

When we judge that something is good or bad, that decision is based at least in part by how our brains work. If we can find these commonalities, then we can know a bit more about why our ethics are as they are and, more importantly, use the instincts to help form a guide for thinking and acting, as well as provide some evidence that can be used in judgments. To distinguish belief intuitions from these more fundamental intuitions, let us call the latter intuition-instincts.

We will begin with intuitions across species. In order to behave in similar ways, then there must be more basic motivations to perform in that manner. We can suppose that human beings feel similarly about certain very general values and principles, such as the idea that pain is to be avoided. Now, how they managed to have those values and principles could be a matter of nurture, nature, self-created, or a combination. That is, their environment taught them to have those values or their very nature provided those ideas as a form of intuition-instinct.

What would add strength to the claim that there is an intuition-instinct given to us through the evolutionary process under which our species has undergone is proof that other species appear to have the same values and principles. Of course, given the intellectual distance between more mature *Homo sapiens* and other species, these values and principles—intuition-instincts—would not be too abstract or complex, such as a consequentialist theory or W.D. Ross's Formalism. However, we should be able to see sufficiently similar value-expressed behavior in situations designed to elicit a certain response if the values and principles are present. For example, altruism requires an individual to risk its self-interest for another individual's interest fulfillment. If we can find evidence of this, then intuitions may merely turn out to be the way that a sufficient number of entities think because of social or evolutionary processes their species have undergone.

In some areas we see similarities between the workings of human and non-human animal brains. First, *Homo sapiens* share 99[20] and 97.7 % of their genes in common with chimpanzees and gorillas (Broom 2003, 31). In addition, Bernard Campbell argues that primates' eye and hand coordination from a large, complex

[20]Bernard Campbell states that humans share 98.4 % of their DNA with chimps (Campbell 1995, 118).

brain gave rise to humanity (Campbell 1995, 119).[21] Therefore, we must have shared genetic-based brain structures.

Despite this, the mere fact that our DNA and brains are remarkably similar does not establish anything about morality or how our morality is possible. What we need is to show that the physical "hardware" of brains creates a platform of some sort for morality by enabling humans to have the morality they do. We can look at emotions, desires, behavior, and other mental states and faculties associated with morality to find similarities.

First, we have some complex emotions in common with some animal species. Empathy is fundamental to our morality; without it we could not have sympathy, and therefore, would have no ability to have morality in the first place. That is, we could not care about others, so what happens to them and their needs would be of no concern to us. There is good reason to believe that we share empathy with other higher primates.

Frans de Waal argues that empathy is "the original, pre-linguistic form of the inter-individual linkage that only secondarily has come under the influence of language and culture." (de Waal 2006, 24). However, he also believes that "emotional contagion", which is "when an emotional state of one individual induces a matching of closely related state in another" (de Waal 2006, 26), will develop into empathy. If correct, then there is a shared emotional-motivational state more fundamental than empathy that we can find in the vast majority of human beings. In addition, De Waal sees evidence of empathy across species boundaries; non-human primates respond to distress and need for help from other individuals by exhibiting the very same behavior, including facial expressions that humans show to other humans in the same situation (de Waal 2006, 25).

Besides empathy, reciprocal altruism is necessary for the existence of morality.[22] It has long been known that there are some behavioral similarities between non-human primates and human beings that cannot be explained as mere accident. Cultural behavior, tool use and creation, ethical behavior, socially coordinated aggression, and infanticide are present in at least simplified forms among apes and chimpanzees (Campbell 1995, 120–1). Singling out ethical behavior, Campbell argues that bio-altruism and a simple ethical system based on punishment, which fosters and maintains a smoother running of social life, can be easily found in many primate populations (Ibid., 121–2). Of the two components of ethical behavior, the latter seems to be learned behavior, which might have its roots in evolution, but the former is considered to be an evolution-generated trait. Bio-altruism is not real altruism in the sense that it merely appears that an individual is sacrificing her interests for the benefit of another individual, but really is not. The apparent self-sacrifice actually will increase her chances of being rewarded in the long term

[21] De Waal argues that other species recognizably share relevant human abilities.

[22] Zamulinski argues that altruism cannot be accounted for by adaptive evolution if individual organisms are the exclusive unit of selection (Zamulinski 2007, 5). He believes that altruism is an evolutionary by-product that is a free-rider on a gene that adaptive evolution can explain (Ibid., 21).

through reciprocal altruism, when the now-benefitted entity returns the "favor." To be able to have true altruism develop, it makes sense to hypothesize that there was at least one intermediary step between pure self-interested behavior and altruistic actions performed by moral agents. Bio-altruism can fill in part of the story on how true altruism came to be a central component in human moral codes.

The mere fact that reciprocal altruism exists entails that there must be other activities in which the brain is engaged, and if the brain for one primate works in this manner for the shared feature of altruism, then we can plausibly assume that other primate brains work in the same manner. In order to have reciprocal altruism, the entity's brain has to function in a certain way; namely, "it is necessary to remember and evaluate who conferred the benefit, what degree of benefit it was, and what would constitute a similar benefit" as well as who should not benefit (Broom 2003, 84). That is, reciprocal altruism would not be reciprocal if there was no ability to return a favor to the entity that performed it, or if the favor was inadequately returned. For example, risking my life to save yours would merit far more than one banana from your pile.

In order to perform reciprocal altruism's necessary functions, the entity would be able to recognize others and discriminate between individuals; have awareness and consciousness, such as perceptual, cognitive, assessment and executive awareness;[23] have feelings and emotions relevant to morality, such as pleasure, pain, fear, thirst, and exhilaration; and have cognitive ability (Broom 2003, Chap. 3). Although this mental function might at first appear to require a very high level of cognitive ability, natural science has shown us that we share many of these brain functions and features with primates and other non-human animals, such as pigs. This common heritage does not entail that pigs think the same way as human beings do, but only that many components of the functioning human mind are shared between animals with sufficiently developed brains, which means that we likely have very similar biological platforms for ethics.

Besides emotions such as sympathy and altruism, we share a significant number of mental or brain traits with other species based upon how our various brains work. Our biases in weighing evidence and decision making, for example, might be the result of shared mental structures, primitive though they are, that are the result of evolution (Santos and Lakshminarayana 2007, 295). "Across a number of domains, people tend to systematically ignore problem-relevant information and systematically overestimate the importance of problem-irrelevant information", such as emotional salience to the person making the decision (Ibid.). To explain why this is the case, especially for those who have relatively little experience, it is suggested that at least some of the components that lead us astray are innately built into our brains. Non-human animals and small children, for example, will tend to make common mistakes based on bias, whereas adult humans with sufficient learning from experience will not.

[23] The latter is generally found only in humans.

Other significant shared traits include cognitive features that require higher-order reasoning, if de Waal is correct:

1. Sympathy-related traits
   (a) Attachment, succorance, and emotional contagion.
   (b) Learned adjustment to and special treatment of the disabled and injured.
   (c) Ability to trade places mentally with others: cognitive empathy.
2. Norm-related characteristics
   (a) Prescriptive social rules.
   (b) Internalization of rules and anticipation of punishment.
3. Reciprocity
   (a) A concept of giving, trading, and revenge.
   (b) Moralistic aggression against violators of reciprocity rules.
4. Getting along
   (a) Peacemaking and avoidance of conflict.
   (b) Community concern and maintenance of good relationships.
   (c) Accommodation of conflicting interests through negotiation (de Waal 1996, 210–1).

Cognitive empathy, internalization of rules and anticipation of punishment, moralistic aggression against violators of reciprocity rules, community concern and maintenance of good relationships, and accommodation of conflicting interests through negotiation require a higher order of thinking than those found in species that merely react to external stimuli. Accommodation, for example, requires a desire to make accommodation and do the necessary work to achieve it; recognition of interests and their conflict (that it is possible to negotiate between the concerned parties); recognition of a starting point to negotiations; and an ability to alter the potential solution based on reactions of the concerned party until a point in which accommodation between all is reached. Since the four features de Wall names are common to all primates, then it follows that they must be part of the basis for our morality, especially since the features are fundamental to what we think our morality is.

Before proceeding further in sketching out the similarities between humans, non-humans and the platform for morality, here is one caveat: although we share fundamental or rudimentary traits associated or fundamental to morality with non-human primates, that does not mean that we can legitimately claim that we share the exact same trait when it comes to degree or brain structure. "[T]he cognitive capacities shared among cetaceans, nonhuman primates, and humans – including self-recognition, symbol-based communication, abstraction, and complex social structures…are associated with markedly different brain features than mere size alone." (Jung and Haier 2007, 136). Hence, the capacities can be shared at some level, but the brain structures for the various species need not be identical, and are not identical, which allows for significant differences in how the cognitive capacities work and are expressed.

Moreover, even if we do share cognitive capacities, they are at best identical in kind but not necessarily in degree. In particular, Campbell has argued that language, religion, ethical behavior, and self-consciousness/awareness are without counterpart in non-human animals (Campbell 1995, 124). For example, elephants appear to mourn, but to equate their mourning with human beings neglects important cognitive and neurophysiological differences. Although animals do have a rudimentary form of language, the difference is degree, if not kind, and can be seen through what language allows *Homo sapiens* to do what animals cannot. Only humans have a:

1. Symbolic environmental reference,
2. Enlarged and classified data base,
3. Precise and rapid data transmission,
4. Thought displacement in time,
5. Discussion, bargaining, planning,
6. Enhanced reciprocity between individuals and groups,
7. Instruction,
8. Complex social organization with demarcation of individual roles,
9. Imagination,
10. Detachment from emotional level of experience, leading to reason and logic (Campbell 1995, 125).

Although it might be tempting to take one characteristic out of the ten listed, such as instruction, to show that non-human animals exhibit that behavior, human beings instantiate each of the ten, sometimes in various combinations. Moreover, not only do *Homo sapiens* have quantity on their side, they also have quality. The depth at which sufficiently mature *Homo sapiens* can bring about all 10 states makes *Homo sapiens* unique among the primates. That is, there are rudimentary or primitive levels of instruction, but they are only at the level we would expect if evolution is true. These innate ideas or structures must have existed as primitives at first, and then evolved over time to the more complicated, higher abilities that they have become. Therefore, we would expect to see common features at primitive levels across the species, especially in species closely related to *Homo sapiens*, but not to the level at which mature humans have them. After all, the capabilities to do all ten rather than fewer and in far greater depth and abstraction is what allows *Homo sapiens* to be moral agents in the first place, whereas animals are not moral agents because they do not have the ability to instantiate one or more of the ten features.

Although we share a great deal in common with animals, especially primates, significant differences remain in how the various species' brains function.[24] For example, how humans interact socially, gather empirical evidence for social

---

[24]Heim argues that imitation is possibly a unique ability in *Homo sapiens*, and might be fundamental to human culture and rationality (Heim 2004, 256).

interactions, and use the information for moral decisions requires mental activity that seems beyond the grasp of minds with less cognitive power, perhaps because such characteristics are emergent (Brown 2010).

> [W]e grasp what is going on in others' minds by a kind of mental mimicry in our own. We reproduce their mental process in our own neurological field, in the same way that, when we observe an action by another, or mirror neurons reproduce the pattern of neurological activity that makes the same motion. There is evidence that such "mirror" effect exists also in regions of the brain that process emotional responses. (Heim 2004, 257)

If the simulation theory of mind reading above is true, then our ability to feel empathy for others is hardwired into our brains. Seeing others in certain situations, such as being in pain, makes us feel for them by, so to speak, putting ourselves in their position. However, a cognitive component separates imitation of this sort from that which animals might do. Mind reading imitation allows humans not only to discern another being's emotional states through mimicry, but also other agent's intentions.[25] Reading another person's intentions involves taking the evidence about them that we observe, and by mirroring it in our own minds, generating the same intentional mental state, although we might not be committed to bringing the intended state about.[26] That is, our mirroring will not be perfect so that our mental states are identical to those of the agent being mirrored, just as we cannot feel their actual pain that generated our empathy, but we can read the other's behavior, which then allows us to know the other's intentions by reflecting them in our own minds.

Given the similarities and differences between *Homo sapiens* and other species' brains, how are we to use the information in building a practical moral code? Great care is needed, especially in cases in which the data and conclusions drawn from them will have a considerable impact on humans and other intrinsically valuable entities. I will consider a variety of issues raised with overreliance on neuroscience and evolutionary explanations below.

Often overlooked is the assumption that the science being done on the brain is performed in a way that provides the most useful or useable data. For example, neuroscience has relied heavily on the average of single cell responses to repeated stimuli. Although data were produced using this method, they were at times incomplete or unclear: "much more knowledge can be extracted using decoding and information-theory methodologies and how information that is ambiguous at the single-cell level can sometimes be clearly interpreted when considering the whole population." (Quiroga and Panzeri 2009, 183). There is a danger of overreliance on a scientific study method that appears adequate, but in reality is unable to clearly and completely capture reality to a sufficient degree. Resulting hypotheses about

---

[25]Laurie R. Santos and Venkat Lakshminarayanan argue that human babies and primate minds are remarkably similar in how they understand simple physical principles, such as "how space and time constrain object motion" and have an appreciation of goals and intentions (Santos and Lakshminarayanan 2007, 294).

[26]Intention, here, is not being used in a philosophically rigorous way.

how minds work and what that entails based on these results, therefore, might be justified by those results, while at the same time they are false.

We should also recognize that there is a danger of drawing unsupported conclusions; it too often becomes more important to make the data fit the hypothesis than for the genetic hypothesis to fit the observations.[27] Although some socio-biologists, such as Richard Dawkins, reject religion, sociobiology becomes like a religion to them. Without scientific evidence, they believe that the evolutionary basis of human behavior can explain everything or virtually everything, and "use a language replete with religious metaphors and concepts such as immortality and essentialism" (Nelkin 2000, 15). For example, the Strategic Learning Model states that it is adaptive for an individual to learn the values and conventions of a group, including the group's morality and language (Sripada 2007, 316–7). We can easily see why that would be beneficial to make one's values and language conform to that of a group with established values and conventions and of which one is a member. However, language and morality would have to be invented and held in common ***before*** an individual could adopt them so that they could be selected for in the first place.

Although we might now have brain structures in place that allow us to have the morality we do, we should not throw out learning as a significant contributor.[28]

> In this emerging neuro-ontology, it is not that human being are brains, but that we have brains…our selves are shaped by our brains but can also shape those brains. (Rose and Abi-Rached 2013, 22)[29]

Science, such as biology, is a very useful tool to explain many things through models based on evolution, but to assume that such models have the power to explain everything requires an illegitimate act of faith that is not based on science or careful reasoning. "In fact, brain-derived data currently add little to nothing to the more ordinary sources of information we rely on to make…determinations [about the mental states of others.]" (Satel and Lilienfeld 2013, 150). Understanding a person's addiction and how to intervene efficaciously, for example, relies far more on the person's mind that "contains the stories" of how the person's mind functions

---

[27] See Gabriel Dover's critique of Richard Dawkins self-imposed "Paradox of the Organism" and the "selfish" gene in "Anti-Dawkins".

[28] Cultures do have a large impact on our thinking about morality, ourselves, and others; people from different cultures have different conceptions of individuals and others (Markus and Kitayama 1991).

[29] There are those who have attempted to explain social learning through adaptive evolution. The Information Learning Model states that social learning is adaptive because the information the individual can obtain through the group is valuable for survival and reproduction. The Strategic Learning Model, on the other hand, is intended to replace the Information Learning Model because the latter cannot explain the social learning of a groups values and conventions if that is not beneficial to the natural selection of the individual. The Strategic Learning Model states "that it's often adaptive to adopt the prevailing practices of the group because the very fact that others engage in these practices makes it the case that doing what they do will be in one's long-term selfish reproductive interests." (Sripada 2007, 313).

than can be found in the person's neural circuitry (Ibid., 70).[30] In addition to the physical parts of the moral story, to learn more about humanity, such as how its language works and how propositions are conveyed through language, non-science sources such as literature, philosophy, ethics, and other information streams have to be used (Heyes 2012; Kaye 1984, 46). Part of the need to go beyond neuroscience —or even to ignore it to understand humanity—is the qualia problem that cannot be explained away in terms of how brains work (Tallis 2011, 293–6). Moreover, attempting to reduce minds to brains, and trying to prove it through neuroscience, is self-defeating: "The very existence of the mind-brain identity theory demonstrates the extent to which the mind transcends, and so is not identical with, activity in the brain." (Ibid., 338).[31] That is, asking the meta-question shows that there is a level of thinking different from the brain that can formulate and answer meta-questions. If we were merely brains, then this would not be possible for we would be able only to understand the machine from inside the machine rather than from outside. Finally, morality could very well be emergent much as language appears to be (Haag 2010). Hence, a reductionist model based on neurophysiology is too barren a theory for the practical work that needs to be done explaining who we are as human persons and moral agents, and why we have the morality we do.

Not only do we get an incomplete picture of humanity through reductionist evolution models, we have to ignore the science that raises legitimate, and at times compelling, critical objections to those models. For instance, many non-genetic factors are conveyed to offspring at or after conception (Mameli 2007, 53; Dover 2000, 64–5). Michael Skinner has shown that epigenetics "provides a molecular mechanism for environment to influence development, program cellular differentiation, and alter the genetic regulation of development." (Skinner 2011, 51). Environmental factors, for example, might cause a gene that would not have expressed itself in later generations without the environmental exposure to express itself because an *ancestor*, such as parent or grandparent, experienced the exposure. Therefore, genes are not the only causal story as to why brains and minds are the way they are, and attempts at pure adaptive or by-product evolutionary explanations of human traits are not possible.

Learning's power on mental abilities and faculties should also not be discounted or dismissed in favor of purely genetic explanations. In rebutting Steven Pinker's mental faculty conception-the Swiss Army Knife Model of human minds-Annette Karmiloff-Smith argues that it is equally plausible to believe that minds do not have innate structures similar to those of complex artifacts. In fact, it is reasonable to believe, based on available evidence, that "cortical micro-circuitry is not innately

[30]Legrenzi and Umilta argue that neurophysiology is not even required for studying or explaining memories, emotions, and other higher-level mental states (Legrenzi and Umilta 2011, 51).

As I have argued, this position seems a bit too extreme because it fails to recognize that we must think in certain ways, which limits what is possible in our morality and why we have the morality we do have. That is, it is part of the story, but not the whole story.

[31]It should be acknowledged that the machinery might require thinking about the machinery in this way.

specified by evolution but is progressively constructed by the postnatal experience of processing different kinds of input". (Karmiloff-Smith 2000, 146). For example, studies of the capacity for face processing in infants show that specialization and localization grow very gradually rather than being in place and fully operational at birth (Ibid., 151). If these are innate abilities or faculties, much like a module from a Swiss army knife, then it makes no sense for these skills to be underdeveloped from the moment of birth or whenever the brain is sufficient to have them. Although it seems too quick to claim that evolution played no role, it is reasonable to assume that learning in the environment also contributed a great deal to the development of various mental capacities and ideas.

In addition, social learning can powerfully contribute to the formation of minds and the development of brains. Even if genetic traits came about because they were favored by evolutionary forces, it does not mean that they are sufficient for why we have the morality that we do. "Even if valuing aggression is genetically determined, how aggression is expressed might be a matter of social choices." (de Melo-Martin 2005, 30). Our society might encourage those with aggression to channel it into useful actions benefiting society, such as competing in the marketplace, rather than trying to destroy all other members of the species who could interfere with the aggressive members' ability to survive and reproduce. Hence, biology provides the platform for morality, but socialization and learning further refine and build upon it (Kitcher 2011, 6). To get a more accurate understanding of human morality and how it affects moral issues of death and dying, then we need a more inclusive, complicated picture of its origins and its components.

1.3.1 What Can Trolley Problems and Neuroscience Say About Morality's Possibility?

While it is clear that neuroscience and other sciences dealing with brains provide data useful for the purpose of finding out how we can have the morality that we do, we cannot expect them to tell the entire story. Joshua D. Greene and others, for example, have performed fMRI studies to attempt to achieve a better understanding of how the human brain functions when it is evaluating moral dilemmas of the Trolley Problem variety. Basically, the researchers wanted to discover if the brain processed information differently when it came to consequentialist versus deontological situations and decision making (Greene et al. 2001, 2004, 2008). What they found is interesting. Deontological claims and thinking are associated with emotional neural processes. Consequentialism is associated with the more cognitive processes of the brain. When a person is confronted with a moral dilemma, these two areas of the brain are activated. The judgment that ensues is determined by which part of the brain is able to "dominate" the other part. If there is a "Me" component in the dilemma, i.e., one that requires the agent to personalize the dilemma rather than to evaluate it impersonally, then the brain makes a deontological decision using the emotional neurological process. If the dilemma is impersonal, then the cognitive portion of the brain performs a dominate cost/benefit analysis to determine a result. This phenomenon partially tells us why people will answer the Trolley Problem in different ways.

From this empirical information about how the brain functions and why ethics is possible for human persons, an overly strong conclusion has been drawn by some. Greene and Peter Singer have argued that consequentialist philosophical intuitions are reliable, while deontological intuitions are not (Singer 2005; Greene 2003, 2008). The crux of their argument is based on the old distinction between the rational and the emotional. Given that consequentialist judgments are the result of thinking in the cognitive portion of the brain, they must be more reliable than those generated by a more emotion-based process. Personal decisions are "morally irrelevant factors and are therefore unlikely to track the moral truth." (Greene 2008, 70). After all, according to this line of justification, the best arguments and decisions use reason and impersonal evidence rather than emotion; hence, deontological conclusions need to be set aside in favor of those that are consequentialist.

However, the data do not support claims that the moral truth is found merely in consequentialism and the cognitive part of the brain. Without adequate evidence, it is premature to draw conclusions as to which part of the brain is higher and which is lower. First, we cannot legitimately assume which function is superior on naturalist grounds because the "lower" emotion-based process is shared in common by the species, whereas the "higher" rational-based process is held by fewer members. That is, what is lower might be more fundamental and closer to the truth than the rational part, which can create false abstractions and theoretical conclusions bearing little on reality. Thus, the lower might actually be the higher-as-more-fundamental if we are talking about accuracies in moral judgments and morality in general.

In fact, going too far with an emphasis on the brain's rational part and consequentialism might be inconsistent with why morality is possible. If the consequentialist maximization of certain values is what is true and nothing else matters in morality, then it would be possible to program a computer to do cost-benefit analysis to identify the right actions in all Trolley Problems or any moral dilemma. In each and every situation, we should maximize the good over the evil, and not concern ourselves with motives, intentions, or anything that cannot be accounted for in the cost-benefit analysis of all alternatives in the situation. Given human moral agents' inability to reduce their emotional basis for morality and problems with biases in working with the numbers, the program might actually be a better decider of what is morally right and wrong than moral agents themselves. After all, many moral agents will not push the fat man in front of the trolley to save the five people on the tracks. Consequentialism here fails to engage the agents' reason centers sufficiently for the agents' rationality to overcome their biases and emotional reactions. The moral agents might even refuse to turn the trolley on to the track that will take the life of the innocent person standing there if the agents found out that the person was someone for whom they cared a great deal, such as their child, and the five other people on the tracks are strangers. Therefore, those who do not do whatever maximizes utility or whatever the consequentialist theory demands in each and every situation is wrong. Hence, the computer is superior to the person.

This is a very strange result that could have only come from allowing the theory to dictate the outcome rather than building the theory to fit the data. Once again, it seems that many of those working in ethics and bioethics use belief intuitions that merely reflect their personal beliefs and biases. Singer, for example, is a consequentialist. Greene appears to be one as well. When they dismiss deontological factors as morally irrelevant, they are using their complex belief intuitions rather than actual evidence to prove that doing so is justified (Singer 2005, 347; Greene 2008, 75).[32]

A more plausible, pragmatic conclusion would state that both the emotional and rational components of the brain involved in making ethical decisions are required; hence, any adequate moral theory must reflect that reality. Both contribute essentially to the process of moral judgments and morality; we most likely could not have the morality we have without both, as Hume and others have recognized, because we would be missing some necessary values and motivations without including the emotion factor and be lacking direction without the reason factor. Moreover, a person is willing to act on a moral belief only when the emotional part of the person's brain is involved in the decision making (Gazzaniga 2005, 167).

Consequentialists building their theories only on genetics or neurology are not alone in using the same belief intuition that says morality can be reduced to one principle.

> Thus as with the Gettier literature in epistemology and the causation and personal identity literatures in metaphysics, increasingly baroque proposals have given way to increasingly complex counter-examples, and though some have continued to struggle with the trolley problem, many others have simply given up and moved on to other topics. (Berker 2009, 328)

The strangeness to these well-meant attempts and surrender of the topic is that there is no need to find the one correct principle that fits these belief intuitions. We want to avoid contradictions or instances in which it is impossible to have at least one action that is morally right, and a principle that generally avoids such situations is intuitively appealing on the grounds of consistency over those that make ethics a whole lot messier. However, the story might be about multiple ethical principles at work that give contradictory results—a form of pluralism is true—or that we should focus on morally right actions without worrying about whether a contradictory

---

[32]Both F.M. Kamm and Selim Berker make similar points. Many of the conclusions drawn by Singer and Greene seem to be based on "arm-chair philosophy" rather than on any empirical data from the human participant studies. Moreover, the conclusions drawn by Singer and Greene go far beyond what the results can bear. (See Berker's "The Normative Insignificance of Neuroscience" and Kamm's "Neuroscience and Moral Reasoning.")

action is automatically morally wrong. A consistent set or even one principle might not consistently do the work in every case in which morality is to be found.

What we need is a lot more argument to get from the descriptive premises or data on what people believe is right and has value to the evaluative conclusion that they should think this way. We would have to show that because we think in a certain way as a result of natural and social influences, then that is the way we should think. Nevertheless, we would need to add some principle or further argument, such as thinking this way is practical for achieving the goals of flourishing, but that is a consideration beyond merely thinking a certain way because of evolutionary processes.[33,34] However, if we can avoid the informal fallacies of absolute authority and Is/Ought, it makes pragmatic sense to use the evidence and argument that the majority of people would give in support of their conclusion rather than using a belief intuition. It seems to be some sort of evidence for how morality is possible for us.

Even with all of the potential problems, we should not lose sight of the fact that the brain thinks in certain ways as a result of evolutionary processes, and that thinking in this manner is something we have to do. That is, we have to think about ethics in the way we do because our brains are somehow hardwired to act in that way. We have the values of X, Y, and Z, and sufficient evidence to think that we have these values hardwired because we share them in common with other sufficiently similar non-human animals and people, or with people in general. Moral principles, values, and faculties based on these values could also be hardwired, but even if they are not, they still make sense in that they seem to have a role to protect that which we think is valuable and avoid that which we think is disvaluable, which is especially useful in death and dying morality. Therefore, if intuitions are these forms of intuition-instincts and ideas derived from the intuition-instincts, then they have an evidentiary role to play in what our moral code should be because they serve as part of the moral platform that allows us to have ethics in the first place.

---

[33]A more useful survey to establish this would ask participants what they thought about people who made decisions in these cases. That is, give the participants the results of the first surveys along with the information that the participants are reasonable people. Then ask the participants of this proposed study if they would blame the other participants for the decisions they made. For those in the minority group, did they do the wrong thing by selecting the minority answer? Are they worthy of blame, although we would not do anything about it, for choosing an alternative they should not have chosen? Given that each person from the first surveys was a reasonable person, it is likely that the answer is that they did not do the wrong thing by making the selection. It is not something the majority of people would select, but there is nothing unethical about it in these extreme situations.

[34]Perhaps we can limit survey respondents only to people who are so intelligent and have enough relevant information that they always get the matter right when they are voting democratically.

1.4 Social and Cultural Intuitions

We should not declare that instinct-intuitions provide greater evidence or explanation than they actually do. It is dangerous to assume that there are more fully developed feelings and instincts, such as altruism, than actually exist. It is one thing to have common values between species, and quite another to have them between members of the same species. The fact that people can articulate what altruism is and understand it more fully in its context is probably part of why people have morality and duties, and non-human animals do not. We could share the same primitive feeling and value and behave accordingly to the instinct—that is, the most fundamental parts of the moral platform are identical—but the values taken in context of non-person thinking and person thinking might very well be essentially different. Moreover, explanations beyond genetics and faculty evolution add to the story of how the morality we have is possible, and we have already seen evidence about the influence that socialization and learning has in explaining why we have the morality that we do. The fact that humans have a value or principle or set of these things in common might be the result of their environment and learning more than any innate functioning or activities of the brain. With that in mind, consider whether some intuitions are the result of socialization or nurture and being a person in that environment, and what evidentiary weight such social intuitions should have in developing a value theory and moral principles.

I will not attempt to find all the social and cultural intuitions that people have, but will consider how such intuitions can be identified and how to evaluate their evidentiary weight.

One might think that social intuitions are veridical for persons with a language and social set of conventions and customs. That is, an intuition is merely a socially constructed instinct about how the world is, which can be determined by careful social scientific testing. Many intuitions seem to be like this. "It is intuitively plausible" is one way in which an argument for some claim is started out, which then proceeds to talk about how people view a particular claim, dilemma, or other moral situation requiring people to make an ethical evaluation. The majority of individuals, or at least a plurality, think this way, and the intuition is merely recognizing the fact that this is the way people currently think, and perhaps think through time. When speaking about the Termination Thesis, which is an interpretation of neo-Epicureans' view on death, Fred Feldman states:

> These examples, ['The fish you eat today, last night slept in Chesapeake Bay'], and many more of the same sort, decisively establish that the termination thesis…runs counter to common sense views[35] about death. We often think and speak of dead things in a way that reveals that we think that dead things formerly lived. (Feldman 1992, 95)

[35]Jeff McMahan also uses the term.

Basically, this is an ordinary language argument for American, or more broadly, western culture. Feldman is relying upon how people communicate their ideas in the western world to establish his claims about identity over time.

This approach can be a useful one, but caution must be maintained. People can use language any way they want, but their usage does not entail that they are correct in asserting factual claims about the world. If they are wrong, say about the definition of death, it does not mean that the convention corresponds to any existing fact other than it is the convention of the society. Furthermore, if intuitions are merely social constructs, then they are only interesting if we are cultural relativists.

Calling something common sense seems to be a way of using an intuition pump, but instead of the intuition's nature or the process by which it is found conferring legitimacy on the intuition, a poll or *vox populi* argument appears to be used. Since most people think this way, then there is legitimacy in thinking this way because it becomes common sense or a widely held belief.

The problem is that poll arguments are general fallacies of reasoning to be avoided whenever possible. After all, the mere fact that a majority of people agree to some rule, definition, code of behavior or other moral factor does not entail that they are correct. In fact, biases in groups and societies often take the upper hand over reason. Too often, we have seen that ordinary people, when in a mob, for example, make decisions that are far more extreme than any one of the mob members would make on their own. Cass Sunstein has argued that the dangers of group think include moderate people in a group moving toward an extreme as their ideas feed off each other in an echo chamber to which no dissenting views are admitted. Information cascades in which one person tells others what resonates with her, and they tell others, who pass it on. In the end, the latter's information is assigned great credibility merely because it appeals to the biases of the hearers as well as being from sources like them, which, in turn, makes the information seem trustworthy (Sunstein 2005, 98).

In social cascades, one member's fear can cascade and build through a group if availability and salience are sufficiently high for the population (Sunstein 2005, 94). Provided that the group's members are adequately unsure in the situation—that is, they have no evidence or beliefs that can ground their beliefs or provide fodder for their decision processes—the more likely it becomes that a group member's fear, no matter how unjustified it is, will be accepted as rational, and then magnified as it progresses through the group. The repeated affirmation of the fear makes it appear justified, which in turn makes it more frightening to the next person, who then legitimizes the increased fear level, and then passes it on. Basically, the chaotic nature of the situation can increase the effects of what should otherwise be dismissed as unfounded or trivial.

We should not always trust poll arguments or the majority's view of a situation because the majority can be a poor arbiter of morality. The force of persuasive arguments, the overconfidence caused by having one's views echoed and magnified, the emotional contagions of groups, and the value of seeing oneself as part of the group all work together to make the group less able to evaluate risks thoughtfully than an individual member would have been (Haidt 2001, 818).

Therefore, there is no guarantee that a group will be better able to discern morality or which values and principles should be used than an individual using careful reflection and critical reasoning.

On the other hand, there is no reason to believe that information that is firmly grounded in reliable evidentiary processes is not useful for making decisions, especially if those choices involve morally valuable entities being significantly affected. Sometimes, the majority or plurality view that is based on fundamental neurophysiology and society's requirements, in general and in particular, can be useful in informing us about how ethics is possible. In order to have a society in the first place, certain universal conditions must be met, such as some form of market to provide necessary goods and services, protection for its citizens, and social rules and institutions that would allow the society to function. How the society is constructed in particular is based on the general necessities, but it can achieve these requirements (determined by the citizens or some authority of the society, such as the society being primarily capitalist or socialist). These shared general and specific social or cultural conventions will also form part of the moral platform that makes ethics possible for us. They will be the social intuitions.

Two benefits from adopting conventional discourse on controversial, emotional subjects should not be overlooked. First, if a coherent, ethical position can be formulated using these conventions that works better in society to promote flourishing, then in the absence of evidence that such a position is false, it is permissible, and perhaps obligatory, to adopt it. Since the position is already partially established in the society, it is more likely to be adopted to everyone's benefit, or at least quite a few would receive something positive from it. A more controversial position that would replace the convention would not fare as well, all things considered, merely because it will be more difficult to accept and could cause unrest. Of course, if the new position is much better for individual and general flourishing than the conventional one, it becomes more difficult to determine which one is permissible. In these circumstances, it would be left to reasonable people to make the decision for themselves, and perhaps for their groups and communities, about what common sense should say.

The second benefit from adopting ordinary language use of terms, or their underlying intuitions, is that it might be easier to formulate a coherent, complex position. The various intuitions are the result of ideas that have been tested over time in the environment in which they are found. By using tested intuitions in a careful way, formulations of ethical codes and decision procedures will benefit the community's experiences and the long-term refinement that the members have performed on the ideas. The ideas, hence, will not be roughhewn material with which to build, but rather more nuanced, finished goods that are better able to apply to the community environment. Since the policies and other structures are identical or similar enough to existing social intuitions, then the former become more likely to be adopted by those who have bought into the community's values and beliefs.

In order to avoid the worst excesses of the ignorance of the majority, it is vital to limit the social intuitions we try to incorporate into the moral framework to only those alternates that are morally permissible, are compatible with the values found on the intuition-instinct level and are selected by the majority. That is, each option on value and rightness open to the group has to be morally justifiable on its own and consistent, as well as being the majority view, in order for it to qualify as common sense or a social intuition we should take as being valuable evidence of morality in general. If an alternate is impermissible (for example, it goes against the universal values found through neurophysiology), then even if the majority selects it, we should not consider this social intuition as valuable.

It is probably a fool's errand to attempt to untie the Gordian knot of primary, efficient, sufficient, and final causes of intuitions because the source and contributors to intuitions will not easily reduce to solely genetic or social explication since we are truly social creatures heavily influenced by evolutionary genetics and epigenetics. However, we can safely claim that intuitions can provide some evidence of how morality is possible and why we must think about morality in the way that we do. Many intuitions are at least partly the products of genetics and socialization, which give us many of the values and moral principles we actually have. In order to build a moral code that will work, we have to incorporate these intuitions into it because they are part of the moral platform that allows us to have morality in the first place. Without them, our moral code can be interesting, but not one that applies to the real world.

1.5 A Very Brief Case for a Pragmatic Approach to Morality

A much better approach to developing normative principles and a value theory, and then a moral code crafted from them, has been given by some pragmatists. If an essential or natural definition of a word, such as hydrogen, is unavailable, instead of allowing intuition pumps to serve as a primary source of evidence, we should figure out what work a discussion, definition, or word is supposed to do, and then build to that end. This approach is standard practice in biology with its 22 operational definitions of species. The appropriate definition is determined by what work is needed in the particular circumstances.

The same approach works well for bioethics, and I will argue for an interdisciplinary, pragmatic approach to bioethics problem solving. "The pragmatic method…is to try to interpret each notion by tracing its respective practical consequences." (James 1963, 23). More clearly, we set a goal, and then, among other things, figure out which means to achieving the goal works best in the situation. If we have ties for the best, then it does not matter which of the tying alternatives we take. However, if there is a clearly better way of doing things in comparison to other live alternatives, and that fact should be obvious to the person, then we are obliged to do what works best to achieve our goals.

Philip Kitcher's pragmatic naturalism shares features with the type of pragmatism that I will develop. According to Kitcher:

> Ethical truths are those acquired in progressive transitions and retained through an indefinite sequence of progressive transitions. Pragmatic naturalism proposes that some ethical statement - typically, vague generalizations, commending honesty and disavowing violence, for example - are true. They owe their truth to the role they play in ethical progress: "truth happens to an idea." (Kitcher 2011, 7)

Pragmatic naturalism is not a relativistic theory in that anything can be morally right or wrong depending on the whim of an individual or the conventions of a society. Instead, morality has to be based on "central human desires and needs, arising from our special type of social existence." (Kitcher 2011, 8). In other words, limitations are already in place that establish the groundwork for how morality is possible and what it will look like. A morality that is based upon peripheral desires or impossible ones, such as desiring pain for its own sake, is not a practical morality, and should be put aside in favor of one that is possible for us to use. Morality is therefore limited by genetics and other forces creating what is possible, but morality also flows from the moral platform that is held in common with all moral agents as human persons.

Given the difficulty of making morality merely a function, and altruism and fixing its limitations as morality's original source, as Kitcher does, we need to have some sort of ultimate end in place. Since flourishing does so much useful work without getting into the difficulties associated with progression prior to truth and ethics as a function, and it seems consistent with evolutionary, social learning, and individual choice—in fact, it incorporates them—we will assign flourishing as the end of morality.[36] For ourselves, other people, and living things, it is a flourishing life relative to what the individual is as *that* particular individual in *those* particular circumstances in which the individual exists. For non-living things, such as societies, it is flourishing as that entity qua that entity in its actual situation.[37]

Incorporating science and other fields' data into understanding morality makes sense for a number of reasons. First, if there is a priori or empirical evidence for a belief, position, or argument, then it must be used. The pragmatic reason for such a conclusion is that if we want to practically flourish and have others do likewise, then positions reflecting reality will be more useful than those of a non-extant universe. The work of socio-biologists, neurophysiologists, and other evolutionary and epigenetic scientists' work fits nicely here. Given that people use both emotion and reason when solving moral dilemmas, then that fact should not be forgotten when developing moral principles or codes. A set of emotions, such as empathy and altruism, are associated with morality. Moreover, actual situations affect reasonable people's reasoning. Sometimes they use more of a consequentialist view than a deontological one; at other times, a deontological view prevails. It also might be the case that both can simultaneously be used in a more nuanced decision procedure

---

[36]Others will have different goals, but what I am striving for here is a plausible goal that other reasonable people can accept as a plausible goal. The aim is not to get universal agreement on what the end is, but rather to get consensus that a particular end is something an entity can reasonably pursue without, in general, being in error morally.

[37]This rather vague notion will receive further explication in the next chapter.

that better captures the complexity of many moral situations, especially those in which other people are harmed in order to save yet other people.

When the available evidence does not yield a convincing result, then an additional criterion can be employed to evaluate whether we may adopt conclusions and beliefs. An argument, for instance, is a strong one if it does the work we want it to do; it is a bad one if it does not.

The measure of whether an argument works well will be that of a reasonable person. A reasonable person would evaluate an argument, criterion, definition, or anything that is being used by how well it works. Although Feldman might reject the use to which I put his criterion to evaluate the acceptability of a criterion, it is valuable for my purposes here.

> A *criterion* is something that is proposed for acceptance; that might be adopted if enough people think it would be useful; that might later be rejected if it proves inadequate or becomes obsolete as a result of technological advances. A criterion is better if it is more useful—easier to apply; more practical; more decisive. It would apply only to human beings. Thus, if enough [people working in the field] chose to adopt [it], it would be the criterion…(until replaced). (Feldman 1992, 58)

What is pragmatic in this approach is that the criterion's acceptability and continued acceptability is dependent on whether there is a better criterion that will be more useful for achieving the goal of flourishing. For the important matters that affect society and for which social policy needs to be made, the polling of experts on the criterion is necessary for public policy or general social interactions. Too many people are being affected to allow each individual to have her way in defining or using criteria. In order for society to run effectively, some uniform standard must apply to all, and should be accepted only upon the best evidence available, which might not be accessible to every person in that society.

On the other hand, if someone is making decisions for her personal use, then a general standard might not be applicable. Given that the standard by which arguments and criteria are evaluated, viz. flourishing of the individual or group, it follows that individual flourishing can be promoted, at times, by allowing idiosyncrasies that do not excessively harm social flourishing. For individuals, the criterion need not be approved by the majority. Instead, a criterion is acceptable for an individual if at least one reasonable person believes that the criterion pragmatically works. If so, then there is *prima facie* reason to adopt it.[38]

This pragmatic approach to approving a criterion allows multiple answers to moral dilemmas to be correct, and should help disabuse the notion that there is only one "True" way of answering certain questions. For example, one can believe that a particular individual is a better person than another based solely on certain features

---

[38]Of course, at times there may be more stringent criteria needed. For example, in public policy or actions affecting a great number of people, then it might be most useful to adopt what the majority of reasonable people believe. If a majority opinion cannot be reached, then a plurality would be sufficient. However, in actions affecting only or primarily the agent of the action, where the consequences are minor to the flourishing of the agent, then the standard may be lower. At least one reasonable person, usually the agent, believing the argument works can be sufficient.

one person has that the other does not share.[39] Suppose that two people are wise and temperate. One also is courageous, while the other is just. If an agent decides that the courageous person is morally superior because of her additional characteristic, then that is a correct choice if it works for the flourishing of the individual and others. If another agent decides that justice is a better characteristic to instantiate, then the just, wise, and temperate person is better than the courageous person within those circumstances. Although this is a form of relativism, if either choice is permissible in the circumstances because the choice works for flourishing and does not go against the available evidence, then making decisions of this type are permissible. Both are right at the same time, but in a slightly different way based on universal values and how they are idiosyncratically weighed.

Definitions have the same evaluative scheme. If they work for the ends of flourishing, then they are acceptable until we find a more efficient definition to use in those contexts.

We might as well begin with why we are interested in death issues in the first place. To define death, we must know what work has to be done. In some cases, medical professionals need to determine when a patient has died or when treatment is futile so that no more therapy is expended on them. Brain death works well for this purpose, but if another definition of death did a better job, then we should be willing to accept it. In other contexts, death could be defined in other ways as well. As long as the end intended for the definition is clear, then we can more easily decide if it is a good one based on the work it does.

1.6 A Pragmatic Method for Developing a Value Theory

Although some entities' intrinsic value is not all that controversial in the philosophical literature and common usage, in part as argued above because they are innate or learned values, others are far more problematic. The problem might lie with how complex the idea is or how far away it is from its supporting foundational innate or learned value. Kant would argue, and many agree, that being a human person is intrinsically valuable, and we could add that how our brains innately and socially think about the issue lets us value human persons as ends in themselves.[40] Less commonly accepted is the notion that being alive confers intrinsic value on the entity that possesses the characteristic. The disagreement is often seen in debates on

[39] There are limitations to what works here. Believing that women are inferior to men is impermissible on the grounds that women and men are persons, and it is personhood that gives moral dignity to a person. Our moral platform and reason prohibit such a belief from being true.

[40] Of course, how people have been defined over time has allowed societies to treat persons who did not fit the social definition in ways that those individuals should not have been treated. However, that is the fault of those who did not use their moral platform and reason in the correct way.

environmental ethics and animal rights in which the flourishing of human persons is in conflict with either that of the environment or animals.

Although it might be possible to find a convincing a priori or a posteriori argument in favor of or against an object's intrinsic value, the fact that so many intelligent people reasonably disagree on the subject provides weighty evidence for the claim that a compelling proof is impossible. Even those things most of us would consider uncontroversial intrinsic worth holders, namely human persons, are not immune from attack from some bio-centrists or eco-centrists. Instead of spending more time and effort on trying to discover a proof capable of convincing all reasonable people of the value or disvalue of an entity instantiating a characteristic, a mental state, or pursuing a telos, it would be best to formulate an argument that most, it not all, reasonable people can accept. Basically, they do not have to agree with the value theory or use it, but they should be able to understand and appreciate the fact it is reasonable for any person to use it in his or her evaluations and decision making. I propose using the Benefit of a Doubt (BD)[41] principle and a ranking system of intrinsic value capturing the moral intuitions or ideals underlying our ability as persons to have morality in the first place (Gert 1998, Chap. 4).

The BD principle assigns intrinsic worth to an object when there is no overriding evidence to prove that the object does not have such value. Instead of believing that something lacks intrinsic value until it is shown to have it, we should adopt the Eastern position of assuming that everything has intrinsic value, until it can be proven not to have it. Put more formally,

> **The Benefit of a Doubt principle (BD) = df.** In cases in which it is an open question to reasonable people acting as reasonable people as to whether an entity has intrinsic value, assuming the entity has intrinsic value is required provided that the assumption does not unjustifiably harm the flourishing of those entities with better evidence supporting their possession of intrinsic value or that of the entity under consideration.

We should give an entity with unknown intrinsic value as high of a status as we can when it is something that is to the benefit of the entity because that will respect her value even if she does not actually satisfy the requisite conditions, as long as entities with better cases for their inherent worth are not illicitly affected. We want to avoid instances in which we do not treat those beings with intrinsic value with the respect their moral worth demands, while at the same time giving credible candidates for such value the benefit of a doubt.

When we are unsure about the actual value, we should also not be hasty in claiming that the entity has such value when doing so will be detrimental to her. In this case, it is her flourishing that also matters. If we cannot tell if the entity is a moral subject or a moral agent, when deciding her fate in regard to punishing or taking other adverse action toward her, then we should take the safer option of treating her as a moral subject. We would make a serious mistake in punishing someone who might not be a moral agent as we would a moral agent because it

[41]BD is based upon Pascal's Wager, which is a cost/benefit analysis of what a rational person should believe to be in her best interest.

would be cruel to hold her to a unobtainable standard. If she is not a moral agent, then she cannot make the proper evaluations of her conduct because she cannot do the proper reasoning required to make her responsible for it. Hence, to be better safe than sorry, we should assign her the value of a moral subject who has the potential to be a person, and punish her according to this level, if at all. The ultimate, pragmatic end would likely be rehabilitation so that she can achieve the goal of being a person who is a good person because that would be good for her flourishing, as well as that of society.

Another immediate concern arises from what might appear to be over-inclusiveness. If people are expending resources on objects that do not have intrinsic value, then the former will be unable to fulfill their duties to those objects that actually are valuable in and of themselves. For example, a person's intrinsic value might not be recognized in the proper way because agents are busy trying to recognize animals' worth in and of itself. A medical experiment that helps save the lives of a small number of human beings could be prevented based on the inability to perform it on animals. If it is the case that animals do not have the value that requires the respect paid them, then the human beings have been treated wrongly, and possibly, abominably.

Over-inclusiveness might also retard the ability of people to respect those things with intrinsic value, which in turn destroys the motivating force required to make people want to be ethical in the first place. If everything that potentially has intrinsic worth is assigned it, then as Janna Thompson claims, the term becomes meaningless. First, if all objects are equal in worth, then there is no reason to select one thing over another (Thompson 2000, 577). Furthermore, if all things are deserving of respect because each is equally intrinsically valuable and there is no reason to select one over the other, then respect becomes an empty term. Suppose that existing is assumed to be intrinsically valuable, then everything that exists must be treated the same way, and we cannot distinguish between the value of one existing thing over another since each entity has existence in equal quantities merely from existing. Hence, instead of being able to show the proper level of respect to those things deserving such treatment, "respect" will mean very little, if anything. Once again, the impetus of intrinsic value on morality will be lost in our attempt to be inclusive.

Although there could be some set of circumstances in which this unfortunate result might occur, its chances are greatly reduced if a hierarchical ranking of intrinsic value is introduced. Some things are more intrinsically valuable than other things. The intrinsically valuable things' ranking is based upon the probability of an object having intrinsic worth. On these grounds, people as people are more intrinsically valuable than plants because there is little controversy over people being valuable in themselves in a way that plants are not. On the other hand, there are those who claim that plants also are intrinsically valuable, but there does not seem to be widespread agreement to it. Therefore, people are more intrinsically valuable than plants. The more likely it is that something has intrinsic value entails that it should be ranked higher than those objects with less probability, until that time in which adequate evidence is provided to change the classification. The result of this practical distinction is reducing both the possibility of treating intrinsically valuable things as they should not be treated and the effort required in respecting intrinsic value.

1.7 An Adequate, Pragmatic Moral Code

With intrinsic value and the flourishing of people and societies in mind, what is necessary for an adequate, pragmatic moral code to achieve the designated end? First, there appear to be universal and absolute moral principles and values: consequentialism's utilitarianism, ethical egoism and deontology. Respect for persons, virtue ethics, and care ethics must be incorporated because they are intimately tied into the moral platform that makes morality possible. No matter how they are used in the moral code, we must keep them sufficiently general so that they apply to many different people and can be used by them to make decisions about ethics that can be understood by other people making similar decisions or thinking about the issue in order to try to understand it. Clarity at this level is useful for us to be more precise without being so precise that the result loses a lot of individuals who have the same general value or idea but do not employ it in the same way, especially when there is a lot riding on the decisions we will be making. General rules and values built on the shared moral platform help us understand each other because we have reference to what they are like from the objective components in ourselves. Our sympathy for others, which is a necessary condition for us to be ethical in the first place, would not exist without an ability to see ourselves in other's shoes. Others matter to us, as they should, because they are sufficiently like us so that we have referent for morality, and we can successfully mirror in our own minds what is happening to them.

Universal and absolute values and principles also allow us to communicate effectively enough that we as individuals or as members of society can make decisions about how we should operate in the real world, as well as allowing us to choose for the flourishing of others and society. Ethics, in fact, can be considered to be akin to communication in a certain way: "Communication is…a goal oriented process" to share information (Rescher 1998, 15). Ethics also is a goal-oriented process that can only be effective if there is a shared understanding among those engaged in it which allows them to share information. Without sufficient referents in common, communication of this sort would be impossible. That is, we need a shared moral language with common meanings and referents that we can grasp, as well as enough relevant data to stimulate us in an appropriate way. Of course, what is being communicated might not be perfectly conveyed by what is said, but as long as it allows us to function well, then that is all we actually need. For example, if we hear from one reliable friend that another friend has died, then we reasonably believe that the latter is dead, and that information makes us feel appropriate emotions such as sadness, grief, and regret. We can converse with our information source friend about the person's death, all the while sharing our feelings through mirroring. That is, the two of us can grieve together and give support to the other by sharing these common feelings being experienced. Therefore, the verbal and non-verbal, and rational and non-rational, communication has been successful in this case.

Although we must have shared values to have effective communications and morality, we must understand, much as Hume did, that we can rank these values in different ways for a variety of reasons, including accepting how our society ranks the virtues and vices. This is relativism of a sort, but one that has a realist base to it. The values and general moral principles are the same, but how they are ranked or developed or used can be based, in part, on the subjectivity of the person's moral code. For example, various moral principles can be used to make a decision, and the decision can be the same but be based or justified on different grounds that work for the individual. In addition, the ethical code of each person can be different in significant ways by what they choose to emphasize from the universal moral platform, and shared values and principles and how they emphasize them.

Although there will be universal general values and rules for all human persons, there is still room for individual and social variations in moral codes, but their use is limited. Relativistic theories such as Divine Command Theory,[42] Existentialism,[43] and Conventionalism[44] may raise the bar for individuals and societies that choose to use them, but the objective principles that are much more closely linked to the shared moral platform set the lowest level for all of humanity as to what is minimal decency. The objective principles provide a universal morality and tell us what is morally required for us to do lest we err objectively in our actions and who we are as persons and individuals. Subjectivism allows us to err, but the error is a result of a possibly changeable standard that we take on ourselves by being a member of a particular society or being a particular person who has made particular choices, including which religion we follow and if there is a divine command theory attached to it that we have chosen to make part of our individual or social moral code.

Relativism allows us to make our moral codes our own by stamping our individual emphasis on the objective principles and values as well as creating new ones. This power enables us to have a code that can be different from every other moral code in existence. Our ability to create our own morality in part gives power to our choices to make our codes ours, and our choices make us more likely to follow our codes because they are ours and reflect our shared and individual values. That is, there is an emotional and intellectual tie to our creations, which are artifacts of who we are as individuals and species beings.

---

[42]For example, an act is morally right if and only if the act is commanded, expressly permitted, or approved by God.

[43]For example, an act is morally right if and only if the agent of the action chooses that the act is morally right.

[44]For example, an act is morally right if and only if the agent of the action does not violate any of the conventions of the society in which the act would be done.

The end result of my pragmatic framework is not simplification, but possibly increasing complexity dependent on the individual as an individual and as a species being in a particular context. That is, individual moral codes could be so extraordinarily complex that they might not be fully explicated, or so involved that they might actually have contradictions in them that only are sorted out when the person stumbles across situations in which the inconsistency becomes apparent, and the person has sufficient motivation to alter the moral code's consistency failures instead of ignoring them. If the inconsistency is ignored, then the person is not acting rationally or reasonably because the path to flourishing needs to be efficient, and the intellectually lazy person should be condemned on objective grounds for a moral failure to improve her morality by improving the decision procedure, including her moral criteria.

Given the potential complexity and idiosyncratic nature of moral codes, we will not be able to formulate a moral code that applies in all situations and works for all people. Despite that, we do not need to do so because we can accomplish the work we need to do without creating or discovering a moral code that provides all sufficient and necessary conditions. We can evaluate individual and shared codes based on their ability to be useful in the situations they encounter, which would include making adequate moral decisions in the situations, being grounded in shared values, and being able to communicate with others about the ethics being used. The reason for the latter is because we often have to justify our decisions to others, and at times, try to persuade them to do likewise. A code that would not allow communications between moral agents would be relatively worthless in the real world in which we interact with other people.

What we can do is to evaluate a moral code for its usefulness for a situation or set of situations. Generally, we should adopt those that are most useful because they more readily allow us to obtain our ultimate end of flourishing. The necessity of this demand for pragmatism hinges on the situation. If a person lives a life in which the person is not challenged morally because there are no real issues confronting her, then the person might not need very detailed moral code from which to operate. Suppose this is a person who has very little contact with others, and lives a solitary, subsistence level of life. She spends most of her time trying to get the physical goods necessary to survive and needs to think little about complex world and social affairs, or even much about what she is doing because it has low impact on other intrinsically valuable things, although it might be of great interest to her. Even with a simpler life, she will have the same moral platform as any other person, but most likely without the external and internal forces that might cause the code to become more complex.

Others would, of necessity, have more complex moral codes. For example, those who have much more involved lives with far greater power over others would need to make finer and finer distinctions between values and principles. A much more developed moral code is required for adequate ethical decision making in this context and this group of individuals. Basically, we are thinking about what skill set is needed for the role being played, and how many roles the person has. A surgeon, city planner, ruler, and master carpenter need much finer tools and the skills to use them efficiently than someone who digs ditches. A person with a very complex social and intellectual life would have more roles than someone with a very simple, isolated existence with few needs and desires.

In the following chapter, I will develop a complex, hierarchical value theory, and then a practical set of moral principles consistent with and incorporating the shared values. The moral code is crafted in this order because values seem to be prior to the existence of principles; principles seem to be developed to benefit positive values and avoid negative values. However, it really does not matter which comes first because neither would be very useful in morality without the other.

References

Barrett, L.F. 2013. Psychological construction: The Darwinian approach to the science of emotion. *Emotion Review* 5(4): 379–389.

Bateson, P. 2000. Taking the stink out of instinct. In *Alas, poor Darwin: Arguments against evolutionary psychology*, ed. H. Rose, and S. Rose, 157–173. London: Jonathan Cape.

Berker, S. 2009. The normative insignificance of neuroscience. *Philosophy & Public Affairs* 37(4): 293–329.

Bloom, P. 2013. *Just babies: the origins of good and evil*. New York: Crown Publishers.

Boehm, C. 2004. Explaining the prosocial side of moral communities. In *Evolution and ethics*, ed. P. Clayton, and J. Schloss, 78–100. Grand Rapids: William B Eerdmans Publishing Company.

Bouchard Jr, T.J. 2007. Genes and human psychological traits. In *The innate mind*, vol. 3, ed. P. Carruthers, S. Laurence, and S. Stich, 69–89. Oxford: Oxford University Press.

Broom, M.D. 2003. *The evolution of morality and religion*. Cambridge: Cambridge University Press.

Brown, W.S. 2010. Nonreductive human uniqueness. In *Human identity at the intersection of science, technology and religion*, ed. N. Murphy, and C.C. Knight, 97–115. Farnham: Ashgate Publishing Limited.

Campbell, B. 1995. How much in the human conditions is without counterpart in other animals? In *Evolution and human values*, ed. R. Wesson, and P.A. Williams, 117–37. Atlanta: Rodopi.

De Melo-Martin, I. 2005. *Taking biology seriously*. Lanham: Rowman & Littlefield Publishers Inc.

De Waal, F. 1996. *Good natured: The origins of right and wrong in humans and other animals*. Cambridge: Harvard University Press.

De Waal, F. 2006. *Primates and philosophers: How morality evolved*. Princeton: Princeton University Press.

Dover, G. 2000. Anti-Dawkins. In *Alas, poor Darwin: Arguments against evolutionary psychology*, ed. H. Rose, and S. Rose, 47–66. London: Jonathan Cape.

Ekman, P. 1957. A methodological discussion of nonverbal behavior. *Journal of Psychology* 43: 141–149.

Feldman, F. 1992. *Confrontations with the reaper*. New York: Oxford University Press.

Gazzaniga, M.S. 2005. *The ethical brain: The science of our moral dilemmas*. New York: Harper Perennial.

Gert, B. 1998. *Morality: Its nature and justification*. New York: Oxford University Press.

Greene, J.D. 2003. From neural 'is' to moral 'ought': what are the moral implications of neuroscientific moral psychology? *Nature Reviews Neuroscience* 4: 847–850.

Greene, J.D. 2008. The secret joke of Kant's soul. In *Moral psychology*, vol. 3, ed. W. Sinnott-Armstrong, 35–79. Cambridge: MIT Press.

Greene, J.D., et al. 2001. An fMRI investigation of emotional engagement in moral judgment. *Science* 293: 2105–2108.

Greene, J.D., et al. 2004. The neural bases of cognitive conflict and control in moral judgment. *Neuron* 44: 389–400.
Greene, J.D., et al. 2008. Cognitive load selectivity interferes with utilitarian moral judgment. *Cognition* 107: 1144–1154.
Haag, J.W. 2010. The emergence of morality. In *Human identity at the intersection of science, technology and religion*, ed. N. Murphy, and C.C. Knight, 131–143. Farnham: Ashgate Publishing Limited.
Haidt, J. 2001. The emotional dot and its rational tail: A social intuitionist approach to moral judgment. *Psychological Review* 108: 814–834.
Hare, R.M. 1981. *Moral thinking: Its levels, method and point*. Oxford: Oxford University Press Inc.
Hare, R.M. 1989. *Essays in ethical theory*. Oxford: Oxford University Press Inc.
Hauser, M. 2006. *Moral minds*. New York: HarperCollins Publishers.
Heim, M.S. 2004. A cross-section of sin: The mimetic character of human nature in biological and theological perspective. In *Evolution and ethics*, ed. P. Clayton, and J. Schloss, 255–272. Grand Rapids: William B. Eerdmans Publishing Company.
Heyes, C. 2012. New thinking: the evolution of human cognition. *Philosophical Transactions of the Royal Society B: Biological Sciences* 367(1599): 2091–2096.
James, W. 1963. *Pragmatism and other essays*. New York: Pocket Books.
Jung, R., and R. Haier. 2007. The parieto-frontal integration theory (P-FIT) of intelligence: Converging neuroimaging evidence. *Behavioral and Brain Sciences* 30(2): 135–154.
Kamm, F.M. 2009. Neuroscience and moral reasoning: a note on recent research. *Philosophy & Public Affairs* 37(4): 330–345.
Karmiloff-Smith, A. 2000. Why babies' brains are not Swiss army knives. In *Alas, poor Darwin: Arguments against evolutionary psychology*, ed. H. Rose, and S. Rose, 144–156. London: Jonathan Cape.
Kaye, H.L. 1984. Cultural being or biological being. In *Evolution and human values*, ed. R. Wesson, and P.A. Williams, 35–48. Amsterdam: Rodopi.
Kitcher, P. 2011. *The ethical project*. Cambridge: Harvard University Press.
Krebs, D.L. 2011. *The origins of morality: An evolutionary account*. Oxford: Oxford University Press Inc.
Legrenzi, P., and C. Umilta. 2011. *Neuromania: On the limits of brain science*. Oxford: Oxford University Press Inc.
Locke, J. 1992. An essay concerning human understanding (abridged). In *Central readings in the history of modern philosophy*, ed. R. Cummins, and D. Owen, 93–271. Belmont: Wadsworth Publishing.
Mameli, M. 2007. Genes, environment, and concepts of biological inheritance. In *The innate mind*, vol. 3, ed. P. Carruthers, S. Laurence, and S. Stich, 37–54. Oxford: Oxford University Press.
Markus, H.R., and S. Kitayama. 1991. Culture and the self: Implications for cognition, emotion, and motivation. *Psychological Review* 98(2): 224–253.
Mill, J.S. 1972. *Utilitarianism, on liberty and considerations on representative government*, ed. B. H. Acton. J.M. Dent and Sons Ltd: London.
Nelkin, D. 2000. Less selfish than sacred? Genes and the religious impulse in evolutionary psychology. In *Alas, poor Darwin: Arguments against evolutionary psychology*, ed. H. Rose, and S. Rose, 14–27. London: Jonathan Cape.
Noe, A. 2009. *Out of our heads*. New York: Hill and Wang.
Peirce, C.S. 1955. *Philosophical writings of Peirce*, ed. J. Buchler. Dover Publications, Inc.: New York
Pinker, S. 1994. *The language of instinct*. New York: William Morrow.
Quiroga, R.Q., and S. Panzeri. 2009. Extracting information from neuronal populations: information theory and decoding approaches. *Nature Reviews Neuroscience* 10: 173–185.
Rescher, N. 1998. *Communicative pragmatism: and other philosophical essays on language*. Lanham: Rowman and Littlefield Publishers Inc.

Richerson, P.J., and R. Boyd. 2004. Darwinian evolutionary ethics: Between patriotism and sympathy. In *Evolution and ethics*, ed. P. Clayton, and J. Schloss, 50–77. Grand Rapids: William B. Eerdmans Publishing Company.

Rose, N., and J.M. Abi-Rached. 2013. *Neuro: The brain sciences and the management of the mind.* Princeton: Princeton University Press.

Santos, L.R., and V. Lakshminarayanan. 2007. Innate constraints on judgment and decision-making? Insights from children and nonhuman primates. In *The innate mind*, vol. 3, ed. P. Carruthers, S. Laurence, and S. Stich, 293–310. Oxford: Oxford University Press.

Satel, S., and O.S. Lilienfeld. 2013. *Brainwashed: The seductive appeal of mindless neuroscience.* New York: Basic Books.

Scarre, G. 2007. *Death.* Montreal: McGill-Queen's University Press.

Segal, G. 2007. Poverty of stimulus arguments concerning language and folk psychology. In *The innate mind*, vol. 3, ed. P. Carruthers, S. Laurence, and S. Stich, 90–105. Oxford: Oxford University Press.

Singer, P. 2005. Ethics and intuitions. *The Journal of Ethics* 9: 331–352.

Skinner, K.M. 2011. Role of epigenetics in developmental biology and transgenerational inheritance. *Birth Defects Research Part C: Embryo Today* 93(1): 51–55.

Sripada, C.S. 2007. Adaptationism, culture, and the malleability of human nature. In *The innate mind*, vol. 3, ed. P. Carruthers, S. Laurence, and S. Stich, 311–329. Oxford: Oxford University Press.

Sterling, C.G. 1994. *Ethical intuitionism and its critics.* New York: Peter Lang Publishing Inc.

Sunstein, C. 2005. *Laws of fear.* New York: Cambridge University Press.

Tallis, R. 2011. *Aping mankind: Neuromania, Darwinitis and the misrepresentation of humanity.* Durham: Acumen Publishing Limited.

Thompson, J.L. 2000. A refutation of environmental ethics. In *Contemporary moral issues*, ed. R.L. Hinman, 576–87. Upper Saddle River: Prentice Hall.

Wertz, A.E., and K. Wynn. 2014. Thyme to touch: Infants possess strategies that protect them from dangers posed by plants. *Cognition* 130(1): 44–49.

Zamulinski, B. 2007. *Evolutionary intuitionism: a theory of the origin and nature of moral facts.* Quebec: McGill Queen's University Press.

Chapter 2
A Pragmatic Framework of Values and Principles: The Beginning

Abstract In order to make moral judgments about death and dying issues in bioethics—or on any moral issue—it is necessary to have a value theory and a set of normative principles that allow us to identify which things have value, what that value is, and to what degree the objects have it, and then to act on that information. Perhaps the most plausible theory is one based on a form of pragmatism that builds its foundation on how we individually and as a community actually perceive morality. The result will not be simple, but it will provide reason to use it and respect other's use of it because it is built on what we already think is true.

2.1 Introduction

Before navigating the more difficult waters encountered when talking about death's values, it is necessary to formalize a pragmatic moral code that includes both normative principles and a value theory. We need both because one without the other is useless. Normative principles without values tell us what to do but provide no motivation as to why we should do it, whereas value theory without normative principles gives us no direction on what to do. Because the issues are grave and urgent, we will discuss, among other things, the value of various lives and whether there is a moral obligation to die; the code must be plausible, if not compelling, to most reasonable people. Universal agreement on such a code is impossible. Given the nature of morality, a universal set of rules and values that everyone shares in all of its aspects does not exist, although there is a universal moral platform that gives us elements in common and a general similarity that allows communication and understanding.[1] Thus, the pragmatic approach is to find a code that allows communication and understanding by building upon the shared moral platform.

[1]Emrys Westacott argues that using phronesis—practical wisdom—is the appropriate approach to morality. Phronesis will find or develop general rules that allow us to factor in and weigh up all sorts of variables, but we are not going to have a universal principle that can adequately handle all cases (Westacott 2012, 188).

D.R. Cooley, *Death's Values and Obligations: A Pragmatic Framework*, International Library of Ethics, Law, and the New Medicine 62, DOI 10.1007/978-94-017-7264-8_2

Among everyone who considers it, there is little doubt that death is a loss. Two of the key disputes, however, are how and to whom it is a loss. Since it is generally a person's death we are most interested in, the first question is, naturally, whether the deceased person is harmed by his new state. Epicureans hasten to argue that injury to the individual is impossible because there is no individual left to help or harm after the person expires. That is, you have to exist to be injured. On the other hand, there are those who believe that death harms the deceased in some way. Priorists, for example, allow harm to stretch back in time with the death causing injury to the individual the moment she made plans that cannot be completed because of her death. Others use Cambridge changes in an attempt to show that death can harm the deceased.[2]

What will put the claims of injury and loss into perspective is a value theory that, as much as is appropriate in developing a practical, consistent theory, reflects pragmatic thinking on harm. Death causes losses in intrinsic value in a variety of ways, including but not limited to the value terminated when the intrinsically valuable individual dies. What is different about this theory is that no claims will be made that it is ***the*** correct theory of value, nor is it asserted that this is the best value theory, although I think it comes rather close. Instead, the claim is that the value theory developed is one any reasonable person can reasonably adopt as her own. Given that any reasonable person could adopt it, then it is understandable to all reasonable people thinking as a reasonable person. Moreover, in bioethical, philosophical and other discussions, although not everyone engaged would agree with the conclusions drawn here on value and what should be done and how to do it, they can understand and appreciate why others think the way they do. In the end, civility between individuals making up society is maintained and enhanced, which should be an ultimate goal of ethics and morality.

I will begin with the practical normative principle before developing the value theory. The order is one of simple expediency—it is easier to develop the principle than the value theory. Of course, the normative principles are rather useless without intrinsic values, but the value theory will spill over into Chap. 3's discussion of what we are as individuals and how we should be valued in the different ways in which we can and are legitimately conceived.

2.2 A Practical Principle

Returning to one of the central questions of this work, is death an injury to the individual? We first need to ask what work we want done when answering the question. What purpose are we trying to fulfill? The main purpose for all ethical actions is, for me, to have a flourishing life, which while it is flourishing, I will call happy in the moment; if it is overall a flourishing life, then it is a happy life. More

[2]These issues are addressed in Chap. 4.

controversially, a happy life is a good life, a life of well-being. Equating all these things together is not something that most would do, but it gets a lot of work done; it not only identifies ultimate values but also provides great motivation to act in certain ways.

In order to evaluate duties to oneself and others and find permissible actions in the absence of duty, it is necessary to have a moral principle or moral code that allows decisions to be made. As in the case of value theories, the idea is not to find ***the*** right moral principle with which we can answer all questions without being open to any reasonable objection.

I propose what I will label the Practical Principle. The Practical Principle is a combination of utilitarian and Kantian requirements which, when satisfied, will produce at least one right answer to a moral question. The principle's broadness, at times, allows multiple solutions, all of which are morally justifiable even if they disagree with each other. In certain situations, such as medical treatments in which both therapies have the same efficacy and difficulty for the patient, and the only difference between them is how they are delivered, then the Practical Principle will classify either choice as morally right, and even be able to tailor the solution to the individual who is making the medical decision. That is, one moral agent might be able to justify only one action, while an equally reasonable moral agent might come to a different conclusion, while using the very same moral principles.

Instead of continuing to vaguely describe the Practical Principle, it might be best to introduce its two conditions. First is Reasonable Person Utilitarianism (RPU), which is a version of act utilitarianism without many of the drawbacks of standard act utilitarianism.[3] RPU is:

> If a reasonable person in the same conditions as the agent making the decision about what to do would reasonably believe that the action will produce at least as much utility as any alternative to the act, then the act is *prima facie* morally right.

Basically, RPU is a variation of the Ideal Person Theory, which bases morality on what an ideal person would do in the situation. Since no one really knows what an ideal person is or whether such a creature can exist, it is reasonable to substitute something that both does exist and can serve as a useful standard. In addition, the relative nature of the situation is incorporated into the condition by making the reasonable person decide the issue under the same conditions as that of the actual moral agent. An ideal person might not be under pressures caused by time stress, lack of information, or any other factor that would make the decision-making process less than ideal. Hence, what an ideal person might choose is irrelevant to what a real person may permissibly do.

RPU makes the morality of actions dependent on a reasonable person standard. A reasonable person is a person who has the two general goals of having a good life and making the world a better place so that others can flourish as well. Within this

---

[3]Standard act utilitarianism states that an action is morally right if and only if there is no alternative to the action with greater utility, where utility is defined as the result of subtracting the amount of evil caused by an act from the amount of good caused by the action.

context, a reasonable person is an individual who understands what morality is, accurately applies her maxims to each situation she encounters, tries to make herself and others better as long as doing so sacrifices nothing of comparable worth, adopts reasonable goals, knows and accepts that reasonable people will have different reasonable views at times, correctly analyzes and uses the data available to her, and is more reluctant to impose risk on others than she is on herself because she respects the others.

The final criterion of a reasonable person is one that brings a central feature of why ethics is possible for people into what it means to be a reasonable person. The main difference between a rational person and a reasonable one is that a reasonable person incorporates all relevant moral considerations into her evaluation processes rather than some smaller set.[4] Suppose that a person has adopted a version of Ayn Rand's ethical egoism as her moral principle. In order to achieve what is required by ethical egoism, the moral agent may act rationally in ways that are morally repulsive to those who care about the intrinsic value of all people. For example, as long as it maximizes her utility, the person can act as Machievelli advised those princes who conquered territory should act—appear to be a good person in public, but be ruthless behind the scenes without being caught out. Assassinations, accusations of treason, and other nefarious actions can be justified on these grounds to the detriment of anyone who is not the agent of the action. Although this approach is rational given the end goals, it is not reasonable. Reasonable people recognize the intrinsic value of other individuals involved, which is a moral consideration that should be taken into account. Hence, some of the worst excesses of pure ethical theory that cause people to strip away their reason, empathy for others, or other morally relevant factors are prevented by the reasonable person standard.

In regard to evaluating a situation, a reasonable person uses appropriate information in the context for what is morally permitted, required, or forbidden. Reasonable people seek out and apply information about the:

1. External world society's rules, practices, and customs.
2. Rules and responsibilities associated with specific roles the agent is playing at the time.
3. Claims that others have on the agent and the agent has on others.
4. Maxims growing out of previous judgments that the agent has made in order to maintain ethical consistency.
5. In conflict situations, what is right on balance.
6. Which consequences are important and their value, as well as the value of other relevant things.
7. Which if any mediated consequences count and which do not (Holmes 2003, 215–7).

[4]Another difference is that rational goals may not be reasonable ones because of the differences in what ethical considerations are included in the decision process.

Given the broadness of the reasonable person definition, it is clear that one agent might come to a legitimate conclusion that is different from another person who has the same information but a different history. In addition, one society might permissibly require certain ethical behavior that other societies forbid. This relativistic circumstance might produce a moral duty for an agent which she would not have had if she had lived in one of the latter societies.

Since a reasonable person can come in many different forms, then there can be diversity in what is right and wrong. Instead of restricting morality only to one right action, and a whole lot of wrong ones, many different actions are ethical, thereby making it easier to do the right thing, at least on the grounds of a wide variety of choices. More importantly, reasonable people are able to understand that other reasonable people can look at the same evidence and come to a different conclusion. Therefore, even if the action is not one a particular reasonable person thinks is likely to maximize utility, she can accept another reasonable person's decision. She can tolerate it in the appropriate way, which will make her life better; therefore, her life thrives more than it might otherwise. Moreover, within groups and communities of reasonable people tolerating each other, flourishing is enhanced by the reduction of internal strife, by fostering positive social relationships and by thriving people.[5]

RPU allows for much greater flexibility, which is required in real life ethical decisions. For example, if a child darts in front of a car, then the driver has a split second to make a decision about what to do. If she has not noticed two children standing on the sidewalk through no fault of her own, then a decision to drive on the sidewalk to avoid the child in the street would be permissible, although tragic. An ideal person would probably say that hitting the one to save the two is required, while a reasonable person understanding the situation's constraints would say either may be done, at the same time recognizing that either alternative is a tragedy, much like Trolley Problems are.

I see two benefits from adopting RPU in place of standard act utilitarianism. First, it does not delink an action's morality and the agent's blameworthiness or praiseworthiness, and it empowers moral agent by allowing them to have sufficient information to be able to choose their actions responsibly. In reality, it may be manifestly impossible to know to any reasonable degree of certainty what the outcomes of various actions are, especially in the case of mediated actions and time and information constraints on the agent's decisions. Hence, under standard act utilitarianism, an agent might do something right or wrong without being aware that the action is right or wrong, respectively. For example, the car driver from the example above might have gone a different way that day, which would have prevented either the one or the two children from being struck by the car. Under standard act utilitarianism, the woman is guilty in the sense of doing the wrong thing, although we might not blame her for it. Under RPU, her tragic action can be

[5]A more developed argument defending RPU can be found in Cooley's *Transgenics, Technology and A Practical Moral Code.*

morally permissible. She did what was right, and because it was right, she deserves no blame for the tragedy.

Second, the emotive appeal of doing the best one can do is preserved by RPU. The agent who attempts to bring about the action that likely is the best she can do can begin to fulfill her obligations to always act right even if the act does not maximize good over evil. In the real world, we want people to try to do their best, but understand when they cannot achieve those goals through unforeseeable circumstances, such as mediated actions.[6]

Trolley Problems are so difficult to solve because two fundamental factors explaining why morality is possible for human persons are in conflict. One is doing the best one can, or the basis of consequentialism, which involves the cognitive parts of the brain. The other is respect for persons, or deontology, which is based more on the lower brain. If the sheer number of people were the only consideration we have for morality, then there should be no difficulty getting universal agreement on what should be done in each variation of the Trolley Problem. However, even when a solution is made to kill one person in order to save several others, there is a residue of negative emotion—generally guilt—that those making the decision have. I contend that it is the residue generated by their realization and appreciation of what happens to the deceased and the intrinsic value lost, even when they do what they think they should do.[7] The decision makers know what they think should be done, but they do not like it. If they were given the option of killing the one and feeling no regret versus killing the one and feeling regret, they would state that the former is superior to the latter. Most likely, they would find someone who did not feel that anything was morally defective because the latter does not seem to understand the moral situation in the first place.

In order to prevent people from being turned into mere numbers in utilitarian calculus, that is, overly privilege the cognitive part to excess, a version of Kantianism is useful. The Quasi-Categorical Imperative, or QCI, is:

> If, in doing the act, the agent does not treat any intrinsically valuable thing, such as a person, as a mere means or as having less value than the agent should think it has, then the action is *prima facie* morally right.

Instead of a detailed examination of what it means to treat something of intrinsic worth as a mere means, I will merely stipulate a reasonable set of requirements. First, to treat something as an end in itself entails that the action's agent must have primarily "pure" intentions and motives. Intentions are the goals of the action, while motives are what cause the agent to choose and perform the action. Both types of mental states are extraordinarily complex, as is what the agent is trying to achieve and why she is trying to achieve it. However, if all the intentions are put together, then a reasonable person would state that the organic whole formed by the primary intentions is good overall, and the organic whole of the motives has the same value.

[6]RPU, QCI, and the Practical Principle are developed and defended in greater depth in my book, *Technology, Transgenics and a Practical Moral Code.*

[7]Appropriate feelings will be discussed in a later chapter.

Second, the agent must have a good attitude to the action. Although the intentions and motives might be good overall, how the agent thinks about the action matters as well. We have a duty to interact with those who love us, and who we should love in return, such as our grandparents. Suppose that the primary intention is to make them happy, and the motive is the desire to do so. At the same time, the agent might resent the whole situation while not appearing to do so. In this situation, the agent has degraded himself by not being a better person and has not respected the grandparents.

Finally, each time an agent acts, she must have a feeling of respect for all those intrinsically valuable things involved in the action that she can be reasonably expected to know about. This feeling recognizes the correct value of personhood for each person as well as the value of those entities that are not persons but deserve respect for being ends in themselves, such as sentient animals. In today's hectic world, for example, many often forget that those who interact with them are people as well. The person who takes your order, cashes out your groceries, or fixes your mechanical items is a person with intrinsic value. Animals who feel pleasure or pain are also intrinsically worthy and should be treated as such. As we will see in the next section, their mental states can and are valuable in and of themselves. An appreciation of that fact is required; otherwise, the true worth of the intrinsically valuable entity is not sufficiently recognized. Although this requirement demands much from us, viz. paying more attention to others, it is difficult, at best, to understand an argument that would say it is generally or always permissible to not do so.

The immediately preceding claim brings up an important issue. Although RPU needs not be satisfied by all morally right actions that exist, QCI is a different case. I contend that QCI is a necessary component for any adequate moral theory on the grounds that rejecting it leads to a theory or principle without practical use. It is, in fact, firmly grounded in our moral platform in which emotions, such as empathy and altruism, are found, and which appears to be the result of how our brains are structured as well as our socialization. As has been shown before, it is necessary to have these emotions for our morality as well as to be motivated to act ethically. A person is willing to act on a moral belief only when the emotional part of the person's brain is involved in the decision making (Gazzaniga 2005, 167).

It is only when we have respect for intrinsic value that we have motivation to act ethically. Of course, this position denies the possibility of the amoral person who understands what ethics requires but is left unmoved by it. With this axiology and QCI, such a person cannot exist because she cannot understand what ethics requires of her—ethics requires her to be moved by it. In the real world, philosophical devices that call into question perfectly useful beliefs, theories and principles should be eschewed until it is shown that such a device is not deeply impossible given the available evidence that it is. Although this issue could use more development, the concession required here is that this axiology and principles are reasonable to adopt and make decisions with. If that is the case, then there is no need to take on all possible objections to whether or not it works in every case.

Finally, the Practical Principle (PP), states that both RPU and QCI must be satisfied in order for an action to be morally right.

> An action is morally right if the agent does not treat anything with intrinsic value as a mere means or as having less worth than the agent should think it has, when performing the action, and a reasonable person in the same situation as the agent would reasonably believe that the action will produce at least as much utility as any alternative to the act.

Although this standard is very high in most real world situations, the difficulty of satisfying both conditions is a guarantee that decisions about death are not treated lightly. If we are discussing the end of life of a person, especially when one person is deciding how and when another will die, then the argument for a particular conclusion should be as strong as possible. These stringent requirements do just that. For a death to be justified, it must be for the reasonable best, and at the same time, no one is treated as a mere means, which eliminates the problem of using people merely to achieve the greatest good for more people.

2.3 A Theory of Value[8]

For William Frankena, a thing is intrinsically valuable if and only if it is good in and of itself or good because of its own intrinsic properties (Frankena 1973, 82). In other words, if something is an end in itself, as Kant would define it, then it is intrinsically valuable. Frankena claims that experiences of pleasure and certain other states of mind are intrinsically worthy when they are instantiated by thinking beings. Fred Feldman recognizes that these sorts of experiences are states of affairs that must incorporate the experiences in some way (Feldman 1986). Pleasure, for instance, is not an object that can exist independently, and we should not act as if it can. If we do, then we fail to recognize the reality that actually counts in ethics. When we are referencing pleasure, then what we actually should mean is that a sentient creature is pleased about something.

The recognition of the complexity of sensory and other experiences will bring in greater complexity in evaluating intrinsic worth. For instance, if the person is rather evil and does not deserve pleasure, then the person receiving it will be less valuable overall than a neutral desert person feeling pleasure (Feldman 1986). Here is where we should begin to see that context matters a great deal in the actual intrinsic value that exists. Moreover, the context we choose to use to understand a situation can also affect the value. For instance, if we are merely looking at someone pleased about something, then anyone pleased about the same thing and in the same way will instantiate an equal intrinsically valuable state of affairs to anyone else with that state. However, if we look at the two people overall, then we can see that the state of affairs in the larger context can alter the overall values significantly, as happens in the example of the evil person being in that state.

[8]Brennan and Lo (2003) provides an excellent overview of value in nature.

Even though it might be incredibly complex—and unable to be described in any satisfactory way to most axiologists—morality requires some sort of end or some intrinsic value; otherwise, it would have no purpose and would not provide motivation to be ethical. Every theory "needs a foundation, something that instrumental goods aim toward." (Russow 2002, 98). If everything is merely an instrumental value without end, then there is no good reason to act or be good, other than to move from one extrinsic value to another without end. Ethics requires intrinsic value so that the principles of right or wrong action and good or bad things have content to them. Intrinsically valuable things help "craft" the theory, although they do not determine its entirety (Ibid.). For example, utilitarianism needs good and bad to do its cost-benefit analysis. If we adopt a value theory that states that the ultimate good is virtue and the ultimate evil is vice of any variety, then what we should do and how we should be are ultimately based on obtaining and maintaining these ends in themselves. If the principle of right and wrong is classic act-utilitarianism, then virtues need to be maximized in some way, perhaps in sheer number or in the strength with which people instantiate them, while vices are minimized in order for an action to be ethical. Hence, both a value theory and normative theory are necessary components to a moral code or overarching ethical theory.

Intrinsic value has another role to play. "It is commonly agreed that something's possession of intrinsic value generates a *prima facie* direct moral duty on the part of moral agents to protect it or at least refrain from damaging it." (Brennan and Lo 2003). That is, the recognition of intrinsic value gives an internal impetus to behave or act in a certain way consistent with the value of the thing. It is the respect we automatically give to those things that have intrinsic worth, which is not simultaneously awarded to extrinsically valuable things, that makes us do what we consider to be ethical.[9] For example, if a person is intrinsically worthy, then all moral agents should recognize that value in the proper way. If a person is merely instrumentally valuable, then that worth can change as the end changes. In this case, the person may be treated as a mere means because that is all she is good for. In fact, she only has value just as long as she is useful for the purpose of achieving some end. Treating any person in this manner, however, is morally repugnant because her intrinsic worth makes the conclusion repugnant; we do not naturally treat things that are valuable in and of themselves without respect. Hence, without intrinsic value, the enterprise of ethics cannot get off the ground.

How to rank the intrinsic value candidates is difficult. Everyone seems to have their own idiosyncratic view about what is intrinsically valuable, why it has that value, and how intrinsically valuable entities, if there are more than one, rank against each other. Hedonists argue the only worthy thing is pleasure, while pain is the sole bearer of intrinsic disvalue. Hastings Rashdall and W.D. Ross propose their own complex, pluralistic theories of worth with multiple things bearing intrinsic

[9]This is one reason that Aldo Leopold argues in the *Sand County Almanac* that the land is alive in the way that a person is. By giving intrinsic value to the environment, then people will automatically show it the respect that they do any person.

value that might or might not be commensurable (Rashdall 1923; Ross 1998). The result is that one value system's foundations could be contradictory to that of another system, and both could be legitimate, as long as they are consistent with the moral platform developed in Chap. 1. Once again, if we think that all normal citizens should be able to communicate with each other on vital issues affecting all, then there have to be some universals on which to base their understandings, discussions, and decision makings.

In order for morality to be possible and practical, certain facts must exist. First, there must be universals such as human sympathy or empathy upon which morality is based (Hume 1990). As we saw in Chap. 1, a moral platform is universal to all moral agents. This platform explains why general moral principles and values look the same wherever they are formulated. For example, Bernard Gert argues that ten moral rules cover all ethical situations (Gert 1998, 20).[10] Gert's assertion is important for two reasons. First, it helps provide good evidence that a commonality exists between all people on basic morality. Second, and tying into the claim for universality of basic moral beliefs or whatever provides the foundation for ethics, these complex rules, such as prohibiting killing, must reflect more basic beliefs about value. For example, killing must be a bad thing for some reason. Perhaps what is disvaluable here is the bad mental state(s) of the agent who would kill, and what it does to the person's flourishing. Perhaps it is the loss of worth that occurs when the thing is killed. It might be the negative effects on the society and the relationships between citizens required to have a flourishing society. It could be a variety of things, but the important point is that there has to be some motivation to say that one should not kill. Something has to underlie each of these rules for them to make sense to us, unless it can be shown that there is a faculty in the human brain that provides us with the complex rules independently of any underlying intrinsic value (Hauser 2006, 43–55).[11] These universalities provide some help in asserting intrinsic value. If vice is universally condemned, as it has always been, then it is a good candidate for an intrinsically disvaluable entity. If something like virtue has been universally applauded, then I will assume it has intrinsic value. In addition, evaluating an intrinsic value candidate's actual worth should be based on whether or not it interferes with flourishing. If it does, then there is at least *prima facie* reason to reject it as intrinsically valuable. If it does not, then it might be neutral or valuable, depending on whether it promotes or sustains flourishing.[12]

---

[10]Each of these rules can be supported by the Practical Principle. What makes PP superior to Gert's 10 rules is that PP is more flexible in capturing all the moral elements of each particular situation.

[11]Since the work in Hauser's book was not accused of being falsified, then it can be used here. Besides, the citation merely illustrates one way of thinking about how ethics is possible.

[12]Those who wish to assert that there are no universals will have a great deal of difficulty criticizing the position presented here. If there are no universals, then we will not be able to have a meeting of the minds to debate the issue. Hence, it is a waste of time to do so. Those who believe that there are universals will be able consistently to question whether I have the right ones.

The approach's inclusiveness could lead to difficulties, but the benefits of its practical nature should more than make up for them. The main problem will be working with so many different values at once, but we think about ethics using multiple ideas, rules, and beliefs. These should be incorporated in any adequate value theory, otherwise we will quickly encounter a number of counter-examples that cast doubt upon the theory. More importantly, if one of morality's ends is individual and societal flourishing, then it is necessary to create an ethical system that better leads to those ends rather than one that is too narrow. An overly circumscribed system will be unappealing to many citizens, in part, because it does not reflect their values. Hence, they are more likely to focus on its shortcomings rather than to identify the relevant moral features and then come to a decision that is reasonable.

What I will be arguing here is a very broad, inclusive axiology. Instead of requiring that anyone proposing that something has a particular value must prove that it has that value, it will be assumed to have it until it is reasonable to assume that it does not have this value (the Benefit of a Doubt principle). Basically, what is being proposed is a rejection of a Western approach to ethics.

> In modern industrialized countries like the United States, the assumption is often that things have only instrumental value unless there is evidence to the contrary. This is in sharp contrast to other cultures that make the opposite assumption; they assume that everything is deserving of respect, and that we ought not to use things as mere means unless there is a good justification for doing so. (Hinman 2000, 563)

One reason for taking this stand is to avoid situations in which entities with intrinsic value are not treated as they should be. The second is to account for the rules in Gert's list. Each of them incorporates assumptions about intrinsic value. How those work out will say something about the value theory detailed in this chapter.

In order to avoid counter-examples in which something that all reasonable people would claim to have intrinsic value is treated equally to an object that only a few would dare argue is an end in itself, a hierarchy of value will be introduced. It is still possible to make distinctions between the worth of various objects; using the approach that a value theory should be useful to individual and societal flourishing requires such discrimination in order to achieve the ends of thriving.

The difficulty, of course, is gaining general agreement as to what has value and to what degree it has value. The most practical way to solve this problem is to rank various items according to the Benefit of the Doubt principle, the moral platform from Chap. 1, including how controversial it is to claim that the entity has intrinsic value. Those less contentious objects are more likely to be ends in themselves; hence, they are ranked as having greater moral worth than those that are more controversial. Human persons as intrinsically valuable are the least divisive contention, thereby making them the category of things with the highest intrinsic value whose interests should generally be pursued before the interests of other intrinsically worthy objects.

Besides establishing which things have intrinsic value and their ranking, it is important to explain why they have the value that they do and how to evaluate it.

As Frankena's intrinsic value definition entails, an object's worth comes from a number of sources, including the teleology of flourishing ***and*** the properties and states of affairs it instantiates. For example, being in a state of pleasure can make a specific person more valuable at that moment in time than an otherwise identical entity, all things considered, but one who is in a neutral mental state or a state of pain. The complex value theory developed below might be guilty of hand-waving, but how it plausibly works in general should be clear by the end of this chapter.

2.4 Beginning to Develop a Theory of Value

For all its positives, the Benefit of the Doubt principle will generate a value theory that will make the task of evaluating actions and moral duties much more difficult than if a simpler axiology were adopted. A more complex value theory requires greater work both for the evaluator and anyone asserting the axiology's appropriateness. The complex, pluralistic value theorist has to first propose the inclusion of each class on the list, and then explain how an object's value is calculated, including, but not limited to, commensurability. For example, W.D. Ross's axiology is, from least to most worthy, desert appropriate pleasure and pain, right opinion and knowledge, and virtuous disposition and action (Ross 1998, 134–141). Ross went on to say that an object from one level is incommensurable to any other object from any other level (Ross 1998, 150). No matter how much desert appropriate pleasure is produced by an action, for example, it can never equal the smallest amount of knowledge or virtue. Other philosophers have also ranked the value of categories based on intrinsic worth and other considerations (Warren 1983; Elliott 2000; Frey 2000). The axiology developed here will follow their lead.

The approach taken in developing this multi-layered theory is to try to start with the smallest components of intrinsic value, and then work toward the larger ones. While doing this, keep in mind that the aspect or context in which the intrinsic value is considered may have an effect on the evaluation. An example might be helpful here. When considering the value of a plant, we might first examine the plant for what it is in and of itself. This would be limited, but it would include its genetics and how it looks, for example. The surrounding environment would not be a consideration because the context lens is restricted to the plant itself. However, we might want to consider the immediate environment of the plant as well as the plant. In plant sciences, a weed is merely a plant out of place. Hence, when considering this plant in this environment, then it might be a weed or not according to its impact on the environment. The plant, therefore, has additional features only because the context has been broadened to the plant in a larger whole. If an even greater aspect is taken into account, then the plant might go from being a weed to being a plant in place given the larger biosystem. Suppose that the plant is actually more beneficial in a larger context than are the plants in the smaller lens environment. Those plants might expend too many resources to make them suitable to producing greater good than the plant that was a weed under one aspect, but is now a plant in place in the wider view.

2.4.1 Mental States

William Frankena claims that a number of properties are intrinsically valuable if they are instantiated in moral agents. The members of Frankena's list are not in themselves ends; rather, "what is intrinsically good is the contemplation or experiencing of them. In themselves, they are inherent rather than intrinsic goods." (Frankena 1973, 89). Frankena has recognized that the basic fact that it is the property's instantiation, not the properties themselves, which has intrinsic worth. Pleasure, for example, subsists only if someone is experiencing it. In other words, there must be a state of affairs in which an entity is pleased about something in order for an intrinsically valuable state of affairs to exist. This is an important point. Although it would be much easier for a cardinal calculus of value for pleasure, pain, and the other types of properties to exist independently of a perceiver, the fact of the matter is that they cannot. Hence, the entity that experiences them and the facts about that creature will have an impact on the intrinsic worth of the mental state experienced as will be seen below, but for now we will consider the smaller components of intrinsic value.

According to Frankena, many different states of affairs types have intrinsic worth. His list of intrinsically valuable things is long, but is a good base built upon the moral platform for a theory of value:

(a) Life, consciousness, and activity.
(b) Health and strength.
(c) Pleasures and satisfactions of all or certain kinds.
(d) Happiness, beatitude, contentment, etc.
(e) Truth.
(f) Knowledge and true opinion of various kinds, understanding, wisdom.
(g) Beauty, harmony, proportion in objects contemplated.
(h) Aesthetic experience.
(i) Morally good dispositions or virtues.
(j) Mutual affection, love, friendship, cooperation.
(k) Just distribution of goods and evils.
(l) Harmony and proportion in one's own life.
(m) Power and experience of achievement.
(n) Self-expression.
(o) Freedom.
(p) Peace, security.
(q) Adventure and novelty.
(r) Good reputation, honor, esteem, etc. (Frankena 1973, 87–8).

A person being pleased about something at some moment of time, therefore, is intrinsically valuable in that specific way. Certain animals may instantiate these experiences, such as having pleasure, as well. Their experiences would then be intrinsically valuable or disvaluable accordingly.

The different states of affairs' intrinsic worth can come in degrees. For example, knowledge is intrinsically valuable for Frankena, but we need not claim that all knowledge has the same worth. The difference in intrinsic value stems from the fact that some pieces of knowledge are more intrinsically important than others (Bradley 1883). Trivial information, such as that of what people eat for breakfast, in most instances, does not have as great an intrinsic worth as general abstract principles about the nature of humanity, ethics, or physics. The measure, once again, is flourishing. We must ask ourselves which pieces of knowledge make a person better just from the possession of them. Although there is some difficulty in distinguishing the extrinsic worth of knowledge from its intrinsic value, e.g., knowing human nature will allow people to make much better choices about their lives, there does seem to be a difference based on the prestige a person has. Philosophy professors are often misunderstood in society. Average citizens have no detailed information about what they do. However, that does not prevent most people from believing that philosophizing is valuable, although it will not normally get someone a job on its own. Hence, there is respect for knowledge even though it might not be thought to be extrinsically useful. The other classes of intrinsically valuable things can be evaluated in a like manner.[13] However, I do think that I have discovered the truth of the matter, but I am limited by the arguments of which I can avail myself. Since the arguments do not have the convincing force of a mathematical proof, I cannot defend them with that level of certainty. But, for the purposes of this dissertation, I will proceed as if I have adequately proven my case.

As can be seen from the list, Frankena does not limit himself to particular instances of mental states, such as being pleased about something at a particular time. Instead, the state of being alive is intrinsically valuable. Furthermore, one's life as a whole can be intrinsically worthy, including, but not limited to, having virtues or dispositions to act ethically in certain ways given certain stimuli.

Positive intrinsic worth does not exhaust the list of what has value in and of itself in the universe. Certain states of affairs are intrinsically disvaluable. If one experiences one of them, then the experience is bad in and of itself:

(a) Sickliness and weakness.
(b) Pains and dissatisfactions of all or certain kinds.
(c) Unhappiness, discontentment, etc.
(d) Falsity.
(e) False opinion of various kinds, lack of understanding.
(f) Ugliness, disharmony, disproportion in objects contemplated.
(g) Unaesthetic experience.

[13]Someone may object that the axiology I have developed is subjective. Based on my background and education, she may argue, I chose to include these things in my value theory rather than others.

(h) Morally bad dispositions or vices.
(i) Mutual dislike, hate, enmity, antagonism.
(j) Unjust distribution of goods and evils.
(k) Disharmony and lack of proportion in one's own life.
(l) Weakness and experience of failure.
(m) Lack of self-expression.
(n) Loss of freedom.
(o) Strife and insecurity.
(p) Sameness.
(q) Bad reputation, dishonor, lack of esteem, etc.

It should be noted that death, unconsciousness, and mental inactivity are not included in the list of intrinsically bad experiences. Since the entity is dead or otherwise mentally inactive, then such states cannot be mental states or other states of affairs instantiated by an experiencing entity, such as a moral agent or animal.

Our ultimate ends and moral platforms support the mental states' intrinsic value and disvalue, especially if we put that platform in the context of flourishing for ourselves and others. In order to have the morality we have, then we must have a particular moral platform built from nurture and nature and the ultimate end of morality is the flourishing of individuals and groups. In our moral platforms, we as individuals and as species members in society have specific feelings and motivations, such as compassion, empathy, fairness, status, punishment, and a natural tendency to favor one's own group and those for whom one cares over others (Bloom 2013). We are altruistic and have other sympathy-related traits such as attachment, succorance, and emotional contagion. We have norm-related characteristics, reciprocity, and getting along characteristics,[14] and whatever else makes our morality possible and universal. This platform is necessary for Frankena's mental states to exist and have the value that they do, for without that platform, our morality would not be possible. Moreover, a person would be hard pressed to instantiate all 18 intrinsically valuable states and not be flourishing in a good life, whereas someone who had only the 17 intrinsically disvaluable mental states for his entire life would have a miserable existence. Hence, the mental states have their independent intrinsic values while simultaneously being necessary for the ultimate end of flourishing.

Some members of Frankena's list are better understood as short-term mental states, while others are associated with longer-term mental states. For instance, aesthetic experiences are generally more momentary than are the mental states of peace and security which usually are based on experiences built up over time. When first seeing a significant painting, there is an aesthetic experience in the moment that cannot be repeated on additional viewings. For instance, the first sight of a Van Gogh landscape can be an extraordinary experience, an awakening that is not and cannot be duplicated when seen again, since the perceiver knows that the

[14]See De Waal (1996, 210–1).

experience is coming and is already building anticipation. Basically, great arts' first impact on people is like being startled, and being startled is not a state that can be maintained.

On the other hand, feelings of peace and security are longer-term mental experiences based on a steady set of experiences that will eventually lead one to feel peace and security. If a person lives in a good neighborhood with little strife or crime, then peace and security are likely. Those background feelings are threatened only when there is sufficient challenge to the experience to cause the agent to question them. For example, one act of terrorism or a series of actions that make one feel as if there is sufficiently increased danger to herself or others might destroy them. If, however, there is one small crime in a span of some years, e.g., one house burglary in a 20-year period, then the agent can still maintain peace and security.

Frankena realizes that a state of affairs' intrinsic worth does not reside solely within a person's immediate mental states and properties. Rather, some intrinsic value arises when the moral agent is in her environmental context. For example, a person feeling peace and security is intrinsically valuable for Frankena. However, it is impossible for a person to feel either without taking into account her surroundings, regardless of whether they have the properties the person ascribes to them. In order for her to feel peaceful and secure, she must think that the entities in the area, including other persons, are of no real threat to her. If she is afraid of bears, for instance, and thinks one is in close proximity, then she can feel neither beneficial state because she feels afraid and insecure. Hence, the agent's assessment of her situation and its effects upon her are integral to her overall intrinsic worth.

Another difference between immediate or short-term mental states and those that are longer held is that some are more vivid or immediate than others. If someone is in pain, for example, then the person is aware of that feeling for every moment the person has it. Other mental states do not lend themselves to that immediate and continuous apprehension. Happiness, for instance, is an extraordinarily complex concept; some even argue that it does not exist. A person may be happy in the moment, i.e., feel happy because something is causing the person to feel that way for that moment, or the person can be happy for a much longer spate of time. The latter is often associated with contentment with one's life. The former is more momentary, but perhaps more noticeable to the entity experiencing the feeling or mental state. It is more intense an experience so that the person realizes that she is in a state of happiness, and it usually allows her to be able to identify the happiness source. On the other hand, the longer-term happiness that is associated with contentment might not be as noticeable unless the agent reflects on a longer period of her life. Since it is constantly present, then it becomes part of the background medium until the agent's mind brings it consciously to the forefront. That does not entail that the agent is not happy when she is not thinking about it, but is not aware it is there, much the same way that we do not realize what happened on a routine drive to work. We filter out the standard so that we can pay closer attention to the unusual or more demanding that requires more of our immediate attention. However, they are all intrinsically valuable states of affairs.

When comparing value for longer-term mental states and those that are shorter, no hard and fast rule exists to establish which has greater worth. Once again, we will rely on reasonable people thinking about these and ranking them according to how reasonable it is to believe that one state is more worthy than another in context of the agent flourishing in the particular environment in which she resides. Those more conducive to the agent flourishing are to be prized over those less likely to do so. For example, suppose that an agent can have a very pleasurable physical experience that has high intensity for about 1 h. On the other hand, the agent can be content or have appropriate self-esteem. The high intensity pleasure has value in and of itself, and as we will later see, alters the intrinsic value of the agent in the moment and in the longer term. This pleasure seems to aid the agent's flourishing at the moment. If it has no longer-term consequences in the person (for instance, if he never thinks about it again), then the experience remains only part of the flourishing in the moment. However, the longer-term mental states of contentment and self-esteem can be felt in the moment, but seem to make more sense when considered in the longer term. In general, these are more worthy than more immediate mental states because of their longer duration and more integral role in an agent's flourishing. Although they might not be as intense, they have greater bearing on who the person is and the value of the person's life when considered on the whole.

From his various works on the subject, it is apparent that Fred Feldman believes many things have intrinsic value. Unfortunately, Feldman is not always clear about what they are. Regarding his complex axiology, Feldman says "that [his] intuitions about what has intrinsic value are not sufficiently firm." (1986, 35).[15]

However, what is made obvious is that Feldman holds that every episode or state of affairs in which someone is pleased about something is intrinsically valuable. On the other hand, all states of affairs in which someone is in pain are disvaluable. Among other things, what makes Feldman's value theory interesting is his recognition that states of affairs can affect other mental states' intrinsic value.

When someone is pleased or pained about something, for example, the intrinsic value of the state can be isolated when we need to talk about it in a very narrow context, but there will be more to the story if we expand the context. More specifically, the value of the overall pleasure or pain state of affairs is not determined solely by pleasure or pain. The intrinsic value of a pleasure or pain must be adjusted according to other states that will affect the overall value, like justice. That is, if someone were to receive what he does not deserve, the value of her receiving it is a worse state than otherwise would have existed. On the other hand, an agent receiving what she deserves is a better state than would have otherwise occurred. Feldman proposes six plausible principles which adjust the values of pleasure and pain states of affairs according to just desert:

(P1) Positive desert enhances the intrinsic goodness of pleasure.
(P2) Negative desert mitigates the intrinsic goodness of pleasure.

[15] I will show later that not providing an adequate catalogue of intrinsically valuable things may pose a problem in regard to Feldman's normative theory.

(P3) Neutral desert neither enhances nor mitigates the intrinsic goodness of pleasure.
(P4) Positive desert aggravates the intrinsic badness of pain.
(P5) Negative desert mitigates the intrinsic badness of pain.
(P6) Neutral desert neither enhances nor mitigates the evil of pain (Feldman 1995).

Principles 1, 2, 4, and 5 are reminiscent of a claim G.E. Moore makes in *Principia Ethica*. Moore contends that adding more pleasure to certain situations can make the overall state worse rather than better (Moore 1998, 210–213).

Examples of P1 and P2 in practice are helpful in understanding how Feldman adjusts utility for justice. Suppose that Faith is a good person who at this particular point of time deserves ten units of pleasure, and that Faith receives what she deserves. According to P1, this positive desert state of affairs' intrinsic value is greater than the value of Faith merely receiving ten units of pleasure. Feldman doubles the original ten units for it (Feldman 1995).[16]

Next, suppose Ami does not deserve ten units of pleasure but receives them anyway. She is a bad person who actually deserves ten units of pain at this time. According to P2, the intrinsic value of Ami receiving ten units of pleasure is mitigated by the fact that she does not deserve them. "In such cases, pleasures—of any intensity—have no intrinsic value." (Feldman 1995). According to Feldman's Graph C, the intrinsic value of such states would always be a small negative number.[17] Principles P4 and P5 work in similar ways affecting states' of affairs intrinsic value.

Frankena's other intrinsically valuable mental states can be affected similarly by the intrinsically disvaluable states of affairs. As we have more and more complicated states of affairs involving more and more intrinsically valuable and disvaluable mental states, we will have more complex evaluations as one mental state of affair might enhance, mitigate, or leave unchanged the values of other states. It should also be noted that other combinations of intrinsically disvaluable or valuable states of affairs can affect value in significant ways. If valuable states of affairs are combined, then in this context, they are more worthy than either one individually. The upper limit seems to be what a person can experience. If the person is at the limit of being able to experience something, then the combination value is determined by what is actually occurring in the person's mind at that time. It might even turn out that the combination becomes a negative experience, as occurs when someone's pleasure centers are overstimulated. On the other hand, combining negative mental states will make the overall experience worse. This will also be limited to what a person can experience.

Given the complex nature of these combinations and the effects of their interactions and interrelations to each other, it would be best to think of them as organic wholes in much the same way that G.E. Moore defines the term. Moore held that a

[16]Of immediate concern is the cardinal nature of the calculus, but that issue may be put safely aside until later in the chapter. These numbers merely indicate that there is a significant alteration in intrinsic value once the scope of what is considered is increased.

[17]Feldman's particular adjustments in the value of his example states of affairs may only apply to them. However, I will use them as guidelines in adjusting utility in other states of affairs.

thing may have little or no intrinsic value on its own, but when part of an organic whole, it may change the value of the whole out of proportion to the value of the part itself (Moore 1998, 187–8). Although each part of the whole is necessary to the existence of the whole, the part's essential natures do not entail that they have intrinsic worth on their own, or that the sum of the values of the parts is proportional to the value of the whole. According to Moore, knowledge has little to no intrinsic value (Ibid., 196), but when it is part of the organic whole of the appreciation of beauty, knowledge makes the whole much more valuable than it otherwise would have been. In this particular case, knowing that the beautiful thing exists creates a better overall situation than if the beautiful thing were merely believed to exist. Hence, organic wholes are entities with supervening properties, which makes them incapable of being reduced to their component parts and retaining the same properties.

2.4.1.1 Comparing Values of Mental States as Organic Wholes

How a reasonable person should go about the business of comparing value needs to be connected with Feldman's mitigation/aggravation modifiers. Although the fundamental principles P1–P6 are correct, how these work out in practice is based on an assumption that is a bit troubling. Feldman assumes that the values can be determined using cardinal calculus. If true, then our contentions that simple and complex states of affairs should be understood as organic wholes, and that intrinsic value should be understood in terms of organic wholes, are mistaken. Instead of the value of the whole not necessarily being equal to or proportional to the sum of the whole's parts, simple cardinal mathematics can determine the worth by doubling or halving the value of the base pleasurable or painful state of affairs.

It is very appealing to reject organic wholes. Instead of relying on the fuzziness of what a reasonable person would determine in the situation, which might be contrary to what another reasonable person thinks, there is an answer that is the same for all reasonable persons. If the value of a state of affairs is X for one person using the objective approach, then it must be X for all people. The agreement between persons on moral values will go a long way in helping make decisions which everyone can use. If the facts are the same for everyone, then it would be irrational to deny them merely because one's opinion on the matter differs. Hence, it could be more pragmatic to use cardinal rather than ordinal calculus to achieve the end that we desire because reasonable people can understand and accept the results; it is the same result for all of them regardless of what they feel about them.

Although what Feldman claims in general about intrinsic value being altered is correct and vital to any adequate theory of value, it neither fully recognizes the impact context has on worth nor recognizes that reasonable people can reasonably disagree about value and still maintain their reasonableness. First, granted that it is simpler to isolate these various states of affairs from each other, the truth of the matter is that doing so may mislead or misrepresent the true value in the situation. Once again, like the plant that would be a weed in one context but at home in another, we can understand worth but only in the context of some scope of which we are aware. We can have the simple state of being in pleasure, and evaluate that, but the overall value of the larger state of affairs might not be apparent until the lens is broadened to include a larger context. If desert can mitigate a pleasurable state of affairs' worth, then it is reasonable to believe that any positive state of affairs may have its intrinsic value altered in more inclusive contexts. For example, having an aesthetic experience is good in and of itself. However, if it is not deserved, the overall value in this context is considerably reduced. Deserving such an experience might increase its overall worth.

Second, in the absence of any hard proof that intrinsic value is based on a cardinal calculus, then it is unjustified to favor it over an ordinal calculus.[18] Although cardinal calculus is useful in its universal objectivity, it still faces the fact that a diversity of opinion can be legitimate in ethics and moral decision making. If the cardinal calculus is true, then there should be no disagreement in the same way that it makes no sense for people to claim that one person plus another person is not equal to two persons. However, reasonable people understand that although morality for all humans is universal on some primitive level, there is room for morally permissible variations between societies and individuals. In the same context, two people might come to contrary opinions and each be morally justified in her conclusions.

A much better approach to determining something's intrinsic value is the Comparison Test (CT). As Moore's Isolation Test does, CT relies on thought experiments in which reasonable people compare objects to each other. Unlike Moore's test, CT requires the comparison be done relative to a stated or understood context. For example, if the scope is that of the universe or world as a whole, then the comparison must be of world pairs. In each world pair, two worlds are similar except for this major difference: one contains the thing whose value is to be determined and the other does not. For example, if considering a mental state of affairs in the universe, then Jack being pleased about X is compared to a world in which Jack has a neutral view of X, or Jack is experiencing a neutral state of affairs.[19] The question is which world is more valuable to a reasonable person. Regardless of the object to be evaluated or the context in which to evaluate it, it is important to treat the thing as it is within the context rather than modifying it in any way. The thing itself may be simple or complex. If the thing is complex, it ***must*** be considered only as a whole within the chosen context. That is, the whole must always be treated as simple because it cannot be reduced to its parts and still maintain its properties.

A thing's intrinsic worth is determined by deciding whether or not the situation-context containing the thing under consideration is better, worse or

[18]People might favor cardinal calculus because it provides clearer answers and fits with the part of the brain that favors consequentialism.

[19]Of course, we could also compare Jack being pleased about X and Jack being pained about Y where X and Y are not identical.

equally valuable to the situation-context that does not include the thing.[20] If the former situation-context is intrinsically better, then the thing under consideration is intrinsically valuable. If the situation-context is intrinsically worse, then the thing is intrinsically disvaluable.

We can use the same test to determine more complex mental states involving more than two simpler intrinsically valuable or disvaluable mental state types, such as an agent receiving pleasure when she deserves an appropriate amount of pain. In this situation, we could compare a situation-context of the person receiving pleasure when she deserves pain to that of a person receiving pleasure when she deserves pleasure or is desert neutral when it comes to pleasure. We would probably rank, from highest to lowest, the intrinsic values of the three possibilities as positive, neutral, and negative desert situation-context. The test becomes far more complicated—thereby making it far less likely to gain consensus—as we increase the number or the differences between the types of mental states involved. It would be difficult, for example, to compare the state of receiving deserved pleasure to that of self-expression through an aesthetic experience. Nevertheless, what we have seems workable for our needs.

A metaphor used to help explain gravitational pull can also be useful to explain the impact that different states' of affairs intrinsic value has on the value of other states of affairs and the whole. First is the actual physics metaphor. Suppose that a flat, taut-looking rubber sheet—space-time—stretches out in its infinite and bound way in front of us. If we press our hand down on it, then the sheet sags around our hand and cups it while stretching out in all directions from the hand. Furthermore, suppose that we have two steel spheres: matter-energy. One is twice the size and twice the weight of the other one; for our purpose, make one five inches and one pound, and the other ten inches and two pounds. If we place the smaller sphere on the sheet, then the sheet acts in much the same way that it did for our hand. If we then place the remaining sphere on the sheet four feet away from the first, then the sheet reacts by becoming tauter in all directions. As a result, the smaller sphere depression in the sheet is decreased. As we move the larger sphere farther away, then the first sphere's depression becomes greater and smaller as the larger sphere is moved toward it, until it rolls into the larger sphere's depression. This, in a way, is how intrinsic value works when we have one intrinsically valuable or disvaluable state of affairs affecting the value of the other and being affected in return.

For intrinsic value evaluation, the sheet and spheres represent two different things. First, the sheet is the context. In this case, it will represent the psychological person's mind that is being evaluated, but in other situations it could represent other contexts.

Second, the spheres are the states of affairs that affect intrinsic value. Those that are essential to the person's identity are placed closer to the "center" of the sheet because that placement will have more effect on the sheet as a whole than locations

[20]Care must be taken here in order to determine whether or not the thing is intrinsically valuable rather than extrinsically, inherently, etc.

further out. Those states of affairs that are important are between the essential states and the trivial states. In addition, the spheres' or states' of affairs distance from each other on the sheet is determined by their effect on each other, although since these are organic wholes, an effect on all the others is assumed even for the smallest, most distant one. Those more intimately tied together, such as desert of pleasure and the pleasure state, will be closer together than less connected states. In addition, their respective weights are determined by the intrinsic value and disvalue of the localized state of affairs, e.g., being pleased about something (the greater the intrinsic value, the heavier the weight to be used). When all of the relevant spheres are assembled properly, then the reasonable person can use the image to compare to another entity.

Although the metaphor has weaknesses, such as not using ordinal mathematics, the metaphor works in showing that a person's mental states and instantiated states of affairs are interconnected in a certain way. They do not change the spheres themselves in most cases, although that can happen when a sphere's size or weight is increased or decreased. When a sphere changes, or is added or deleted, then all the others change simultaneously. Perhaps the change is more noticeable when a large positive sphere is added, but it might be minor as when a distant small state is eradicated. The point is that each one is dependent in a way on every other one. This claim is vital when considering the value of people in the context of their society and in their world. If the person has many relationships or has a great deal of impact, then she is a large, heavy sphere that will have a great deal of influence. If the agent is isolated and alone, then her influence will be small but will still exist. These ideas will be useful when defining persons and distributed reified personas considered in the next chapters.

Any thoughtful person reading about the Comparison Test will realize that it is not without its difficulties, but I do not see how to make it strong enough to answer any challenge posed to it. I attribute this problem to the difficulty of establishing a value theory. As Moore said, the best that one can do is to present one's arguments and then let each person decide whether or not she is convinced by them. If she is not convinced, then there is no further recourse to take. Since there do not seem to be any logical proofs to determine the value of a thing, perhaps it actually is impossible to logically prove which things have what value. The best one can do is to evaluate the test on the grounds of reasonableness. That is, can one or more reasonable people use the test to make good decisions about flourishing? Moreover, can the test be used to make other reasonable people at least understand why the original agent came to the decision she did and why it is reasonable for that person to come to her conclusion? If both these measures can be met, then it would be difficult, at best, to show why the test is not a legitimate method of intrinsic value estimation. Perhaps there is an evaluation scheme that is more useful than this one to such a degree that no reasonable person would reject that method in CT's favor. It also might be the case that there is a yet undiscovered or underappreciated test that is capable of establishing a competing criterion as the one that all reasonable people must use to make ethical decisions, or else they lose their classification of being rational people as well.

Until that moment in time in which a better criterion is established, I will use the Comparison Test to determine an object's value in comparison to others, especially because the scheme works so well in cases in which there are combined mental states.

2.4.1.2 Virtues and Vices

Virtues and vices can come in two varieties of mental states. Like pleasure and pain, they can be immediate as when they are being used for some purpose by the agent, e.g., acting wisely. Virtues and vices can also be dispositions to behave in a certain way; they can exist without being used or noticed by the agent who has them. For example, a person might be brave but need not be acting bravely at every moment to be brave.

In order to show the complexity of Frankena's mental states of affairs as intrinsically valuable and disvaluable, I'll briefly consider how virtues and vices work as values. Virtues and vices, or to connect them to mental states of affairs or the instantiation of them in a person's character, are intrinsically valuable or disvaluable as well. There are two ways to confer value on the entity possessing them. First, an agent might have an episode of virtuous or vicious consciousness, much like a person may be in a pleasurable or painful state. While performing a particular action, for example, the person might exhibit courage, yet not be courageous overall because he has yet to acquire the habit of acting courageously in situations in which it is appropriate. Second, having the disposition to act in a certain way, either virtuously or viciously, can enhance or mitigate the intrinsic value of a person who has that particular trait.

Virtuous consciousness, as I use it, not only includes certain feelings and attitudes that the agent has, but the agent's intentions and motives as well. Roughly, virtuous consciousness may be defined as:

> **Virtuous consciousness** = df. Any particular feeling, attitude, motive, or mental state of affairs that is caused by at least one virtue or is part of having or acquiring at least one virtue.
>
> **Vicious consciousness** = df. Any particular feeling, attitude, motive, or mental state of affairs that is caused by at least one vice or is part of having or acquiring at least one vice.

If an agent, at a particular instance, intends to do what is just, then she has a virtuous consciousness of justice. Of course, this does not entail that she has the virtue of justice in any of its forms. Rather, we realize that virtues are generally acquired over time; a virtue can be acquired by an agent through repetition until the action becomes habit. On the other hand, the agent might not need the repetition if she can become virtuous merely from seeing the value of having the virtue, and then through whatever reasoning and other mental processes she has, adopting the characteristic. However, it is the former manner of character alteration that is relevant for virtuous consciousness. In order to be courageous, then the actions that make the person courageous must themselves be courageous. That is, the person must do what she thinks is the right thing despite her fear of doing it. The courage shown by the individual in the situation is in fact intrinsically valuable, and helps make the action a courageous action.

Intentions have an impact on the intrinsic value of worlds based on the mere fact that more goodness tends to make things better and more badness makes things worse. The presence of good intentions can increase the intrinsic value of a world as a whole as well as that of a complex mental state including them and the action that is performed. The same is true for good motives on the agent's part. On the other hand, the absence of any of this good consciousness makes the value of a world less than it would otherwise have been. The presence of vicious consciousness lowers the world's value even further. Several examples help to better explain what these types of consciousness are and what intrinsic values they have.

Suppose that ten-year-old John has a choice between two alternative worlds. In the first, he finds a wallet filled with money and returns it to its owner as he believes he should. In the second world, John discovers the same wallet, but he intentionally keeps it. Further suppose that John wants the money to buy a bicycle, the wallet's owner needs the money more than John does, John knows who the owner is, and he knows she needs the money. No other considerations are morally relevant.

By using CT properly, the world in which John returns the wallet is better than the world in which he keeps it. This result is due in part to the virtuous consciousness that exists only in the first world.[21] John returns the wallet and its contents because he intends to do a good thing by returning what is not his and believes that one should not gain from another's misfortune. Other good intentions will also tend to improve the overall value of the world or mental state or situation, but possibly in different amounts depending on what the intention is and how it fits into the overall context with its complex mental states.

In addition to intentions, motivations can affect the value of worlds. Once again, consider the world where John finds the wallet and returns it. Suppose that instead of only one world, there are two worlds open to John where he returns the wallet. In the first, John returns the wallet because he thinks such an action is good, and his motivation to be good is what causes him to return the wallet. In the second, John returns the wallet because he expects a reward. The bad motive in the second world decreases that world's intrinsic value because it brings evil into that world that would not have existed had John had a morally good or neutral motivation. John's motive is bad due to the fact he is returning the wallet merely to gain a monetary reward; his motive is one of greed. On the other hand, the good motive or virtuous consciousness that John has in the first world makes the value of that particular world better than it would have been if John had no motive or a bad one in returning the wallet. Everything else being equal, the better world is one that contains more virtuous consciousness. The value of the other virtuous and vicious consciousness can be shown in a similar manner.

[21] I am leaving aside considerations of organic unity.

Another type of vicious and virtuous consciousness is having inappropriate or appropriate emotions, feelings, or attitudes about a situation. A person has vicious consciousness if he brings inappropriate feelings or attitudes into a situation. Appropriate emotions, feelings, and attitudes are virtuous consciousness because appropriateness is a virtue. If the virtues are excellences that help a person thrive, in part because to have them is to flourish, and the virtues assist the person in finding happiness and being happy, then having appropriate emotions, feelings, and attitudes in the right way and to the right degree must also be virtuous. Conversely, inappropriate emotions show a defect in a person's character that will not allow the person to thrive as she otherwise would.

Although it might appear to be strange initially, negative emotions can enhance the intrinsic value of a situation if such negatives are appropriate, while generally positive feelings can mitigate positive intrinsic value to a point in which the disvalue of the situation is enhanced. For example, it is a virtuous consciousness to feel regret when regret is appropriate.[22,23]

In addition to the intrinsic value and disvalue of virtuous and vicious consciousness, respectively, possessing a virtue is also intrinsically valuable. If we grant that virtues are excellences, as Aristotle contended that they are, then having one makes the person a more excellent being than she otherwise would be *ceteris paribus*. On the other hand, having a vice makes the person less valuable than if she merely possessed intrinsically neutral habits of behavior.

Just how a virtue's possession will affect the intrinsic value of the person who has it is not terribly clear. Using the Comparison Test, a reasonable person would have to take into consideration a variety of factors. If we are comparing two otherwise identical people with the only difference being that one has a particular virtue while the other does not, then it is reasonable to conclude that the former is more intrinsically valuable than the latter on the grounds, once again, that having more excellences should make one better. On the grounds of sheer quantity, then the more virtues one has, the more valuable one is. On the other hand, it follows that the more vices a person has as part of her character, the less valuable she is to the point where she might be overall disvaluable, as happened in the case of Hitler and other extremely evil people.

However, evaluations of intrinsic worth based on virtues become difficult quickly and might lead us into using intuitions that do not reflect the universal moral platform. First, although we can talk about one person being identical to another save the possession or lack of virtue, assuming that this can happen might be a bit too glib. The quantity and quality of the different experiences required to

[22]I define regret as the feeling of doing something that is normally morally prohibited but is, in these unusual circumstances, ethically required. One would rather not do it, but understands that it is a duty that cannot be shirked.

[23]Of course, this entire discussion of vicious and virtuous consciousness presupposes that the agent has control over her intentions, motives, emotions, etc. The appropriate emotional consciousness issue will be developed in much greater detail in Chaps. 5 and 6.

create a virtue or vice and their effects might make the two entities have different identities rather than being the same person with one uncommon virtue.

In order to have a virtue, a number of other conditions might be required in order for the person to have that disposition to act. Consider courage versus having cowardice or a neutral characteristic of neither courage nor cowardice. Clearly, a considerable number of mental states would have to change in the agent to move her from being a coward to being courageous. These altered mental states might in turn affect other mental states. After all, if someone is courageous, then she has to act courageously a sufficient number of times to acquire the characteristic of doing what she believes is the right thing even though she is afraid to do so.

Each act of courage might very well affect other reactions to situations of which courage is part or might be part. That is, acting courageously might make the person bolder, view the world in a different way, or alter other mental states as a result. The impact would be considerable because the more changes to important traits such as virtues or vices, then the more likely it becomes that there will be different people with different identities instead of one person with a trait and without a trait. Hence, there is no simple removal and insertion of courage in a person the way one might replace an air filter in a motor vehicle. Moving from a neutral state to one in which one has courage would have a lesser impact than that of replacing cowardice with courage because the person has to instantiate less of a change in the former, but the same complexity of change holds in this situation as well.

Second, the degree to which one holds a vice or virtue will also affect intrinsic value. If someone is more courageous than another, *ceteris paribus*, then the former is more intrinsically valuable for that reason alone. Vices work much the same way. The stronger the vice, the more disvaluable the person. Once again, the measure will be that of a reasonable person comparing entities that are as similar as is reasonable to make them.

Third, vices and virtues should not be considered in isolation. Much like pleasure needing to be considered in light of desert or other intrinsically valuable and disvaluable things relevant to the mental state, the value of a person having a particular vice or virtue should be considered in light of what other vices or virtues she possesses. For example, a person might be courageous but at the same time be willfully ignorant of basic ideas about common decency. Suppose that the person is a courageous racist. The community proscribes racism and racialism and visits harsh punishment in the form of disapprobation on anyone transgressing its conventions. The racist acts as a racist in public as a protest. This individual is clearly courageous, but horribly mistaken about what is right in the situation. Hence, the value of the courage and other virtues must be reduced by the related vices, and the related vices' disvalue mitigated by that of the virtues associated with them.

To reduce the complexity involved in evaluating the value of vices and virtues and the resultant discord that will erupt because of the inevitable disagreement about what is more valuable and what is not, a reasonable person standard and context will have to be used again. When it is useful to discuss a bare vice or virtue, which merely means that we are considering a person having the vice or virtue and nothing else in comparison to another bare vice or virtue, then we can evaluate bare virtues for this purpose. Perhaps we are trying to figure out if moderation is more valuable than justice. If we need to consider the value of one person against another, then all the vices and virtues of each person must be considered. In most situations, it will either be clear which one is better because her combination of virtues and vices are intrinsically more valuable than that of a rival, or it does not matter which is better or worse. More precisely, in the latter case, if a reasonable person says one is better, then that is an acceptable answer to use for decision making for that person and those who agree with her. If another reasonable person makes a contradictory claim, then that is permissible as well. This form of relativism is not only acceptable, but it should be encouraged on the grounds that ethics is about making good decisions on how to live our social and individual lives. As long as reasonable people can understand why a person evaluates as she does, then her answer is legitimate.

2.4.2 Relationships

The final component of this section deals with relationships, whether they have intrinsic value, and if they do, what value to assign to them. A relationship is a relational property whose definition must involve reference to at least two things, generally in a comparison. A simple relationship is between two objects, such as A and B. We can say that A is taller than B, B is hungrier than A, A is wealthier than B, and so on. Most relationships, whether they be simple or complex, have no impact on intrinsic value because they have little, if anything, to do with persons or other intrinsically valuable entities. However, if B becomes upset that A is taller than B, then the relationship might be part of the causal change that causes someone to have an intrinsically valuable or disvaluable state of affairs, but is not in and of itself valuable.

Value-bearing relationships at their very core have intrinsic worth because they involve intrinsically valuable or disvaluable mental states of affairs in at least one of the parties. More complex and generally more valuable or disvaluable relationships involve some form of cooperation or emotional interaction-tie between the objects. If John and Mary love each other, for example, then there is a relationship between John and Mary. John and Mary are in a reciprocal emotional relationship; Mary is the object of John's affection; hence, the essential definition of the relationship must include Mary. However, a reciprocal relationship cannot be a simple relationship with the emotion running in one direction. What makes it an emotional relationship is the fact the objects are emotionally tied together, viz, John loves Mary and Mary has to have some sort of reciprocal emotional tie to John.

Without the tie, there would be no reciprocal relationship as I am using the term. If Mary has no idea who John is, then John is not in a reciprocal emotional relationship with Mary, although he can have many relations with Mary based on loving her, e.g., being her admirer or stalker. These unidirectional emotional relationships can also have intrinsic value. Being Mary's admirer would generally be a positive, whereas being Mary's stalker would be disvaluable.

If Mary is in a reciprocal emotional relationship with John, then the relationship can be described in such a way that it captures the essence of it. It is a love-hate, love-love, love-indifferent, or love-colleague relationship. Whatever emotional relationship it will be, there must be some sort of interaction or exchange between the two to create the relations with Mary as the object/subject, respectively, and John as the subject/object, respectively, that link them in this particular way.

Of course, in most relationships, the objects are emotionally tied to each other in some way. That is, there has to be some sort of linkage between the set of two or more relationships. Suppose that Mary loves John and John loves Mary, but they have never met. We could say that they love each other, but that claim would not capture what is usually meant when mutual love is asserted. They, in fact, do have a loving feeling for each other, but the mutual feeling must be expressed so that each is aware of the relation of the other and then can reciprocate. That is, Mary must know, in some way, that John loves her, and then react accordingly with John; John must know that Mary loves him, and then reflect that back to Mary accordingly for there to be a relationship implied by 'John and Mary love each other'. Once again, the simulation theory of mind reading seems to be at work.

Relationships can be significant in several ways. If it is important to the person's identity and affects her flourishing by altering her intrinsic worth, then the relationship can be valuable or disvaluable in and of itself. Note that the claim here is not that the relationships are necessary in all cases to an individual's nature, but only that they can be important. Defining what makes a relationship important rather than having some other status is a difficult task. First, we can say that all necessary relationships that are part of a person's essential nature are important, but not all important relationships are necessary. Second, an important relationship is not a trivial one, although it might be an accidental one. Perhaps the best way to characterize an important relationship is to state that it is one that is a non-trivial part of the identity of the person because the relationship has a great deal of impact on the person's flourishing. Although the person need not essentially change if the relationship is altered, the person's nature is significantly different when the relationship begins or ceases to exist.

Marriage, for example, can be a trivial or essential relationship. Obviously, to be married, one has to be married to someone else. In the developed world, if a woman is married, then an important (not necessary) element of her identity is the fact that she is married. If it is a marriage in name only, then her unmarried status will not alter her essential identity in any way because if she was no longer married, she would be the same person. However, it is an important relationship because it does affect important decisions she can make about her life, especially those involving her flourishing. If she is married to one person, but wants to marry another, then she cannot do so. Suppose her first husband is no longer the love of her life, but her new business partner is. Because she cannot marry her business partner as soon as she wishes, her flourishing is reduced significantly in comparison to what it would have been had she been free to marry at will. Hence, these relationships, relations, and those of a similar nature are important to who the person is.

Marriage as a more Kantian ideal is an essential relationship. Kant argues that sex can only be a morally legitimate activity inside a marriage because it damages one's human dignity otherwise.

> But if I yield myself completely to another and obtain the person of the other in return, I win myself back; I have given myself up as the property of another, but in turn I take that other as my property, and so win myself back again in winning the person whose property I have become. In this way the two persons become a unity of will. (Kant 1963, 167)

Of course, Kant seems primarily concerned with showing why having sex does not necessarily degrade each person to a moral status below that of an animal, but his claim about the organic unity of marriage makes a great deal of sense. When two people are married in the romantic sense of the word, they create a new entity, which is the married couple. Although we can identify each person as an individual, and the psychological identity of each remains firmly ensconced in the mind of each moral agent, the couple is something above and beyond the mere addition of each couple member. In important ways, the full identity of each is no longer distinct but rather shared with the other. Each component has unique features, but when the individual is by herself, then she no longer feels whole. This emptiness might simply be a case of wanting to share with the other person. For example, if one spouse is in a museum looking at a painting the other spouse would find aesthetically pleasing, and the latter is not there, then the former feels a want in the experience. The experience of the painting would have been much different if the other had been there to share it with. This feeling of incompleteness cannot be accounted for merely by saying the present spouse would have had a different pleasurable state had the missing spouse been there as well (living, in part, through the other). Although it seems terribly vague and full of hand-waving, when both spouses are present, their pleasure and experiences affect each other in a way much like that found in Chaos Theory. The experience cannot be the sum of each one experiencing the painting added together; there is more than that at work.

What happens to people in a relationship could very well be based on the simulation theory of mind reading. As discussed in Chap. 1, our ability to feel empathy for others is based upon the simulation theory. We can read others' minds to feel what they are feeling and intend what they are intending, although we do not feel or intend in the exact same way.. Rather, we imitate what they experience, which allows us to understand what they are experiencing. The reciprocal nature of a relationship and mirroring, which allows the relationship to exist and develop, seems to use this mental ability. Therefore, what is happening is not merely that two different people are having independent mental states, but rather there is an organic whole developed of the two people acting and interacting both internally and externally. The more intimate the relationship, the better able each person is to read the other's mind through simulation because the intimacy of the relationship requires a great deal of understanding of the essence and important characteristics of the other person, which each shares with the other.

When a spouse dies, the survivor often states that part of him has also perished. This is generally true for couples who share a deep commitment, love, or whatever positive emotional relationship they have, or for couples who have been together a very long time. Their individual identities as separate entities has been surrendered a long time ago. They have become a couple who make decisions for the flourishing of the couple which might be detrimental to either individual separately. For example, they might save for retirement, which prevents one member from buying a boat when she wants to do so. Since it is for the good of the couple, each is willing to subvert her self-interests to that of the couple.

A clear difference between couples and other types of groups creates organic wholes and cannot be reduced to the members or parts of the whole. Businesses are moral entities that can have unique duties that no individual member of the business has. For example, if Milton Friedman is correct, the company has a duty to make as much profits for its stockholders as it permissibly can within the confines of the law. However, someone who works for the business does not have the same duty. An employee has an obligation to fulfill her job's requirements, which might or might not be consistent with maximizing profit for the business as a whole. In fact, it is often not in the individual's interest to maximize stockholder profit because that will reduce her salary accordingly. That is, by reducing her salary, the company can increase its profitability. As seen above, couples can have the inconsistent duties based on duty for the individual versus duty as the couple. The same fact applies to groups versus members of the group.

The morally significant difference between businesses and marriages is the emotional commitment in each. A true romantic marriage is based on love, self-respect, respect for others, and all the other common or necessary relations between two people that make the relationship a marriage. Businesses require loyalty, cooperation, and the emotional relations one needs to pursue the company's flourishing, but there is no expectation of the emotional commitment required of marriages. Therefore, marriage is generally more central to a person's identity than is the person's job.[24] If the marriage status changes, then the identity of each individual changes simultaneously. If a person loses or gains a job, on the other hand, the relationship is important but not essential.

Given that these positive relations and relationships are part of a person's identity and fundamental to flourishing, then they can be intrinsically valuable or disvaluable and affect an organic whole's worth. A married couple who deeply love each other are more intrinsically valuable than a couple who feels less than that for each other. The intrinsic worth comes not only from the actual feeling of being in love that each member of the relationship has, but also the reciprocal nature of the relationship that creates the new entity of a married couple. Friendships and other positive, nurturing relationships can work in the same way.

Hate and other disvaluable relationships can also have an impact on the value of a person and other intrinsic value bearers. If a formerly happy married couple now

---

[24]Of course, some jobs are careers, which will be part of a person's identity.

loathes one another, then that hate-hate relationship is an intrinsic evil. Not only are the hate mental states intrinsically disvaluable, but being in such a relationship is an evil in and of itself. It cannot possibly be connected to their individual flourishing or flourishing as a couple.

It should now be clear that death can cause a loss in a few different ways. First, dying when an entity is experiencing an intrinsically valuable or disvaluable mental state makes that state cease to exist. For example, if someone is having a pleasant experience, and that experience is cut short by the death, then the full pleasant experience is lost. Second, and more importantly, the death of an individual with knowledge, virtues, and other long-term mental states eliminates each of these mental states. The intrinsic value of the world, in general, is lowered, whereas the death of a person instantiating vice, ignorance, and other intrinsically disvaluable mental states would more likely increase the world's value as these negative mental states cease to exist. Finally, when one member of a relationship ceases to exist, then so too does the relationship. If it is a positive relationship, such as marriage or friendship, it is not simply that half of the positive is gone. What is lost is the relationship that had value as well as the mental states of one of the entities involved in that relationship. How much this is worth depends on a variety of factors, and should not be thought to be proportional or equal to the sum of its parts because certain relationships create new entities that cannot exist without one or more of its parts.

Now we know a bit about death's values as well as a value theory that can be useful in making decisions about death and dying when combined with the Practical Principle. Despite this, we still do not have all the tools we need to begin applying the practical moral code to issues about death.

At this stage of the development of the value theory it is most readily apparent that the concept of a person plays a very large role in the axiology. We can discuss mental and other states of affairs, such as relationships, separately, but what can never be eliminated is that fact that each of the intrinsically valuable things involve people, or at the very least, sentient creatures in some way. It is impossible, for example, to have a mental state without a mind that instantiates the state. The relationships that are intrinsically valuable generally require objects that are people. Hence, it is vital to begin discussing the various ways an entity can be a person and how intrinsic value can be affected or understood in each case.[25]

References

Bloom, P. 2013. *Just babies: the origins of good and evil*. New York: Crown Publishers.
Bradley, F.H. 1883. *The principles of logic*. London: Oxford University Press.

[25]PP and the value theory here are used to develop applied ethics positions on war, physician-assisted suicide, the death penalty, and a number of other controversial issues in death and dying in Steffen and Cooley's *The ethics of death*.

Brennan, A and Y-S, Lo. 2003. Environmental ethics. In Stanford Encyclopedia of Philosophy, Stanford Center for the Study of Language and Information. http://plato.stanford.edu/entries/ethics-environmental/. Accessed 27 August 2014.
De Waal, F. 1996. *Good natured: The origins of right and wrong in humans and other animals*. Cambridge: Harvard University Press.
Elliott, R. 2000. Environmental ethics. In *Contemporary moral issues*, ed. R.L. Hinman, 567–576. Upper Saddle River: Prentice Hall.
Feldman, F. 1986. *Doing the best we can*. Boston: D. Reidel Publishing Company.
Feldman, F. 1995. Adjusting utility for justice: A consequentialist reply to the objections from. Justice. *Philosophy and Phenomenological Research,* 55: 567–85.
Frankena, W. 1973. *Ethics*. Engelwood Cliffs: Prentice Hall Inc.
Frey, R.G. 2000. Rights, interests, desires, and beliefs. In *Living well*, ed. S. Luper, 436–442. Fort Worth: Harcourt Brace College Publishers.
Gazzaniga, S.M. 2005. *The ethical brain: the science of our moral dilemmas*. New York: Harper Perennial.
Gert, B. 1998. *Morality: Its nature and justification*. New York: Oxford University Press.
Hauser, M. 2006. *Moral minds*. New York: Harper Collins Publishers.
Hinman, R.L. 2000. An introduction to the ethical issues in our relationships with the environment. In *Contemporary moral issues*, ed. R.L. Hinman, R.L., 559–66 Upper Saddle River: Prentice Hall.
Holmes, R.L. 2003. *Basic moral philosophy*. Belmont: Thomson Wadsworth.
Hume, D. 1990. *A treatise of human nature*. Oxford: Oxford University Press.
Kant, I. 1963. Duties towards the body in respect of sexual impulse. In *Lectures on Ethics*, ed. L. W. Beck, Trans. L. Infield, 162–8. Indianapolis, Hackett Publishing Company.
Moore, G.E. 1998. *Principia ethica*. Buffalo: Prometheus Books.
Rashdall, H. 1923. *The theory of good and evil, vols. I and II*. London: Oxford University Press.
Ross, W.D. 1988. *The right and the good*. Indianapolis: Hackett Publishing Company.
Russow, L. 2002. Environment. In *Life science ethics*, ed. G.L. Comstock, 93–123. Ames: Iowa State Press.
Steffen, S., and D. Cooley. 2014. *The Ethics of death*. Minneapolis: Fortress Press.
Warren, M.A. 1983. The rights of the nonhuman world. In *Environmental philosophy*, ed. R. Elliott, and A. Gare, 109–133. University Park: Pennsylvania State University Press.
Westacott, E. 2012. *The virtues of our vices*. Princeton: Princeton University Press.

Chapter 3
Defining and Valuing Properties and Individuals

Abstract One of the largest controversies in death and dying focuses on what dies and what is lost when that entity is dead. For the most part, those who work in the field are interested in the death of people, but it is a reasonable position that people are not the only things that matter in the discussion. Properties other than being a human person can confer intrinsic value on those beings that instantiate them. For example, in the animal rights debate, some argue that anything that is living deserves moral consideration based upon that fact. Others draw the line among the entities we must respect somewhere between those beings with the ability to feel pleasure or pain and those who can feel neither. Other folks have a different standard that will create bigger or smaller groups depending on what properties they think are relevant. In this chapter, I will argue that an adequate, pragmatic value theory must be very, very complex. First, it should view value as hierarchical. Some properties being instantiated matter more than others, although each is worthy of consideration in its own right. In addition, how we value life depends on the context we have carved out to consider the living thing. Sometimes, we focus on the general properties it has, whereas in other cases, its individuality should be of main concern.

3.1 Introduction

Death is not merely a cessation of existence, but rather the cessation of life. If it were the former, then any time we destroy an object, such as grinding a rock into dust and then making it part of a concrete wall, it would make sense for us to say that the rock died. It might also make sense to claim that we killed the rock, if we were the sufficient and actual cause of its demise. We don't think about the world in this manner, nor do we need to. We already have a perfectly useful conception of life that does all the work we need it to do.

In what follows, I will not focus not on what life is because it seems to be another one of our complex ideas with multiple definitions; there is little

D.R. Cooley, *Death's Values and Obligations: A Pragmatic Framework*,
International Library of Ethics, Law, and the New Medicine 62,
DOI 10.1007/978-94-017-7264-8_3

controversy over what is claimed to be alive below being alive, but there is controversy over what can die and how to appraise it. I will present a hierarchical value theory. At the very lowest level of intrinsic worth is an organism. From that base of intrinsically valuable entities, we will climb upward through biomes, biological bodies, biological people, moral subjects, psychological people, embodied people, and embedded people.

Part of discovering an individual's worth is determining what it is. A mistake is to try to find a "common sense" natural definition for any of these terms with necessary and sufficient conditions.[1] For example, the seemingly simple statement "I will die" can mean many different things based on to what the words in the sentence refer. A list of different interpretations is shown below, but the compilation is not exhaustive; the first interpretation alone could mean different things according to how the speaker defines the self (Murphy 1993, 54).

1. *I* am an individual with a life of my own, a personal existence.
2. I belong to a *class* of beings, one of whose attributes is mortality.
3. Using the intellectual process of logical deduction, I arrive at the conclusion that my personal death is a *certainty*.
4. There are *many possible causes* of my death, and these causes might operate in many different combinations. Although I might evade or escape one particular cause, *I cannot evade all causes*.
5. My death will occur in the *future*. By future, I mean a time-to-live that has not yet elapsed.
6. I do not know *when* in the future my death will occur. The event is certain, the timing is uncertain.
7. Death is a *final* event. My life ceases. This means that I will never again experience, think, or act, at least as a human being on this Earth.
8. Accordingly, death is the *ultimate separation* of myself from the world (Kastenbaum and Aisenberg 1976, 376).

Although reduction to one definition for each word would save a great deal of resources and settle issues instead of encouraging intuition-pump duels, we want to capture reality as much as we can so that we can do the work we want to do.[2]

> The object of the present study…[is] to free the language of everyday life from confusion. It is essential, for this purpose, to realize that everyday life presents us not with "a" usage of terms, but many different usages. Nor is this a phenomenon which arises from trivial causes, which analysis can easily offset. Ambiguity will always persist in common discourse; hence, the practical task lies not in seeking to eliminate it, but rather in seeking to make its presence evident, and by careful study of its origins and functions, to render it no longer a source of error. (Stevenson 1944, 34)

[1]Lloyd finds a similar problem in defining species of animal and plants. He argues that different criteria can be used depending on assumptions made earlier in the taxonomy (Lloyd 2007, 40).

[2]I will limit myself here to a discussion of human persons and human personal identity, although what is said about them will work well for many other entities that can fulfill the requirements of how personhood or personal identity is used in at least one circumstance.

Once again, we should look at the purpose of our pragmatic enterprise and then find the set of definitions that can do work in the circumstances we encounter. We understand what it is to be a person in different ways not because we are terribly confused about the whole thing, although we often could do with greater clarity, but because the various definitions of personhood and individual identity do useful business in helping us understand who we and others are, how things work in our world, and how we and others should be, act, and interact. Which definition is true for a particular set of circumstances is determined by which one works best or well enough, as measured by need in that situation, to satisfy our immediate and long-term goals of flourishing. Of course, there is a limit to what is an acceptable definition, since "works" implies that the definitions should reflect, in general, what is real, but it need not be totally reducible to a natural definition in which the term's essence is complete.

With this in mind, what works in our reality? We, as *Homo sapiens*, classify things, as our near and distant ancestors did, to survive. This is evolutionarily based, the result of evolutionary forces working on our ancestors, and still retained by us. An evolutionary adaption hypothesis would explain this property's existence as the result of a fight or flight response rate that was central to how species' members survived in an unfamiliar situation. Those species' members who were slower at making correct classifications of the situation were less efficient than those who could make rapid and correct categorizations.

What helps separate *Homo sapiens* and moral agents from other animals is our ability to classify the universe and its components in ever-finer ways, as long as that was a sufficiently valuable way to do things. The fight or flight classification coarsely divides the world into two classes of thing, thereby ignoring the safe option mentioned above, which would create a world separated into three different classes. However, this three-category system is inefficient. These categories are still far too large and unhelpful for the vast majority of sentient creatures to use in making decisions affecting lives. For example, we have to place other *Homo sapiens* into roles so that we can interact appropriately with them. A stranger in the dark needs to be treated in one way, while a co-worker or spouse requires other forms of interactions and feelings. The better we are at identifying what others are, in general, the better decisions we make about them and the better we interact with them.

Although the world would be a less contentious place if each class of thing had a natural definition that could be objectively discovered, and then captured in the essential and sufficient conditions for a thing to be a member of the class, reality does not work out that way.[3] The reason, in part, is that we actually help create reality through our classifications or what exists and how we interact with it. Classifications run from purely natural kinds to purely artificial kinds. For example, there can be triangle shaped objects, but the mathematical concept of a triangle—

[3] Lloyd asserts that there are no cross-cultural universal definitions of terms such as self, agency, and causation (Lloyd 2007, 108–9).

that is, its essence—is a purely artificial kind of thing. Of course, this denies Plato's view that the forms have eternal, independent existence. However, the essence of a triangle is pure mathematics because such a figure cannot exist as a figure in the real world. "Triangles" are merely our way of figuring out how the world works through a form of logic, viz. mathematics.

Natural kinds of things exist independently of human perceptions and are capable of having natural definitions, which can capture the object's essence without vagueness. Hydrogen, for instance, is a natural, objective thing that was scientifically discovered. We can define it by saying that hydrogen is this element and nothing else. Against Darwin's belief that species are a subjective, artificial classification, and biology's 22 pragmatic definitions, Gould argues that species are natural kinds like branches on a bush (Gould 2003, 466). The objective division between the stem and the branch, according to Gould, is clear and obvious to anyone who is examining it, although whether the branch collar is stem or branch is a matter of debate.[4] Most definitions of persons, death, and other related concepts and classes tend to portray themselves as objective divisions which can be naturally defined so that all necessary and sufficient conditions can be captured in the definition. In other words, the definitions are descriptive and complete.

Many classifications, such as personhood, death, and personal identity over time, fall somewhere between purely natural kinds and purely artificial ones. That is, they are part discoveries and part human creations. Bioethics and ethics are another example. They have an objective, natural reality as a component, but part of it is subjective, and therefore, artificial and vague. C.L. Stevenson captures this, in part, with his definition of the good, which has both a descriptive and emotional component to it (Stevenson 1944, 207).

To determine if a definition is adequate for a term, we should discover what it is that we want to define. If it is a natural kind, then the definition should be natural. If it is artificial, then an artificial meaning should be sought. If it falls somewhere in between, then many definitions of the artificial or subjective component might be involved. The vagueness and ambiguity caused by subjectivity should not surprise us; even the definition of red, which would seem to be one of the simplest of definitions to grasp, is ambiguous. Some things are clearly red and other things are clearly not red, and a whole range of items could go either way, such as near-orange (Stevenson 1944, 35). Instead of declaring that red has a natural definition clearly demarcating what is and is not in the class, we should create definitions to fit the needs we have in the situation. Sometimes, something will be red, and other times,

[4] I think that Gould makes a mistake here. He is correct that the branch and stem clearly are different, but only if our perspective comes from far enough, but not too far, away from the bush. That is, if we are very close to the join, then we see an area in which we can claim both stem and branch. Too far away, and we cannot see either, but merely a bush. Vagueness and subjectivity is always present based upon the context. Even in the case of species, there isn't one definition that works, although there is one that we use most of the time because it generally works, just as Newtonian physics is good enough in most cases unless we need to do finer work.

it will not be red depending on the purpose we have in broadening or narrowing the definition.

Pragmatism, however, does not collapse into relativism. Our adequate classifications are objectively limited by two things. First is the natural or objective reality within the identity/definition. That is, if we want to define something that has an objective element to it, then we must build that element into our definition in some manner. A person, for example, is not a rock. Rocks are not alive, have never been alive,[5] and will not be alive. Therefore, life in some way is a necessary characteristic of being a person. Let us make this the first objective step in the process of defining terms because it helps us to capture reality, which is generally useful, and it allows us to have common understanding of the terms. We need people to be rational so that they can be reasonable, and recognition of objective reality is eminently rational; hence, we must incorporate objective reality in our definitions. As I argued previously, one purpose of ethics was to help us make decisions about our flourishing and the flourishing of others, which will require interactions. A definition based on objective reality is better suited to achieving those goals.

The reason that objective reality must come first is that we can eliminate non-contenders for an adequate definition from our class of all possible definitions. We can see that these work for us. Generally, the better a definition represents objective reality, the fewer mistakes we make in classifications. As stated above, persons are not rocks, so we eliminate such obviously wrong definitions from consideration before we waste any real time on them. At times, reason can take us all the way we need to go because the evidence is readily available to us. In chemistry, for example, we can know the truth of a scientific proposition based on abductive reasoning without having to introduce subjectivity to any significant degree beyond what is required to do the science. That is, we can determine which hypothesis is better supported by the data, and then draw a conclusion based on that objective evidence.

Of course, definitions will be in the grey area where we do not know if they fit, but these will not be obviously excluded as some non-contenders are. There are instances in which reason cannot give us the answer because of insufficient evidence for reason to do it on its own. Basically, the argument here mirrors that of Kierkegaard's Leap of Faith. Reason can exclude gods that make no sense based on the evidence, such as what happens in the Problem of Evil argument and a deity with infinite power and goodness, but reason cannot tell us whether it is rational to believe or not to believe in far more plausible gods.[6] When there is such equipoise in the evidence, then emotions, such as faith, can be permissibly brought into play, as long as the resulting belief does not interfere with the individual's and other's ability to flourish. To believe is to believe non-rationally rather than irrationally, which is a rejection of what the evidence compels a rational person to believe, or

[5]I am aware of the existence of fossils and petrified organisms, but these entities were not alive as rocks.

[6]These divine entities are merely causally, rather than metaphysically, impossible.

rationally, which is to believe whatever proposition has sufficient evidence in the situation.

Returning to defining concepts and classes, there are times in which objective reality cannot tell us what a complete definition should be because part of the concept is subjective. Being a person, for instance, has a partly subjective and partly objective meaning. Despite this, we do not want to introduce rampant relativism in our definitions when objective evidence is lacking. Such a practice is not conducive to making the world a better place or pursuing our or others' flourishing. Therefore, we need a second criterion of creating a definition and evaluating its adequacy as a definition.

The second criterion requires us to evaluate the contending definitions that were not eliminated by objective reality by examining how well each works given the context of the situation, although we do not have to be explicitly conscious that it is useful. That is, determining if something works adequately requires that it works in a particular situation's context to achieve some stipulated goal. The situation's context is determined by whom the definition will directly or indirectly affect. It is the goal of flourishing for those in the situation and how to practically achieve it that should tell us which definition works adequately for that goal, defined as good enough or best depending on how much intrinsic value is at stake. For instance, we should use only the best definition when a great deal of intrinsic value is in the balance, e.g., life and death situations.

I think that the Practical Principle can be useful in deciding which of the many legitimate competing definitions it is permissible to use in the situation. If adopting a definition in the situation respects all those intrinsically valuable entities affected by it, as well as being something that a reasonable person would reasonably believe would maximize utility, then that definition works in the context. Consider the unethical definition of person as a white male, which excluded all women and people of color. The Practical Principle would prohibit this definition on the grounds that being white or male does not treat all intrinsically valuable entities as they should be. However, definitions of psychological persons and organisms as people would not automatically suffer the same fate as long as we focus on what is relevant to being a person rather than introducing illicit bias that destroys the ability to flourish to those in the situation.

The best way to understand the entities is by using an analogy from physics. Picture one element in a complex molecule that, if removed, will essentially alter the molecule into something new; it is an atomic entity. The person is as distinct, essential, and atomic as that element in the complex. To help understand what the element is in its environment, one must also comprehend the complex structure of the molecule as well as how the element functions within it, even though it cannot really be separated and the molecule will remain the same. Moreover, for an even fuller understanding, we need to picture the molecule in the various contexts we find useful to talk about, such as being in a solution or a bigger environment. The point is that the element's internal and external contexts play enormous roles in our grasping what the element is within those contexts.

The same can be said about how we envision people and individuals. For its usefulness, we often treat people as distinct from their surroundings—basically, as islands isolated from all other things. However, although the mental lives as minds, brains, or other types of psychological selves are, in fact, private, even they cannot be understood in isolation from all other objects in the universe they inhabit, much as an island cannot be an island unless it is fully surrounded by water or some other substance sufficiently different from the material which composes the island. I contend that whichever definition of personhood, moral subject, or feeling entity one uses, it should be understood as an organic whole and an atomic entity located within greater organic wholes that are also unities. The person-as-organic-whole or greater organic whole we address in a particular situation is decided by the contextual need one has to make good decisions or communicate effectively in the situation.

3.2 What Are We in General and What Is that Worth?

The saying that a coward dies a thousand deaths is closer to the truth of who we are as individuals and persons in general—and our ends—than is asserting that a hero dies but just once. We have many deaths because there are many types of living that we can instantiate and then lose. In this section, I will consider what it means to be a person as well as other living states we can instantiate.

In ordinary language, "person" is used to refer to a variety of objects that fall under the generic category of being a person or having personhood. This interchangeability, ambiguity, and vagueness often lead to a great deal of misunderstanding and rancor. For example, in the abortion debate, "person" could mean, among other things, conceptus, fetus, bearer of a soul, or moral agent. The conceptus and fetus are clearly not moral agents, but many who oppose abortion claim that we have the same duties to the entity created by conception that we have toward moral agents because they share a morally relevant trait, whereas those who support choice make the strong claim that the fetus or conceptus is essentially ethically different from a moral agent, and therefore, the moral duties are not the same.

To avoid some confusion, I will stipulate different definitions of terms to be employed throughout this work. Each has its own controversies, possibly because it cannot be complete and fit everyone's intuitions, but each is the most useful definition in at least one plausible context. The following definitions are sufficiently broad that they can include a number of more rigorous definitions, but we should never forget that each of the following does not include all the ways that the term could be used legitimately. However, the goal to develop a position that a reasonable person might reasonably adopt to pursue flourishing in an appropriate way remains the main measure by which a step, process, or theory is being judged.

Let us begin with those that struggle to show that they are persons for very good reasons, but are often the focus of argument when discussing what dies: organisms and biomes. An ***organism*** is an individual living entity that can reproduce, grow, react to stimulation, and maintain its internal conditions so that they are sufficiently stable and constant. Of these characteristics, the most important appears to be the maintenance of stable and constant internal functioning. The organism is unhealthy, and possibly dying, when it cannot maintain its internal conditions, yet there are still internal functions that entail that life is present. If there is no internal functioning, then the organism is no longer an organism but a dead entity that used to be an organism.[7] In fact, when an organism is incapable of fulfilling one or more of its four traits, then it ceases to exist as an organism and therefore dies. There are those who define death based on the permanent cessation of life for an organism as being the only death that human beings/organisms encounter (Belshaw 2009, 1).

Of course, most organisms are not the same thing as a human organism. Plants, fungi, viruses, and non-human animals are organisms, but we would rarely characterize many of them as persons of any type because they lack one or more essential or sufficient properties we would use to include them in the class of persons. In general, organisms will not be classified as persons unless there is an additional characteristic added, viz. the organism must also belong to *Homo sapiens* or have genetic material that allows it to fall under the heading of a species that is also made up of persons.

As for organisms' intrinsic value, being alive as an organism can be considered to be intrinsically worthy by reasonable people, especially those who have a bio-centric, egalitarian approach to values (Taylor 1986). Such individuals could plausibly and persuasively argue that when we consider cases in which there is unnecessary death of even the most insignificant organism, as determined in comparison to our own standing as human persons embedded in the world, then we tend to disfavor, if not outright condemn, the waste of life. For example, if a child were killing plants and non-sentient animals, then many people would consider it to be a teaching moment on respecting life. This is not a matter of educating the child to respect others' property, for in our example, let us assume that the plants and animals belong to no one. Rather, the teacher is saying to the child that there is permissible and impermissible behavior toward living things. If there is neither need to kill an entity because one is not in danger nor need to eat the former organism to survive or flourish, then it is generally a bad thing and wrong to kill it. Although not everyone would feel the same way about life, those who are reasonable people can see the reasonableness of the teacher's actions toward the child, which is all that is required to invoke the Benefit of a Doubt Principle. Until we know that life does not have intrinsic value, we will assume it, and if we need more evidence than that, we will continue to assume such worth because there are reasonable people who believe life to have it.

---

[7]I put aside the issue of whether the cessation of life needs to be permanent. Although permanent cessation is clearly death, it is sometimes useful to talk about people dying and being brought back to life. For the sake of clarity, we could distinguish between permanent death and temporary death, although the latter might be an odd term in most contexts.

Human organisms as a type of person might be an acceptable position to those individuals who believe that a person exists from the moment of human conception.[8] First, the being is alive in a manner that the gametes that will create the fertilized eggs are not. The gametes are not organisms capable of reproduction; they are part of the reproduction process, but they are not reproducing, nor are they able to grow on their own. Second, although all organisms are identical as instantiating the property of being an organism, and therefore have the same value for being a member of the class, we could stipulate that a human organism is privileged over other organisms. Most of those who believe that personhood attaches at conception do not believe that such a property is universal to all fertilized eggs, only those from *Homo sapiens*. Hence, being human enhances the value of being an organism, or instantiating the two characteristics make the organism more valuable because each property is intrinsically worthy in its own right.

Although human organisms would appear to be unhelpful in discussing the nature of personhood because they are too different from our normal understanding of what a person is, organisms do offer some value in understanding more complex conceptions of personhood if we think of organisms as organic unities or wholes.[9] No organic unity's activities can be reduced to the sum of the activities of the parts (Woodger 1948, 289–90). Hence, a human organism, if it is an organic unity, could have activities that cannot be reduced to its component parts' activities. Part of those activities might be, in some way, being a person, although that might be stretching the definition of a person beyond a point that is reasonable.

It is far more problematic to claim that being a human person entails that a soul attaches at conception, and it is the possession of a soul that allows us to call each human organism a person. Of course, there is a difficulty in establishing a soul's existence, what such a thing would be, and the relationship between it and the fertilized egg, which we cannot perceive to exist in the way that we can see that something is an organism. Since souls and their relationships with physical objects seem based upon theological considerations and pure faith rather than evidence, I will not spend much time on them. Even the Benefit of a Doubt principle would struggle to show that it is permissible to believe that such entities exist because there is no useful evidence to justify such a claim and we can accomplish everything we need to without referring to souls. For example, it is a bad thing, although not necessarily wrong, to destroy human fertilized eggs because they are alive and have potential to become moral agents, both of which are properties that, when instantiated, seem to be intrinsically valuable in and of themselves. The evidence for that conclusion is found in how most people value life itself, reject unnecessary ending of life, and work diligently to preserve life even when doing so will cost

[8]This is not always true given that 1 % of the population is an identical twin, triplet, quadruplet, and so on (Sagan and Singer 2014, 122).

[9]The grand siècle thought it could reduce, if not eliminate, the unknown by reducing it to a mathematical formula. The same idea is at the root of much of the discussion about death. There is an assumption that if we reduce things to their parts, then we can fully understand what is happening.

them something that would serve their best interests far better than keeping the living thing in that condition. In addition, if there is a potential to be a human person, then there is even more incentive to preserve the life. Hence, even if there are souls, we need not make reference to them to make decisions and evaluations that are defensible to other reasonable people.

Another difficulty is that we have been wrong for some time about what a human being is. Human organisms are actually ***human microbiomes*** when their functioning in the world is understood more clearly. A biome is generally an ecosystem in which plants, animals, and other living things in a contiguous geographic and climatic area have complex interdependent and interrelated relationships with each other and the environment. A human biome is similar although it is clearly not an ecosystem per se. The Human Microbiome Project has found that:

> The human body contains trillions of microorganisms - outnumbering human cells by 10 to 1. Because of their small size, however, microorganisms make up only about 1 to 3 percent of the body's mass (in a 200-pound adult, that's 2 to 6 pounds of bacteria), but play a vital role in human health. (NIH 2012)[10]

Although not fully understood, it is safe to say that without these bacteria, we would be unable to digest certain types of food and absorb nutrients we need because we lack the relevant enzymes to do it ourselves. Our microorganisms also produce beneficial compounds, such as anti-inflammatories and vitamins that are necessary for our healthy functioning as a living body (Ibid.). Hence, in some ways, human organisms are useful fictions created to help us understand our world.

The idea that we are organisms can work for us in various ways, but so can the notion that we are biomes. Recall that an organism is an individual living entity that can reproduce, grow, react to stimulation, and maintain its internal conditions so that they are sufficiently stable and constant. In reality, we cannot maintain our internal conditions without the microorganisms that inhabit our skin, gut, mouths and other portions of the *Homo sapiens* bodies we have or are. In addition, those microorganisms, especially given the fact that they exist in the vagina and help with function and immune regulation, might also affect the other three characteristic functions of an organism (Witkin et al. 2007, 347). Hence, we are not fully contained beings as one-celled organisms are, but rather far more complex systems in which a variety of organisms interact in interdependent, interconnected ways.

Defining humans as biomes raises questions with regard to life and death issues. First, given the large number of organisms involved in the human biome,[11] then it becomes a much more difficult matter to determine the intrinsic value of a human biome than it does for an organism. Each organism has its own worth because it has life, which was acknowledged above. Given the sheer number of organisms composing the biome, we have to take trillions of organisms into account in value calculations. Making matters more difficult is if we adopt, as I have argued for, the view that we are organic unities or wholes. Our value as an organic whole might

[10]For more information on human biomes, see HMPC (2012).

[11]Or any individual animal biome, for that matter.

neither equal nor be proportional to the sum of its parts. In the case of a human biome, there are a number of complex, interdependent, and interconnected relationships that will help give rise to an emergent entity with its own unique worth far beyond that of a human organism.

Second, we as biomes have more cells of bacteria than *Homo sapiens* cells, although the weight of the former cells ranges from only two to six pounds in adults. That means when *Homo sapiens* organisms die, most of what was alive as organisms remains alive, if we limit ourselves to the number of living cells. Hence, in the biological sense, there is a problem arguing that we, as biomes, die when our *Homo sapiens* portions cease to live. Most of what was alive in the biome remains alive to begin the bodily decay process of the former *Homo sapiens* organism. The new biome, if it can still be called that because of the interrelatedness and interdependency loss, remains intact for some period of time and continues to function as a living thing.

A plausible approach to this problem is to say that the *Homo sapiens* parts are the only ones that interest us, but this raises immediate questions. Why this is the case? Why are *Homo sapien* cells privileged over those of the micro-organisms? It cannot be because of sheer weight or mass of the human component of the microbiome because those measures would be as arbitrary as counting the number of organisms.

It is not arbitrary to focus on *Homo sapiens*'s components just because it is natural for us to do so. We want to know about them because they matter to us in a way that the surviving organisms of the body do not. When the *Homo sapiens* components die, then our concern is not with the biome's remaining parts. We no longer consider it to be a biome because one of its essential components and the relationships between it and the other organisms composing the biome has ceased to exist. What we are interested in morally is what was lost in the human body's death that was a necessary part of the human biome, and as we are the products of evolutionary biology that favored those beings that worked together to survive and have empathy, we remain interested in those beings that are like us.

Compare how we understand human organisms and biomes to how we understand ***biological persons***, which are developed members of the biological species of *Homo sapiens* (Feldman 1992, 101). These entities can die yet remain biological people for survivors. The difference between a biological person's ability to retain its membership in the species even if it is dead and the failure of an organism and biome to continue on as an organism or biome, respectively, is based in each definition's foundation. For the latter two, it is a matter of being alive. A human organism and biome are different from a human body because life is an essential trait of organisms and biomes. When the human organism ceases to exist, then so too does the biome, although parts of the biome not only survive but also thrive.

A biological person, on the other hand, merely has been alive at one time and retain a sufficient number and quality of the physical features required to be a species' member, be they a certain genetic code or bodily arrangement. If the human body dies, then it still has those physical features until they have been altered sufficiently to render the entity no longer a species member. For example, a corpse of a few moments is still a biological species member, but once it is cremated, it loses that status.

We should not dismiss all dead bodies as being incapable of being persons in certain contexts. A dead body is often the focus of grief and other feelings of loss associated with death. When attending a funeral, we rightly say that S is in the coffin even when the psychological person S no longer exists. We know that the psychological person is gone because we would never treat the body the way that we will do, viz. cremating, mummifying, burying, or doing to it whatever the death ritual requires; it would be morally repugnant to treat the person in that way (Glahn 2009, 34). However, the recognizable body of S is still S in a way that makes sense when the survivors use that term.

In addition, the body is what we immediately perceive and becomes intrinsically, and perhaps essentially, linked to the conception of the person by those the person encounters and, at times, even the person herself. This is one of the reasons that some people so vehemently oppose cremation because it destroys the body, which is what they think is intimately linked with, if not the actual identity of, the person. They realize that the psychological person no longer exists, but the body is what we believe to be an essential part of the person. There still seem to be duties toward it, such as respect and giving it dignity, that exist for persons, which do not exist for mere organisms or biomes.[12]

It is here that we begin to see that being a person can be a social construction at times, which entails that personhood is external to the individual in an essential way. It still makes sense to call the dead body a person; therefore, other properties have to be involved other than it being alive. Besides it having been a person at one time, the only other real difference is that others recognize the individual as a person based on their connection between the dead body and the person who existed, regardless of whether they actually knew that living individual. For example, those who interact with the human body of a stranger treat it with respect and dignity that is not accorded to all dead things. If social recognition constructs a form of personhood for dead bodies, then it seems even more likely to be true that living persons can be ***socially constructed persons*** who exist only if society recognizes them as people.[13]

Lloyd Steffen's work on the concept of moral deaths is helpful in illustrating how others can create personhood. Steffen argues that people are morally alive with those with whom they have relationships until both the survivors declare the other dead and the other is dead psychologically or physically (Steffen 2012, 133). For Steffen, it is the external act of the other that is a necessary cause of moral death, which he illustrates with the Terry Schiavo case. Schiavo suffered a severe brain injury that left her in a permanent vegetative state; in other words, she, as a

[12]Julia Glahn argues that a dead human body has dignity because it is still part of the human community, although it does not have the dignity that a living body does (Glahn 2009, 38).

[13]This notion of social construction will play a very large role in the next section on socially embedded persons and personal identity.

psychological person, had ceased to exist. After a number of years, her husband tried to have her feeding tubes removed so that her body would die, but her parents and brother fought in the courts of public opinion and law for her to remain fed and hydrated. Schiavo had undergone moral death for her husband, but not for her other family members. It was only after her body had died that her family members finally were able to give her the various moral deaths she underwent.

Socially constructed persons can do some of the nuanced work we may need it to do when talking of cases of this type. Although it can be abused when it comes to individuals who are persons but not recognized by some historical societies, e.g., what happened and continues to happen to women and minorities, it will be added to the list of things that fall under the larger category of personhood.

Although it is useful at times to speak of our personhood as being equivalent to a physical body, there is clearly a problem when it comes to trying to value such an entity in a reasonable way. Of course, a living body has intrinsic value because it is living, but seeing worth in a dead body is far more difficult. The mere fact that it is a human corpse in and of itself does not seem to confer intrinsic worth.[14] There is no potential to become a human or have a trait that will make it intrinsically valuable for that reason. It is merely dead flesh that will be disposed of. The only place it might gain value is through how other people view it. As Glahn mentioned above, we still afford the body dignity, which does not mean it has dignity for what it is *qua* body but what it represents to other people. Therefore, the best we might do with valuing the dead body is as a link to our humanity as persons.

There is a significant issue about when an individual changes from being a ***psychological subject*** to being a psychological person. A psychological subject is an entity that is capable of mental experience, but lacks the high-level mental capacity or experiences required to be psychological persons. ***Sentient creatures***, for example, are psychological subjects because they are capable of experiencing pleasure or pain. Even if an individual is not sentient, it can still have mental experiences as long as it has a properly working brain.

Sentient creatures and psychological subjects are intrinsically valuable because they are moral subjects, which means that we have *prima facie* moral duties toward them since they are worthy of respect for their intrinsic value. Frankena argued that it was the mental experiences themselves that had the value, but that fails to acknowledge that such experiences are impossible unless an entity exists that has those experiences. Once again, we need to return to states of affairs discussed previously. A pleasurable state of affairs exists when ***someone or something*** is pleased about something, and this state can be mitigated or enhanced according to whether the person deserved such pleasure. Of course, the value of the experience might say nothing about its component parts given the theory of organic unities, but once again, I will rely on the Benefit of a Doubt principle and what reasonable people think.

---

[14]This might be too Western of a view. Some societies attribute agency and personhood to the dead (Lloyd 2007, 110).

Are sentient creatures and psychological subjects more intrinsically worthy than living things that cannot have mental experiences of any kind? Is there greater value in being a dog than there is in being a plant, *ceteris paribus*? I think the answers to the questions are justified privileging of sentient creatures over living things as can be seen by our laws on animal abuse. We are not entitled to harm sentient animals even if we own them, but we can neglect plants without too much censure because the latter are thought of more as mere property than deserving far greater respect. Now, the difference in levels of respect merited here is correct, although we need to be careful about that of living things. Overall, to be alive is far more valuable than to be dead, all things considered. Thus, even though plants are property under the law and morality, to do what is right and be a good person we need to respect the value of the living thing, but not as much as the sentient being. Finally, since all entities that instantiate the property of being a sentient creature or psychological subject do it equally, then all entities with that quality should be respected equally when thinking about this quality. Hence, a dog is equal to a human in value when all we are considering is their sentience.

Although all psychological subjects and sentient creatures are moral subjects, not all moral subjects can be found in those two classes of things. Those individuals who have been psychological persons but who have lost consciousness temporarily or permanently are moral subjects. Of course, they cannot be psychological persons because they lack one or more of the required characteristics of that classification. Despite this, they are not merely organisms or biomes or physical bodies. Their history of being a psychological person is important because it conveys increased intrinsic worth for these moral subjects that moral subjects who could never be psychological persons do not receive.

The difference in our evaluations is shown by the negative sorrow or regret we feel when a psychological person permanently loses his or her psychological personhood. Those in permanent vegetative states have enhanced intrinsic value because they are moral subjects, but not in the same way as those with potential to reclaim or gain their psychological personhood do. As was briefly seen when it came to giving bodies of people special respect because of the fact that the bodies were intimately part of who a person was, moral subjects with permanent loss of consciousness are accorded the same value. Their worth is enhanced when we consider them as organic unities that have the properties of being alive, being an organism (although one that might need external assistance to keep its internal functioning running), being a biome, and once again, being a human body that is the center of a more complex organic whole. This valuation might be socially constructed in basically the same manner as that of the dead body, but it still exists.

In addition, for those who have temporarily lost their consciousness, their potential[15] to become psychological persons is intrinsically valuable, and makes

[15]I am using Norman Kretzmann's definition of natural potentiality here. An object has natural potential to become X if the object has the capacity to become X (Kretzmann 1999, 39). This is not a mere metaphysical possibility, as found in Tooley's wonder-cat example. "[I]f no one

them worthy of greater respect than that conferred by the lower states of value in and of itself. The same sort of consideration applies to entities that are likely given causal possibilities and environmental factors to become psychological persons, even though they have only been psychological subjects up to this point in time.[16]

It is worth some space to examine a standard problem found in many of the writings on potentiality. Too often, there is an assumption that having intrinsic value entails that each thing with such worth shares it in equal amounts with all other things with intrinsic value, or that we cannot assign it an intrinsic worth because that leads to absurd conclusions. For the former, Jeff McMahan states "For if it's being a potential person makes it worthy of respect, killing it must be as seriously wrong as killing an innocent person, if other things are equal." (McMahan 2014, 148).[17] Patrick Lee and Robert George have argued that anything with "a rational nature," which for them includes embryos, has rights and deserves full moral respect (Lee and George 2006).

Michael Tooley has argued that taking the position that potential personhood is intrinsically worthy leads to the absurd conclusion that we must not refrain from producing those entities that have a capacity for rational awareness (Tooley 1983). Agata Sagan and Peter Singer claim that "if new life is precious and entitled to protection, it seems that a minimal implication of this is that we should, if other things are equal, protect the new life from destruction and do our best to enable it to develop." (Sagan and Singer 2014, 125). They then conclude that we would not be able to use embryos and stem cells in research based on this alleged obligation.

A problem with attempts to show absurd results if one assumes intrinsic value is that they fail to recognize the distinction between obligations to respect value and moral duties to act in certain ways. Something can be morally valuable, yet we do not have an actual duty to protect or bring it about. However, we do have an actual obligation to respect that value for what it is. Of course, there might be very rare instances in which we have a duty to perform acts of commission in which a potential person is protected or nurtured so that it can become an actual person, but in the real world that requires us to respect intrinsic value as we should, we will be too busy pursuing the flourishing of those entities with greater intrinsic worth to put more effort into potential persons than merely respecting their value. That is, conclusions that moral worth entail a duty to do more than respect assume that we have nothing else to do except protect or nurture these beings, which, in the vast majority of conceivable circumstances, never arise when we look at how we actually lead our lives fulfilling our obligations to entities with greater moral worth.

(Footnote 15 continued)

deliberately intervenes to prevent it from happening, it will, in the vast majority of cases, happen." (Feinberg 2014, 70).

[16]This claim is argued for more extensively in *Technology, Transgenics, and a Practical Moral Code*.

[17]McMahan argues that only a developed fetus has the potential to become a person, which he defines as an embodied mind (McMahan 2014).

Hence, the absurdities only happen in a non-existent world or for people with very little else to do in their lives.

Both sets of objections fail to recognize that there are different value levels based on a variety of factors, including which properties are instantiated. However, as Margaret Olivia Little argues, there is a progression of value as the entity moves from human organism to human being to self to agent (Little 2014, 182). According to Little, "status is something that emerges and increases as the balance between potential and actualization shifts." (Ibid, 180). We become more valuable over time as we instantiate properties that make us more worthy.[18] Therefore, saying that potential people have intrinsic value is not equivalent to saying, nor does it entail, that all intrinsically valuable entities have the same worth, that intrinsically valuable entities should be treated the same way unless their values are equal, or that we have a duty to protect and nurture all intrinsically valuable entities in the same way or even at all.[19] A moral saint, for instance, is of greater intrinsic worth than a sinner. However, we, as moral agents, have a duty to respect every intrinsically valuable thing for the value it has, whatever that value might be. Therefore, the moral saint receives more respect from us than we should give to the sinner.

When it comes to discussions about death and dying in bioethics, the main type of person is the ***psychological person***. A psychological person is roughly defined as an entity that "has self-conscious intelligence…is capable of purposive action… instantiates a sufficiently rich psychological profile." (Feldman 1992, 119).[20] Simon Evnine's necessary conditions of personhood somewhat fleshes out this concept by proposing that all entities with psychological personhood have[21]:

---

[18]I think that Little's view is correct for potential versus actualized properties. However, there should be concern about thinking about an individual as either a human being or self. It is sometimes required to think about a moral agent as a human being in order to compare her value to that of other human beings. In addition, we can think of an entity becoming more valuable as it instantiates more properties that increase its intrinsic worth.

[19]James Walters takes the position that moral values can alter as a developing individual draws closer to or moves away from "the threshold of indisputably personal life, the life of the normal adult in any society" (Walters 1997, 63). However, he is worried about social policy based on the concept of personhood because it would struggle to get it right in marginal cases (Ibid.).

[20]Various philosophers have agreed that full moral standing requires a sufficiently high level of cognitive functioning Engelhardt's (1986). Kantian position is consistent in large part with Feldman's. McCormick (1974) argues that personhood belongs to those individuals who can enter into meaningful relationships. Fletcher (1997) proposes what looks to be the lowest standard of these types with his claim that any entity that has at least an IQ in the 20–40 range is a person. McMahan states that, roughly, the entity must have the capacity for self-consciousness (McMahan 2002, 6). McMahan's definition implies that an entity need not be self-conscious in order to be a person.

[21]Christian Smith claims that persons are centers of subjective experience, durable identity, moral commitment, and social communication (Smith 2010, 68). However, requiring moral commitment makes the psychological person a moral agent, which is not identical to being a psychological person. In addition, one can be a person without durable identity; the latter is about personal identity over time rather than personhood.

1. Finitude—spatiotemporally located, causally efficacious particulars,
2. Belief—necessarily possess concepts and have beliefs,
3. Agency—not only the performance of the individual intentional actions but also engaging in relatively long-term plans and projects and deliberating about actions, plans, and projects,
4. Second-Ordinality—the ability to have beliefs about beliefs, both one's own and other peoples (Evnine 2008, 10–5).

The final two necessary features of personhood exclude most, if not all, animals other than humans, as well as a number of human subjects, including but not limited to infants and any unconscious human being.

Although the above roughly define psychological personhood, it is clear that individual personalities are not identical to psychological personhood. The latter seem to be the essential trait that makes all person *qua* persons psychological persons, but does not explain what an individual person is, although Feldman does state that some entities can be more of a psychological person than others (Feldman 1992, 120). I assume that some baseline must be crossed in order for an entity to be a psychological person, but that baseline might be exceeded. For example, someone with severe brain damage can be a psychological person if certain states of affairs exist with regard to her mental abilities, but she will not be as much of a psychological person as someone with typical *Homo sapiens* species' functioning in the situation. When the abilities cease to exist for any person, then she stops existing as a psychological person (Feldman 1992, 102).[22] In fact, I contend that she dies as a psychological person even though her body might remain alive as an organism or biome.

The organic whole that I claim is that the ***generic psychological person*** is equivalent to Feldman's psychological person. Although the definition is rather vague and ambiguous, it shows a high level of intellectual ability that allows the entity to be a rational being and a moral agent. Feldman might disagree with my position, but this is the entity that is the core person without the more comprehensive identity of the person as a whole. Identified by Descartes as the mind, it ceases to exist when the entity can no longer fulfill this type of personhood's essential characteristics. Generally, when talking about a person's death, this is the entity that comes to the forefront. It is us as thinking things, and because it has such centrality in who we are as persons and moral agents—because we do our thinking, remembering, and other mental activity here—then it has a much higher value than our physical body, regardless of whether the body is living or dead.[23]

[22]Feldman contends that psychological persons are not essential to identity. He writes that someone with a degenerative mental disease, such as Alzheimer's, can lose her status as a psychological person yet still be the same entity (Feldman 1992, 103). My organic whole approach to identity would make such a result impossible.

[23]It might not be possible to avoid the internalist-externalist debate with this type of psychological person. An internalist believes that "all the conditions that constitute a person's thought and sensations are internal to their skin and contemporaneous, inside and now." (Mendola 2008, 1). On the other hand, an externalist believes that "features of a person's external environment or history are part of what constitutes that person's beliefs, desires, and sensations." (Ibid.). Given our

Moral agents are different from psychological persons even though psychological personhood is necessary to moral agency. A reasonable, working definition of a person as a moral agent is:

> Moral agent = df. Any being that possesses those capacities by virtue of which it can act morally or immorally, can have duties and responsibilities, and can be held accountable for what it does. (Taylor 1986, 14)

What follows from this definition is that the entity must be able to make judgments; have the ability to engage in moral deliberations and then make choices based upon the deliberations; have the resolve and will-power to implement the choice; and finally, have the capacity to hold herself accountable to others for failing to carry out her choices (Ibid.). In addition, moral agents can be good or bad depending on what virtues and vices they have. Of course, having these high-order faculties entails that there are lower level abilities, including but not limited to possessing and using an abstract language, reasoning theoretically, remembering and using her history, recognizing her identity, and having a moral code to apply in deliberations. Furthermore, persons as moral agents are beings capable of self-fulfillment and happiness. They are creative, active community members, capable of shaping their own destiny in a social context and having a wide range of skills and knowledge relevant for the purpose of flourishing.

Given the characteristics of what it is to be a psychological person and moral agent, there is no reason to think that moral agency and personhood are limited to the set of human beings. It is perfectly plausible to believe that there could be other entities in the universe possessing these abilities and having the same intrinsic worth as another person qua psychological person. If it were true that primates or other Earth creatures perform the relevant functions, then they too would be persons of equal moral value to human persons. Less obvious would be primates or other creatures having a sufficiently developed moral code to make them moral agents, but that possibility is not inherently ludicrous to consider.

A psychological person's intrinsic value is less than that of a moral agent. A psychological person need not be identical to a moral agent, although all moral agents are psychological persons. In addition to children who have not quite reached a certain level of moral maturity and other psychological and moral subjects, there are individuals who are amoralists. Psychological reasons, such as being incapable of empathy, render these individuals unable to understand any moral code or have a set of values and rules of ethical conduct. Given the fact that moral agents are psychological persons who are also capable of morality—and perhaps necessary to morality's existences—whereas psychological persons need not be, then it is reasonable to believe that moral agents are more intrinsically valuable than psychological people.

(Footnote 23 continued)

constitution as interrelated and interdependent components in larger organic wholes, externalism seems to have greater explanatory power when considering who we are as persons in those larger contexts.

In addition, not all moral agency is the same. A moral agent's definition is any being that possesses those capacities by virtue of which it can act morally or immorally, can have duties and responsibilities, and can be held accountable for what it does. The definition sets relatively low standards for human beings, but possibly unobtainable ones for most animals. A 10-year-old might be a moral agent in certain situations, but fails to be one in other circumstances. This does not mean that she is not a moral agent at all times on the whole, but she might be such that particular information cannot be known by her, or she has too few tools with which to deal with the situation, that she no longer has duties and responsibilities in the situation. For example, if the 10-year-old knows why stealing is wrong and is not compelled to do it, then she has a duty not to steal and therefore can be held accountable for failing in her duty. However, if the situation is one of intentionally murdering another person, then the 10-year-old is not a moral agent even if she commits homicide. Since she is unable to understand the consequences of the action or even the morality of her choice, then she is not a moral agent in this regard in this situation, although she is still a moral agent on the whole. On the other hand, a normal 50-year-old is still a moral agent in this situation because she does have the basic requisites to be able to make a decision that will make her accountable for murder.

As a result, much as in the case of psychological persons, we could evaluate moral agency's intrinsic value in two ways depending on context. First, if we were merely discussing moral agency for moral agents in general, then the value would be the same for each entity who fulfilled the conditions to be a moral agent. However, if we were trying to figure out individual value per moral agents, then we can use a sliding scale as to intrinsic worth. Those entities that have greater capacities, for example, would have greater intrinsic worth in these circumstances. The same sort of conclusions can be drawn for psychological persons who are equal in one aspect but can be vastly dissimilar in a different aspect.

Moving up the hierarchical scale of intrinsic worth is a combination of two previous entries: physical and psychological persons. The ***embodied person*** will signify the entity, organic whole or not, that results when the overall psychological person and physical subject are a unity. Although this stipulation might entail some sort of Cartesian dualism, it need not. In fact, the term "embodied person" is unfortunate because it makes it appear that the mind is somehow "poured into" or otherwise fitted into a properly functioning body, when that is far from the case. The overall embodied person might be a simple or atomic thing that cannot be deconstructed or separated into mind and body that exists in a particular time and place. Talking about individual identity for persons, for example, Jeff McMahan states "there need be only enough physical and functional continuity to preserve certain basic psychological capacities, particularly the capacity for consciousness." (McMahan 2002, 69). Given how the mind requires the body for proper functioning, and it is likely that the mind supervenes on a properly working brain, there appears rather good evidence that the two should be thought of as an organic unity or whole in some circumstances.

Oddly enough, the embodied person concept is much more nebulous than that of the psychological person because it includes the psychological person's complexities and adds more to it. First, each psychological person we will discuss is "embedded" in some way in a body. The most plausible way of thinking about this is a form of 20th Century Interactionism described in Douglas Hofstadter's work.[24] Hofstadter claims that the "'vortex' of self is responsible for the tangledness, for the Godelian-ness, of mental processes." (Hofstadter 1979, 714). The mind—the psychological person or psychological subject—supervenes on lower level brain or mind activities:

> It is irrelevant whether the system is running deterministically; what makes us call it a "choice maker" is *whether we can identify with a high-level description of the process which takes place when the program runs.* On a low (machine language) level, the program looks like any other program; on a high (chunked) level, qualities such as "will", "intuition", "creativity", and "consciousness" can emerge. (Ibid., 713–714)

Hence, although the mind is dependent on the many layers between it and a properly functioning brain, as well as the brain itself, the mind is not reducible to the brain (Ibid., 300). Christian Smith argues that persons are "higher-order, emergent beings existing through the interactive functioning of specific, lower-order, human causal capacities." (Smith 2010, 59–60). The causal capacities range from subconscious being at the very lowest order of existence capacity to interpersonal communion and love at the highest order. In between the two are primary experience, secondary experience, and creating capacities, such as volition, interest formation, and identity formation, respectively (Ibid., 54). We able to explicate fully what the psychological person is as a thinking thing because, as Hofstadter argues, we can have either a complete or consistent system, but not both. Hence, it is rational for us to pursue the consistent, incomplete system with its ineffability and inexplicability so that we do not have to deal with the contradictions generated by completeness when there is no need to do so.[25]

Although we discuss and treat the psychological person as if it is somehow capable of being separated from the body, that feat is impossible. We know that the two cannot be separated for the reasons stated by Hofstadter. In addition, the body constantly affects the mind even though we might not be fully aware of it happening. For instance, when someone has short-term or long-term sensory deprivation, there is not an automatic communing in the mind as the Cartesian pure thinking suggests or the Platonic purely rational contemplator of the Forms. Instead, as the result of faulty source monitoring, the mind hallucinates. When the mind is deprived of sensory data, then it cannot identify the origin of the experience—that

[24]Antonio Damasio (2011) has an updated version of this in *Self Comes to Mind*. For an extensive, in-depth development of the emergent-supervening property view of the mind using a very large number of psychological studies, see *Irreducible Mind* by Kelly et al. (2009).

[25]I have argued elsewhere that the moral platform might create inconsistencies because it is built to have different values and principles that might not always be in accord with each other. In general, if we can, we should resolve the inconsistencies by modifying the moral code that supervenes from the moral platform.

is, whether it is being generated from the mind or from outside sources. In fact, if there is an absence of sense experience data, the mind will naturally impose its own patterns; this creates hallucinations that appear to be veridical (Mason and Brady 2009, 783). Hence, the psychological person and moral agent as a rational agent cannot be separated from a properly functioning brain, which helps produce it, and the body sending the mind sensory data; otherwise, the mind ceases to be rational.

At times, though it is useful to talk about the psychological person or physical subject in a way that strongly implies that they are separable, they need not be so. However, given the definitions of personhood so far developed, this narrative of identity, personhood, and existence allows for the separation, at least when we are communicating with each other so that a form of dualism can be adopted when doing so is useful to our needs.

3.3 Individual or Personal Identity or Personality

When discussing how we talk about psychological persons in general and in particular, I have introduced a number of definitions of properties, such as personhood, because each does work in ways that we legitimately associate with morality. If we are talking about general values and duties, then we can use the general concepts when thinking about people and other intrinsically valuable entities. For example, the medical field, business world, societies, communities, and other entities comprised of large groups of people loosely organized for some loosely understood common goal have to address a wide variety of situations involving a very large number of people interacting with each other and other intrinsically valuable entities. To write and defend effective policy in these areas, general concepts of personhood and other properties conferring intrinsic worth are appropriate because we cannot individually tailor laws and policies to specific particulars.

However, there are other legitimate instances in which we need to look at individual people in a variety of situations. For example, I can have an obligation to a particular person that I do not have to the general population of persons. I have the duty because of who she is as an individual person and not merely because she is some type of person in general. Hence, individual identity is an important concept in death's values that needs further exploration.

The use of "person" and "identity" are interchangeable at times. A ***person*** is any object that fulfills any of the definitions of personhood. As illustrated earlier, personhood can be a relatively impoverished concept, but it also can include very complex, numerous, and rigorous requirements. For example, all rational creatures are persons merely because they instantiate the complex property of being a rational creature, which entails being a psychological person, and might also entail being a moral agent. A person might also refer to something much more specific. Don is a person according to at least one of the general definitions of personhood, but Don is also a particular person with a personality. Don, the specific person, is who he is;

that is, it is his numerical identity as a personality. Hence, when talking about Don's identity, we could be discussing Don as a person in general or Don as a particular person or personality who exists over time, or both. Since we are figuring out death's values, then we need to understand its impact in general and specifically for the various ways we understand life, persons, and personal identity.

Given the pluralistic way of defining personhood, it is clear that personal identity is going to be a much more complex concept than personhood. Personal identity, if we are limiting ourselves to persons and not any being, must first satisfy the conditions for being a person in some way.[26] What makes the issue much more intricate is that additional characteristics have to be added in order to distinguish one person from another or to meaningfully say that one person is identical to another beyond the fact that they are both persons. Hence, we can talk about generic personhood and what that means in discussions about death and dying, but we can also think about specific individual organisms, biomes, psychological persons, physical persons, and so on. For example, the death of a person might mean one thing, especially if we are thinking about intrinsic worth, but the death of a particular person can mean another. For the former, all entities that are people have identical intrinsic value as persons in general, but they have individual intrinsic worth based on who they are as individuals. Their values could also change as we alter the context, much as a plant can become a weed by merely changing the context from the plant isolated, to the plant in a smaller environment, to the plant in a much larger biome. Therefore, when considering death's values, we have to be concerned with context that will determine how the values are expressed. For personal identity, they can be expressed as the individual organism, biome, et al.

Personal identity includes far more entities than does personhood because we can consider individuals in far more contexts than those limited to the generic type of personhood. That is, we can consider Carl as a generic psychological person and also Carl as a particular psychological person.

In addition, personal identity does not have to be fully internal (Lindemann 2014). Much as it is impossible for the psychological person to exist separately in the real world from the body, individual people are not in actuality distinct in a way that we can totally isolate them from everything else that exists, or even a great deal of it, and then treat them at all times merely as a psychological or embodied person. Let us call this the ***embedded person***, which stretches through time and is inside and outside the internal psychological person and embodied person.[27]

> [A] person and a person's mind are not confined to particular spatio-temporal coordinates, but consist of a spread of biographical events and memories of events, and a dispersed category of material objects, traces, and leavings, which can be attributed to a person and

[26]Personal identity might also by synonymous with individual identity, which applies to any existing object. It could also be used to refer to the identity of living things over time.

[27]The embedded person can be the immediate, intermediate, and overall psychological person. The decision process or needs of the moment should determine to which entity the conversation refers.

> which, in aggregate, testify to agency and [personhood] during a biographical career which may, indeed prolong itself long after biological death. (Gell 1998, 222)[28]

An embedded person is the organic whole that is the embodied or psychological person in her network of relationships in the larger organic whole of being a family, group of friends, personal community, or environment, but it is more than that. The person extends outward from the limitation of the body so that properties we would normally think external to it, i.e., if we are following Descartes, are part of what it is to be a person and to be this particular individual person. The artifacts made by the embodied person are part of the embedded person. Artifacts made by others that include the embodied person are part of the embedded person. Mental states such as memories and feelings about the embodied person are also components of the embedded person. In addition, the objectively real embedded person is that entity in relationship to everything else in the universe that has, is, or will exist. At times, though, it will be useful to have some contextual limitations on identity.

A particular ***socially embedded person*** is defined as the individual organic whole that is a psychological person or embodied person as the focus or center[29] in relation to all other organic wholes with whom the psychological person has significant relationships, where a significant relationship is one having bearing on who the individual person is as an individual person.[30] Charles Taylor argues for something like this with his view that identity is the self in moral space delineated "by the commitments and identifications which provide the frame or horizon... within which I am capable of making a stand." (Taylor 1989, 27). Instead of being solely internal to the person, the embedded person's identity is also determined by external relationships and social constructions of that identity.[31] For example, being engaged would be a significant, if not essential, relationship with another individual person. It helps make each engaged person that particular being (Helm 2010).[32] Being loathed by a powerful rival is a significant relationship whose power is determined by how that relationship affects the organic whole that is the person who is loathed. It also helps create the individual but might not be essential to who the

[28]Gell contends that the externalist theory of agency-attribution is the correct way of understanding what a person is. This theory rejects the mind as being a set of internal experiences, instead claiming that the mind is "in the public domain, as language, practices, routines, rules of the game, etc." (Gell 1998, 126).

[29]Christian Smith contends that the core of a person is "a centering, interior focal point of personal being, consciousness, and activity." (Smith 2010, 62).

[30]Norman Care claims that persons can be social and historical particulars, which is a vastly more expansive concept than the one developed here. "Consider what persons are in respect of their actual involvement in the circumstances and activities of everyday life." (Care 1996, 44).

[31]I am basing this position, in part, on the distributed processes of social perception and cognition that denies that cognition and perception must be internal. Hence, if cognition can be internal and external, then so can personhood. See Smith and Collins' (2009) "Contextualizing Person Perception: Distributed Social Cognition.".

[32]Bennett Helm also argues that friendship is a "kind of plural agency." (Helm 2010, Chap. 8).

person is. That is, the person could be the person without the antagonistic relationship but could not be that specific person without being married.

From the above, we see that there are different relationships with different powers to determine individual identity. If altering the relationship will necessarily change the person, then the relationship is an essential part of the person. If the relationship is merely important, then altering it will not essentially change the organic whole that is the person as an individual, or even the person as a whole, but will make it more likely that essential changes will occur. A trivial relationship may be altered without the organic whole undergoing significant alteration. Given the changing nature of relationships, such as their intensity and type, as well as the quantity of relationships, the embedded person can vary its identity. Regardless of the relationship, who the person is as a person is affected by not only what happens internally in the embodied person, but also externally as well. This type of person takes seriously the notion that part of being a socially constructed person, in at least one sense, is being identified as a person by others, especially since most of us are social beings.

The many distinctions made in this approach are right and useful in at least one way. When we want to think about personal identity and who the person intrinsically or otherwise is, then we can practically locate the person's identity completely within the node or point in the overall organic whole which is the group, society, or other complex thing that serves as the atomic unit of the context. That is, the entity that is the person, regardless of which type of general or individual personhood is being used, is an identifiable component of the larger whole, although we might be unable to clearly and distinctly delineate the component's boundaries. Consider, for instance, the person's emotional relationships being contained wholly within the physical subject or psychological person. A socially embedded person internally contains all the memories, self-concept, rules, rationality, and other necessary features for not only being a person in general, but also a particular person. At the same time, the socially embedded person's identity external to the psychological person contains all the relationships and beliefs and mental states about her that are in the minds of other psychological subjects that do not exist within the psychological or embodied persons. However, even if it does not capture the greater objective reality, it is often convenient to treat people as isolatable nodes that are wholly distinct from each other so that we can talk about their actions and duties, and judge them. In other words, we have to consider people in different aspects when considering them in the different contexts we create as we pragmatically delineate reality in the situation.

Before moving into a more controversial area, it is worth the effort to discuss the problem of identity over time. Although we act as if we can isolate embodied and other types of people as individuals to talk about their identity over time, such an endeavor is always suspect. Have we ever truly and completely captured the individual object to which we want to refer when talking about a particular individual? This is unlikely because we aim too low. The problem comes about because the only objectively real identity over time is that of the universe as a whole. No matter what happens to it, since there is only one universe in which we exist, it is always numerically the same universe from beginning to end—if there is a beginning or end to it. Of course, conditions within the universe can change—and greatly so—but the universe is numerically identical to the universe and cannot be confused with any other object because there are no other objects with which to confuse it.

Things get much messier if we look at the universe's components in a different context. If we try to understand the universe's "internal" objects and relationships, or parts of the universe rather than the universe as a whole, then we are going to have to be relativists in the physicists' sense of that term. We always must first subjectively create X,Y,Z coordinates of 0,0,0 as a reference framework, and then describe the rest of the universe or its parts relative to this one point. For example, to say that a galaxy is moving away or toward another object requires some foundational point that is not the galaxy one is claiming to be moving. If an observer is standing in the galaxy, then neither is moving relative to them. All other objects in the universe are moving relative to the observer, so the other objects are receding or gaining while the observer remains at 0,0,0 coordinates.

If we are talking about organisms or other living entities in the universe, then we often have to bring even more relativity, subjectivity, and artificiality into the identity of parts of the universe. Once again, we begin with the 0,0,0 coordinates for the object we wish to single out, but at times and according to our pragmatic needs, we expand these outward from a mathematical point occupying no actual space and time to something that is extended in space and time, such as a body. This often pragmatic but subjective expansion of the boundaries of the object can be much larger, as when considering a socially embedded person, and is determined by the needs of what is trying to be accomplished, such as figuring out individual worth in a specific context. The relationships of such an embedded entity, as well as the mental states those in relationships have about the person, are part of its identity, which entail that we must incorporate them into our object occupying the center of the universe position. This makes our mathematical point in space and time a very large and fuzzy-edged object.

Assuming that all things in the universe are matter-energy, we can consider what would appear to be an uncontroversially, objectively grasped entity. Assume that we have a certain amount of matter-energy. When we first encounter it, the matter-energy is in the form of a granite rock. The next moment, the granite rock has decomposed into granite dust. In one significant way, the before and after objects are the same because they are numerically identical matter-energy, as long as the matter-energy is conserved.[33] This is a context that makes sense at times.

Someone might also reasonably say that the rock and dust on the macro-level are numerically identical as long as we add that both are granite. The latter property is sufficient to maintain identity in this macro-level context.

[33]In reality, the transformations would require or release energy, which would alter the amount of matter-energy. For the sake of the argument, assume some sort of Dyson sphere is operating here.

We would also be justified in saying that identity was not preserved because the rock and dust are of different forms or types. We might claim that the change in form is sufficient to alter the identity of the former so that the dust can no longer be called the same as the rock.[34] The delineated context here would be on shape rather than conservation of matter-energy.

It is clear, therefore, that the context we select in the situation determines which conclusion or conclusions are true. In the first context above, we can say that the two are totally identical. In the second, they are numerically identical according to macro-level matter. In the third, they are essentially different because they are far too dissimilar to be numerically identical. Finally, when we combine different properties and levels, some of which are identical and some of which are not, then we have to use our practical judgment as to what works in the situation. If we continued to change the matter-energy to different forms, such as chicken, tree, or water, while conserving the matter-energy composing the macro-level objects, we still can maintain that they are numerically identical on the micro-level, while at the same time truthfully saying they are not numerically identical on the macro-level. Our subjective selection of various aspects of the one reality can yield different understandings as well as truths of propositions.

If we consider what we are as living things in all of our various aspects, then identity over time becomes even more of a difficult story to tell than we may want it to be. First, we must select the general context level we want to use. Physics will give us a story far different from chemistry. Biology gives us yet another level of attempting to understand identity over time. I have been arguing that there is yet another level beyond biology that cannot be explained in biological terms, much as biology cannot be explained by chemistry and physics, and chemistry by physics. Second, once we have the level, then we might need to look at variations on the level. For example, in physics, we can understand physical objects on the micro-level using quantum physics and on the macro-level using relativity. Biology can be divided into micro-biology and macro-biology, and so on. Each of these levels and variations can be used as legitimate contexts in which to think about an object's identity over time.

Finally, even if we remain on the same level, then we encounter different types of people, each of whom has its own definition and context that might be vastly different from another type. What we say truly about physical people, for example, may not hold for psychological people, as in the case of massive trauma to the brain. The psychological person ceases to exist because the psychological capabilities that compose the psychological person no longer exist, which entails that the psychological person no longer exists. However, the physical person of the body may persist even in these trying circumstances. The entity might be a psychological or moral subject, or merely be a living body in a permanent vegetative state, but it is still numerically identical to the body that existed prior to the brain trauma.

---

[34] The decay of radioactive elements would be another way of understanding this point.

Beyond instantiating the general characteristic of being a person or living thing that all persons and living things instantiate, what makes each of us an individual that is different from all other individuals and that persists over time? I will begin with the persistence issue, not because it is an easier question to answer, but because I do not have much to add to the standard answers to it. To restate the question, what are the necessary and sufficient conditions for identity over time, that is, for people at two different times to be numerically identical?

John Locke's version of the psychological account of personal identity is, "a thinking intelligent being, that has reason and reflection, and can consider itself as itself, the same thinking thing, in different times and places." (Locke 1992, 170). For a Lockean, the individual has to subjectively conclude that the individual that exists at this moment is identical to the individual that existed but a moment ago or even further back in time.

Of course, Locke's version faces the problem of what happens to the entity when it cannot reason, reflect, or consider itself as itself because the entity has insufficient mental activity. Perhaps the individual is temporarily unconscious. Does that mean that personal identity ceases to exist for the entity from the very moment of unconsciousness until she regains her ability to perform the proper mental functioning? Most of the time, we want to say that even when the person is unconscious, it is still the same person, but under Locke's view, the person ceases to exist the moment she is not conscious in an adequate way.

A different way to understand the psychological account basically states that a person at $time_{t+n}$ has persisted over time if and only if the person has inherited the mental features of a person at $time_t$.[35] The question then becomes, what mental features have to be inherited in order to have persistence over time? We cannot use memory because of the obvious problem that people forget. For example, if Rose has a stroke that causes her to lose all of her memories of her life between the ages of 22 and 45, then we would have to say that Rose after the stroke is not the same person as Rose between the ages of 22 and 45. Making things odder is the fact that Rose after the stroke would be the same person as Rose before the age of 22 and after the age of 45.

Perhaps the most plausible version of the psychological approach can be found in Sydney Shoemaker's work. Shoemaker argues that instead of memory being the shared mental feature, psychological connectedness and continuity establishes identity over time (Shoemaker 1984, 89ff).[36] Psychological connectedness happens

---

[35]See Garrett (1998), Noonan (2003), and Parfit (1984).

[36]Barry Dainton argues that phenomenal continuity in which we are systems of experiential capacities is the best account and sufficient for persistence (Dainton 2008, 21 and 112–3).

Although Dainton intimately links phenomenal continuity with psychological persistence, my concern about his approach is that two entities might have the very same phenomenal continuity but not have the same psychological continuity. Massive alterations in one's psychology entail the same person as long as phenomenal continuity is maintained, but there are legitimate reasons in many circumstances to say that the vastly changed entity is not numerically identical to that of the old even though Dainton's view would say they are numerically identical.

when a person having psychological states at time$_{t+n}$ has those states as a result of the psychological states a person at time$_t$ had, and the psychological states at the latter time are causally connected through a chain of psychological connections, which makes them psychologically continuous. A person having psychological states at time$_{t+n}$ is identical to the person at time$_t$ if and only if the psychological states at time$_{t+n}$ are psychologically continuous with the psychological states of the person at time$_t$.

Although the split-brain objection to the psychological approach has been raised, which is also a Fission argument variation, it is not a serious impediment to thinking of personal identity being based on mental features, regardless of whether the individual identifies with a previous individual, if the current person has memories that are sufficiently similar to those of the previous individual. The split-brain criticism is supposed to show that, "after the split-brain operation [a cerebral commissurotomy] the patient is literally two persons who share the same body." (Moor 1982, 91). As James Moor has persuasively argued, though, the various two person interpretations are either conceptually impossible, as in the sophistication and separation arguments, or empirically impossible, as in the autonomy argument. For example, the scientific reality of cerebral commissurotomy is often misunderstood by those arguing that there are two minds instead of one:

> First, it is a mistake to think that there is a quantum jump with regard to separation of consciousness between split-brain patients and normal individuals. The split-brain spectrum makes it clear there is a gradual separation of spheres of consciousness as more of the commissures are sectioned…Second, it is a mistake to think that patients with complete commissurotomies have spheres of consciousness that are totally separated. In this regard the label "complete" is misleading. A patient with a complete commissurotomy still has a highly integrated nervous system. (Moor 1982, 94)

The patient is, in actuality, still one person with one mind with a disunified consciousness. Writers often go astray in not recognizing that there is a split-brain spectrum in which kind and degree of disunity vary, thereby not making it an either/or situation of being this person or someone else the moment the operation is completed. (Ibid., 102)

The fission objection does illustrate an important point that I have been making throughout this work. In order to answer the big questions about how to think about our lives, who we are, what we should be and how to be, we should be practical. If an objection or thought experiment would require a magic wand to work, then it is not terribly helpful in trying to figure out what is casually possible. For example, one variation of the fission objection has a person's right and left hemispheres transplanted into two different heads. The question is then asked if there are two of the person or two people instead of just one (Noonan 2003, 139–42). Such an operation would not be successful if we measure results based on whether the patient will die; these operations just cannot work. Therefore, in order to answer these questions and find useful information, we need to be very careful that metaphysical possibilities are not being privileged over casual impossibilities.[37]

Unlike the psychological approach to personal identity, the somatic approach focuses on the physical body, usually a living one, sometimes treated as a computer program (Kagan 2012, 168; Ledoux 2002, 302, respectively).[38] A person X is identical to a person Y if and only if X and Y have the same body or are the same organism. Since we have the same body or basic form as we did when we were children—although with many, many physical changes—we are, in one sense, the same person we were when we first became a body or an organism, and we will continue to exist as this person until the body or organism is sufficiently altered. For example, if the organism dies, then by necessity, the living thing cannot be the same object as the dead entity. A body, on the other hand, can persist in much the same way that Feldman claims that personhood remains for a time after the person's body has died.

Although there are situations in which the somatic approach is useful, there is an interesting problem with it. It deals with the question of how much of the physical body has to change before personal identity is lost. This is not an issue of transplanting a brain into a new body, and then transplanting someone else's brain into the old body, although that also questions quantity and quality of bodily matter required to preserve identity.[39] Instead, suppose that a person has died. If we bisect the corpse from head to toe, then are the two equal pieces each the person? It seems as if both are, or neither one is. On the other hand, if we cut instead across the waist, do we have the same person? It would seem that most individuals using their intuitions would privilege the top over the bottom half when it comes to identifying it as the person.

My guess is that we would apportion percentages in cases in which the body is divided. If there is a portion of the body with an insufficient percentage from the original mass, and it seems as if it would be a very high percentage based on how we identify the person, then we would think that personal identity qua the body would persist. Grandma without her finger would be Grandma, and her finger would not be considered to be her, although it is her finger. Grandma's head or

[37]Perhaps a better way to describe this problem is to keep a person's brain intact inside her head, and then to wonder if we can duplicate that brain and its emergent properties through artificial intelligence. To have a second person, we would most likely need something along the lines of Searle's synthetic brain that exactly duplicates every relevant property of the original brain, but if we could do it—and there is no reason to think we can—then we would have two different entities with the very same psychological states. In this case, the two would be the same person in the short term, and we should treat them the same because they are identical in a very significant sense of the term if one is not a person who adopts the somatic approach to individual identity. As new experiences and thinking happen and alter each, the two will become different persons.

[38]Searle rejects Artificial Intelligence on the grounds that it is about programs and not about the machines. Moreover, "Whatever else intentionality is, it is a biological phenomenon". (Searle 1981, 372).

[39]I do not spend much time with this thought experiment because it is causally impossible and likely to remain so.

brain without her body would be a far harder case to call. Most identify Grandma by her body or head, as they have been doing ever since she was involved with them as two embedded people in relationships. On the other hand, we know that Grandma's brain was the platform that allowed her to exist as a psychological person. Thus, we also might consider her brain to be Grandma in the flesh, so to speak, because of what it did in the past. The rest of her body or head might be considered a shell by which we recognized her, but the real, physical Grandma who cared for us was her brain.

In cases such as these, it probably does not matter if we identify Grandma as the body and as the brain even though it might appear to be a contradiction. That is, it makes sense to say that the brain is Grandma and the body is Grandma. By making this identification, we would be able to achieve certain practical goals, such as making claim to both physical substances so that we can dispose of them properly. If it were Grandma's body and finger, then we would not be justified in claiming that both are Grandma because the finger is insufficiently similar to Grandma to be her in any significant way.

A promising, although very subjective, approach to identity that seems to be based on Locke's internal confirmation of persistence is that of narration. By allowing her to craft who she was, is, and will be, narrative identity gives each individual power over her personal identity, although the power is constrained by the world and context (McAdams and McLean 2013, 233; Bamberg 2011, 4–5).

> When we give an account of ourselves we provide a narrative: a report from a first-personal perspective to a second-person (implicitly or explicitly) that links together third-personal objects, actors, times, and places in a temporally extended, causally related coherent whole, with a beginning, a middle, and an end. (Atkins 2008, 76)

Numerical identity of one individual at one time and another individual at another time is created by the individual who selects continuity by saying that this earlier entity is numerically identical to the one who exists now and the ones who will exist at different times.

One significant problem with narrative identity is truth. Instead of a correspondence theory of truth being used, narrations are relativistic because each individual creates her own reality to suit her own needs and desires regardless of whether that narration is a false story. For example, because of self-deception based on egoistic and moralistic biases, most people are often inaccurate when evaluating their own moral characters or actions. Most of us think that we are far better people than we actually are, or justify our actions as being permissible, and even necessary, when we do not make the same evaluations for the characters and actions of others. Hence, many narratives do not correspond to the facts in the world and are, therefore, false narratives (Paulhus and John 1998). Given the moral requirement to contribute to our and others' flourishing, false narrative identities can actually hamper, if not destroy, our ability to fulfill our overall duty to thrive. If we do not evaluate ourselves truthfully, we do not know what we need to improve or how we can understand our value individually and as part of greater organic wholes.

I think that employing the correspondence theory of truth within the identified context being used can solve this problem. If we are talking about the person within a society, for example, then the person's narrative can be wrong given facts in that context. The person, for instance, might incorrectly think herself to be a morally superior person in her society, when empirically, that belief is false. There are better people than her in her society, which is a context she selected by her judgment. Even in those instances in which she is not implicitly drawing conclusions about entities in the external world, which therefore allows a broader context and additional set of facts, the correspondence theory is still practical. Suppose the individual thinks that she need not improve herself because she believes herself as good as she can be and that her life is thriving at its maximum potential. In this situation, we can compare the narrative to the actuality of the individual's existence and its potential within the circumstances in which she lives. If the narrative does not reflect the facts, then it is false. Hence, for situations in which there are facts, then the narrative must fit the objective facts of the context, but if there are no facts, then the individual's subjective narrative holds sway.

The narrative approach is in keeping with my position that we, as people trying to understand reality, have to select smaller contexts from the one reality of the universe as an organic whole. Our actual identity is that of a three-dimensional axis point of 0, 0, 0 coordinates in relation to everything else in the universe as it exists in the moment, everything in the past, and everything in the future. However, thinking of ourselves in this manner is not only unwieldy most of the time, but nearly impossible given our finite minds' inability to grasp such an enormous amount of information. To think and practically interact in the world, we must carve out more manageable chunks of reality[40] by stipulating smaller contexts, as we will see in the next section. Which chunks we should carve out are determined by our goals of pursuing our flourishing and that of others.

Even though none of the solutions to personal identity works in every case, I do not want to throw out Locke's other view or memory as a candidate for the inherited mental feature, Shoemaker's psychological continuity approach, or the somatic approach because they can be useful in at least one plausible context. I have to admit that I find it rather odd to dismiss the value of something merely because it does not work all the time. My hammer is useless when I am trying to turn a screw, but I do not discard the hammer merely because it is not useful in all tool situations. The same approach should be taken to personal identity, as we have been using for our pluralistic approach to the definitions of terms, which assumes that the definition that is relevant in the situation is one that works well enough in the situation to make the definition relevant. Sometimes it is the somatic approach, and

[40]Their reality should be understood as dependent, in part, on the content and the minds of those involved in the situation. Perhaps the best way to think about it is along the lines of George Berkeley's "Esse est percipi" with regard to the context, but the individual objects within the context might be independent in significant ways.

sometimes it is another that is justified according to the situation and its context. This inclusiveness will also be used for other approaches that can be valuable in at least one plausible context.

3.4 Individual Psychological Identity Over Time

There are many types of personhood and many types of individual identities within the context of each of those personhoods (Ouellette 2014, 7). I will focus on moral agents to think about what exists over time. What is said about them below is true in most cases for the other types of persons and the individuals instantiating those properties.

An ***individual moral agent*** is equivalent to the psychological person who makes decisions and acts using a moral code consisting of a value theory and normative principles. Individual moral agents may be discussed in three different, sometimes vague ways: immediate, intermediate, and overall moral agent. The first refers to the person at a particular moment or otherwise clearly delineated time segments—for example, the specific person who chose to do X at time Y. Which time periods qualify for being immediate will be determined by those needing to do efficient, effective work in the situation's context. In certain situations, we need to talk about the immediate person from moment to moment—the person before and after a major accident, for instance. In other circumstances, we will need to extend the time periods—the person has not been herself for the last month, but this month has returned to her normal behavior.

It is useful to have a concept of immediate psychological people because we sometimes say that a person is not the same as she was only one hour ago. If, for instance, she had been in great pain from a kidney stone, then in one intelligible context, she is not the same person. The pain severely affected her ability to make medical decisions and otherwise function, which altered her from her normal self to someone we might not listen to if her ability to reason is too greatly compromised, e.g., in deciding her care. We could not permissibly treat her the same way once her mental faculties are back to normal as measured against her nominal identity for longer periods of time. There are instances, therefore, when people can be meaningfully different people in relatively short periods of time—in this case, moving from being a moral agent to a psychological subject.

Of course, in order to be a person, the individual has a history that helped form who she is, unless we are talking about some causally impossible world in which we can download "memory" data into a mind. We can consider individuals in isolation because at times that is useful to what we are doing, but often we will need to understand them as part of a greater whole that has existed over a far longer stretch of time. Hence, there are plausible circumstances in which an immediate person must also be understood in context of the overall moral agent. For example, a 50 year-old woman is probably the same person at the beginning of the calendar year as she is at the end of it.[41] If there have been no events or series of events that have changed her essential identity, or altered who she is as an organic whole sufficiently so that she is a new organic whole, then she must be the same.

On the other hand, younger children who are moral agents might not be the same immediate moral agents from one year to the next. Because they have greater potentiality and malleability than do those whose personalities have been set and re-enforced over time, changes that would not alter a more mature person's personal identity could affect who the younger person is, especially around the time of puberty when the latter's world is in greater turmoil. That is, their ideologies and foundational beliefs are not hardened into being unchallengeable and privileged, nor have they developed a set of beliefs as those who have been thinking and experiencing for a far longer time period, which is used to evaluate new beliefs for acceptability. Hence, it is easier to add to or change their belief sets than for those who are older.[42]

The intermediate moral agent is an entity that exists from the time she starts having the mental states and capacities that give the person a particular identity to the time when she ceases to have those mental states and capacities that give her that identity. No 50-year-old individual is essentially the same moral agent she was when she was 10 years old. The former entity ceased to exist because the 50-year-old has sufficiently different mental characteristics from the same moral agent as the younger entity. The 50-year-old is much better at making decisions, has different interests, and has a history that has altered her thinking processes. The older entity is a moral agent and individual in a way that the other is not. At the other end of life, if a person undergoes a mentally disabling disease, she gradually becomes another person, or no person at all. In this case, the older entity and the individual she was cease to be a moral agent.

Of course, the exact moment when the entity ceases having mental states is difficult to identity for both the individual and other psychological entities. If someone is in a deep enough coma without mental states, then the intermediate moral agent has ceased to exist the moment one of her essential mental states did or when a sufficient number of her mental states ended so that there was an essential change in who she is. Mental death, which might be brain death, or the result of brain damage, is also the death of the person. Making things more complex is the fact that different moral agents and psychological persons can come and go over time for the same physical person, which means that an overall person experiences many deaths.

The intermediate person type, as well as those not immediate persons, raises an interesting ontological issue that will affect discussions of death. For Essentialism,

---

[41]However, it is clear that adult personality varies across age. Middle age is the most stable time period (Milojev and Sibley 2014, 29).

[42]For example, unlike adults, infants do not bind beliefs to the belief holders (Kampis et al. 2013, 232).

the very instant that a necessary feature is altered is the precise moment in time that the object becomes a different thing. That is, the identity essentially alters at that instant, but this supposes that each change is instantaneous, when they need not be, as can be illustrated with the Problem of the Heap.

If adding salt crystals to a flat table, the question is when the amount of salt on the table becomes a heap and not merely a bunch of salt crystals scattered on the table.[43] Of course, if there are no crystals or only one, then there is no heap. If we pile salt crystal after salt crystal on the table to form a mound that no one in his right mind would deny is a heap, then we clearly have a heap. However, when in that process does the object go from being a grouping of salt crystals in relation to each other to being a heap of salt crystals? Unlike switching on a working light, there is no exact and clear moment in which one more crystal will create that heap. The fact that there is a Problem of the Heap shows that there is indeterminacy in the real world.

The lack of clear alteration from non-heap to heap—and from life to death and existence to non-existence—opens up a number of possibilities in explaining the situation and when someone becomes a new person. If there is a length of time in which we cannot tell when the salt is a heap or not, then the same might apply to alterations in identity and personhood. There is no exact moment when a child becomes an adult. The society through laws, social institutions through conventions, and individuals through their own belief systems arbitrarily set a time, but social constructs are irrelevant as to whether the entity has actually passed from one state to another, especially to adulthood's moral agency contrary to the general rule. The entity might have achieved that state far earlier, later, or never at all. Therefore, the transition periods between intermediate persons are not to be drawn in solid lines on a timeline with bold markers. At best, we can say that a transition seems to have begun around here and ended around there, and even then an unanimous agreement with our conclusions is unlikely. We would all recognize that the entities before and after are not the same in at least one essential or significant way, but the time between when the entity was obviously X and the new being was obviously not X is not something we can identify with any certainty. The same claim holds true for acquiring or losing virtues or vices, dying, decomposing bodies, changes in relationships, and anything else that is complex and takes time to change through tiny incremental alterations.

This transition time is a very interesting one to try to understand because it will have bearing on the discussion about whether a person can be harmed by his own death, as well as the controversy over declaring someone brain dead. First, during the transition from one object to another, there cannot be yet a third object between them. For example, if the transition is from the entity being a psychological subject to a moral agent, there is no psychological entity that comes into existence during

[43]Earl Conee has argued that it happens the moment that one crystal is placed on top of another. This view can be useful in certain contexts, but not the vast majority of instances when a heap is something we need to think or do something about.

that time that is its own unique self. If there were such a thing, then we would face Aristotle's Third Man Argument, in which we have an infinite number of transitional entities that are unique, or merely be able to make clear distinctions by taking smaller and smaller segments of time until we have an instant switch from the entity being the psychological subject to it being another unique entity. As we have already seen, the latter was not possible with the Problem of the Heap. The transition does not lend itself to instantaneous cessation of existence to something new coming into existence.

Second, it might be that in the transition both objects are present in different ways. When undergoing the death process, part of the person who is dying is still there, perhaps by remembering events that happened to her. Simultaneously, other mental states have slipped away. These were important to who she was because they affected perceptions of herself, and her actions, but they were not essential to her immediate, intermediate, or overall personhood. The new characteristics, however, are essential to the new being. The benefits reflect the gradual change of the actual process of dying and ceasing to exist, as well as living and coming into being, and allow us to plausibly talk about one object being harmed by the other because both are in existence simultaneously. The fact that one is transitioning means that one is still alive during the process, and is being harmed from step-to-step in that process because the person is becoming worse and worse off as she goes along.

However, there is a severe drawback to this view. Since the necessary features are still being instantiated by the individual, we are not actually describing anything that shows the Problem of the Heap argument for the transformation of essential or significant qualities for the involved organic wholes. As each necessary component of the person ceases to exist as a new entity comes into being, the person is no longer the original entity and cannot be harmed by the new being.

The third option is disturbing for Leibniz and the Principle of Contradiction, but might reflect what actually occurs, even though it might offend our desire to never have contradictions instantiated because of their negative effect on rationality. In the transition, it might be the case that both entities are present and not present in the same way. Although Schrödinger's cat in a box thought experiment was designed to show a truth in the quantum realm, it can provide illumination in the macro-level as well. Schrödinger was trying to argue against a certain physics' claim by showing that it would lead to a ridiculous conclusion; instead,, he actually helped establish the claim's validity. Schrödinger postulated that we could make a box containing a bottle of cyanide and having a hole through which a sub-atomic particle could break said bottle should it enter the box. Next, the box would have a mechanism that would break the bottle if we lifted the box's lid after we sealed the box. Finally, we could add a cat to the box with the unbroken bottle of cyanide, and then seal the box closed. Contrary to our intuitions and psychological need for something to be in one state or its opposite, the moment after we put the cat into the box, the cat would be both alive and dead. Without anyone observing to collapse the wave function, the feline could simultaneously be in both states. The problem, of course, is that contradictions undermine our reason because anything can be proven through a contradiction, whether or not it is causally impossible.[44]

Although we would like clear and distinct transitions, answers to questions, definitions, and knowledge, it could very well be the case that such things are causally impossible. We know that there is no moment when one goes from living to dead, at least in most, if not all, cases, unless we have a spaceship that we can use to instantaneously vaporize a person in the sun. Even someone who shoots her brains out might have some transition time, although the action is too quick for her to perceive it fully. If we adopt the position that an entity can be dead and alive at the same time, then we can escape Epicurus' argument and satisfy the intuitions of a number of people.[45] The death of a person might harm her because she is both alive to be harmed and dead simultaneously. Just as someone has and does not have a virtue—another problem of the heap—so, too, may these transitions exist.

Returning to the discussing of immediate, intermediate, and overall persons, the distinction between immediate and intermediate people will be useful for a variety of reasons. First, it enables us to account for the otherwise odd saying that a person is not acting as herself on a particular day. She has to be herself if what is being referenced is the immediate person because that is the immediate person acting. However, the immediate person might not be identical to the intermediate person who can have different qualities. A person who is not acting as herself is merely not acting as the intermediate person would usually act, but is doing what the immediate person would do. Second, we can value the immediate person having a mental state, such as pleasure, and then compare the intrinsic value as to whether or not that pleasure is deserved, which might require us to reference the intermediate person and what has occurred in that person's span of time. John-at-the-moment-of-having-a-pleasurable-state has a certain intrinsic value. It might even be that in that particular time slice, he deserves the pleasure because he has been especially good recently. However, if he has an ethical debt he created for himself earlier in his life that is not overcome by the good, then John as an intermediate person having that pleasurable state has a reduced intrinsic value in comparison to John as the immediate person having the exact same state.

In addition to immediate and intermediate persons, the overall moral agent or person is roughly the unity of all the immediate moral agents or intermediate persons that make up the person's life.[46] The reason that this is merely a rough

---

[44]Of course, if there are contradictory states of existence in reality, then we would have to rethink a great deal of our views about rationality. However, as I established in Chap. 1, our theory should never force reality to conform to it; the theory should conform to reality. To address the issue, we might have to create a more complex view that delineates various forms of rationality that are relative to context.

[45]Satisfying people's intuitions is not necessarily a pragmatic good.

[46]I am concerned that the combination of the different components will lead to a form of reductionism that misses the fact that the whole might not be equal to the sum of the parts. Identities through narration are one useful way of thinking about how this composition works. Hilde Lindemann states that narrative identities consist of "stories [that] display the various facets of who

definition is, once again, the difficulty of capturing such a complex concept in words. First, each immediate moral agent is its own organic whole in a much greater series of organic whole culminating in the universe organic whole. Hence, merely treating each as a discrete entity and then summing them up cardinally might not recognize true intrinsic value in the situation. Second, it is reasonable and useful to discuss overall moral agents while they are still in existence or at a moment in their development. Of course, as Aristotle recognized, evaluating the happiness of a person's life must wait until after the person is dead, but sometimes we need to do it prior to death. For example, when working with our parents to make medical decisions, we can think about them as overall moral agents ***to this point in time*** when we help them decide what medical care they need and want.[47] Therefore, not only the needs and wants of the immediate moral agent are incorporated into the decision process, but also those of the overall person to this time.

In these decisions, it can also be permissible to consider the overall moral agent for periods of time that are in the future. When a parent has Alzheimer's, for instance, the diseases' early stages do not greatly alter the immediate or intermediate persons. Over time, however, both the immediate and intermediate persons begin to fade away into possibly a new immediate or intermediate person or from personhood into the realm of moral subject and psychological subject. Those future people and beings have to be considered as well in planning, especially in light of limited resources for them to use. Do we transfer the house to the healthy spouse now, or will that future person want to retain it? Do the current person's needs and wants defeat those of the future person or subject so that we can spend assets at this time to try to preserve the current entity, which might then reduce resources the future person or being would need? These hard questions show that what an overall person is should be flexible enough to do all the work we need to do in answering questions about end of life decisions, medical care, and other important areas of an overall person's life. The sacrifice for adopting a vague concept is precision, which will affect deadlines on when decisions have to be made and implemented as well as other activities. However, morality does not require precision when many moral situations have fuzzy edges by their very nature.

(Footnote 46 continued)

the person is." (Lindemann 2014, 4). Betty Flowers claims that our personal stories shape us (Flowers 1998, 51). Both recognize that external facts constrain the stories, but that there is still power for an individual to create her own identity and individual personhood. Returning to the irreducible organic whole, David Vellman makes claims that there are two values, momentary and overall, that are sympathetic to my understanding of organic wholes and context creating a form of reality. According to Vellman, momentary and overall value do not reduce to each other, nor can they be analyzed in terms of each other. The overall value of our lives is determined not by summing up all the momentary values, but by the narrative of the whole and how those momentary values are "stitched" together and how they relate to each other and to the whole (Vellman 1991).

[47]Most of us do not know our parents well enough to work with overall moral agents, so we have to make do with intermediate persons.

3.5 Socially Embedded People and Individual Identity

A different way to consider individual identity is to understand an individual as a socially embedded person, which, as defined above, requires us to understand personal identity as not being totally located within the psychological or embodied person. Rather, an individual is actually a larger organic whole in which the embodied person is the central nexus in relation to all other nodes within that whole. Therefore, who the person is turns out to be internal, as in the embodied person, and external, as in the relationships and the mental states incorporating her that others have. Arthur Gell, for instance, makes a distinction between primary and secondary agents, in which the primary agent is the psychological person, moral agent, or embodied person, and the secondary agent is the primary agent's artefacts "through which primary agents distribute their agency in the causal milieu, and thus render their agency effective." (Gell 1998, 20). This is one type of socially embedded person, but there are far richer ways a person can be embedded, including her artifacts, personal relationships, how people view and think about her, and relationships with other entities in the organic whole we are using to provide context in the situation.

Socially embedded people are, in part, the result of their interactions with the other nodes in the organic whole complex. Hence, part of their identity is the result of those interactions. Some of the results rest wholly within the psychological person; for example, if a person is mugged, then the fear the person has from being mugged is part of who the person is. She is the only one who could have experienced it in the unique way that she did. The mental states become part of her psychological history, which will influence her future decisions. Over time, these decisions can lead to important, if not essential, alterations in her identity.

The mugging and other experiences the person has experienced are also part of the person's embedded identity. How others identify her is part of who she is as a particular person. It is a significant component of her objective identity because we can now identify her mentally and physically, in part, as the person who was mugged at this place and at this time. We might feel pity for her because of her ordeal, which will be part of how we identify who the person is. If an aggressive, macho character believes that the person is a coward for allowing herself to be attacked and robbed, then that is part of who the embedded person actually is.- not that she needs to be so as a psychological person, physical subject, or embodied person. How others view her in regards to those entities is external to her, so they are not relevant per se to her internal nature.

However, for an embedded person, the organic whole containing her as an organic whole is the context in which we have to understand her identity, especially when thinking about who the person is in a group, such as society. Within that embedded whole person are all the other psychological individuals who make up a significant part of the organic whole. Hence, their thinking is internal to the whole, although external to the embodied person who is the nexus of the larger whole. Therefore, just as individual thoughts, virtues, and mental states are part of a psychological person's identity, so too are the thoughts of all the non-nexus persons of an organic whole that incorporate them. A component of an embedded person's identity is out of her control because the relationships she has depend upon others, and how they think about those relationships and the people involved in them are under the former's control. In the case of the mugging victim, the victim actually is the individual who is viewed by a particular aggressive, macho character as a coward for allowing herself to be attacked and robbed; hence, her identity can be altered by external agents and forces.

The concept of identity that exceeds the boundaries of a particular psychological person and yet is part of who the embedded person is needs further explication. In reality, the identity of a person cannot be limited solely to the psychological person, although the latter plays a significant, essential role in overall identity. Relationships with other entities are necessarily not internal to the psychological entity, but part of the identity of the overall person or embedded person. For example, suppose that John is the father of Jack. Although John experiences all the internal mental states associated with being the father of Jack, such as pleasure in his son's accomplishments, pain at his son's failures, and so on, the relationship is not wholly contained within those experiences. For John to be the father of Jack, there must be a Jack and a specific relationship between them such that Jack's father is John. Moreover, being a father has more content to it than, for example, standing to the left of Jack. There is an emotional and reciprocal interaction between the two that would not exist if one of the two experiencing subjects did not exist. John's personhood, therefore, is not merely his psychological states, but the relationships he has with others, as well as their thoughts, feelings, and other mental states about him. Unlike the case of John and Mary as a married couple in which there is a new entity that internally includes those relationships,(which makes the case stronger for relationships being part of identity), the existence of a unity along these lines is difficult, at best, to see in the intimate relationship between father and son. There isn't the level of commitment and reciprocity that exists with a married couple. This does not mean that the embedded person's identity in non-essential relationships lacks these features, only that they tend to be less important to who the embedded person is.

Furthermore, given the organic whole context, the memories and perceptions people have of a psychological person in a physical body, the overall person, is part of the person's embedded identity. When describing John, how Jack perceives him, his experiences with John, and Jack's other mental states of affairs in which John figures are all part of the description of John as an embedded entity or person. If Jack suddenly realizes that John never truly loved him as a father should—John was merely behaving in a way that was appropriate to being a father but felt nothing for his biological offspring—then John's identity changes for the worse. Of course, the psychological person that is John does not change with the realization; that was done long ago when John was able to feel nothing. However, John's embedded personhood is now less intrinsically worthy for a number of reasons. The negative emotional states generated by Jack's epiphany, the lack of trust in the relationship, and other intrinsically disvaluable states of affairs reified makes the organic whole of the relationship between John and Jack less worthy because of what it did to Jack and the relationship between the two individuals. In addition, the various organic wholes in which John and Jack are component nodes suffer decreased value, unless the alteration somehow changes other states of affairs in such a way that the organic whole's ordinal worth is enhanced. It might be the case that John deserves to be treated this way; hence, the state is better than when Jack blindly loved him. It might also be the case that the organic whole's value increases in the long run as Jack focuses more of his time on people who deserve and benefit from his attention.

Some nodes and the relationships and interactions between the nexus overall person are more important, such as those generated by peers and relatives, while others have little to no impact on the nexus that is noticeable. The importance of the relationships and interactions is determined in different ways, each of which counts in determining who the person is. First, the stronger the relationship, the greater role it plays in who the embedded person is. If Janice's career is the most important feature in her life, then the relationships she has with her fellow workers will be more central and more influential to her identity than those relationships she has with those less important to whom she is. Second, the overall person is also affected by those relationships and interactions. The internal mental states of the psychological person, such as Janice being happy because she is doing what she loves to do, will compose part of who the overall person is, and hence, who the embedded person is. Interacting with her coworkers will give her greater pleasure and fulfillment than other types of interactions. Having those relationships are also more fulfilling, pleasurable, or productive of intrinsically valuable mental states of affairs than are encounters with acquaintances.

We must not too quickly claim that there is no influence on identity even in what appears to be innocuous relationships and interactions. After all, our habits, such as virtues and vices, are built up from even the most trivial of actions repeated often enough. Although the person may no longer be able to summon a memory of a seemingly meaningless encounter, it could still be held subconsciously, or have altered other mental states or faculties on its way to oblivion. Even the smallest mental state of affairs has to be part of the embedded person's identity when considering the embedded person in her entirety as an organic whole in the universe and in smaller contexts, although we might conclude that on its own it is both accidental and trivial.

3.6 Evaluating a Person's Intrinsic Value

The only fully and objectively real organic whole is the universe in its entirety as an atomic thing. That is, if we wanted to be as accurate as possible about intrinsic value of each node and the whole, then we would have to discuss the universe as a whole. Since it contains organic wholes within it, it too must be thought of as an organic whole. The sum of its parts might equal or be proportionate to the sum of its parts, but that is not necessarily the case. A useful phenomenon in the universe helps illustrate why thinking about it as an organic whole is important and plausible. When one electron changes in the universe, all other electrons in the universe change simultaneously (Francis 2012). The same entanglement result could be thought to apply to persons within the universe, especially since we are working with organic wholes. The universe has an infinite amount of relationships and objects with relationships to each other and to each person. When the person changes, then the entire organic whole that is the universe alters as well. I might move in one direction by one inch, but my relationships to all other things in the universe are now different by that one inch.

When an organism or person dies, then many organic wholes can cease to exist. First, the death of the organism is the death of the biome. Second, the psychological person is gone if its foundational structure is gone. Third, the embodied person ceases to exist if either or both of its essential components dies. Finally, the person as an embedded organic whole in all of their forms dies. Of course, death can happen at a higher level and those entities at a lower level can continue to live, but the farther down the hierarchy one goes, the more entities cease to exist because of the lower level entity's death.

Embedded organic wholes alter essentially when the central node ceases to exist. That entails that all relationships with the embedded person change simultaneously as well. More precisely, the embedded person's relationships are terminated because the embedded person has ceased to exist. However, there is a new entity in the place of the old one—the physical subject—and, I will later contend, the historical or reified distributed persona. When an embedded person dies and is replaced by the reified distributed persona, the substitution does not preserve the status quo. If thinking about the rubber sheet metaphor, the place that the person filled is no longer the same shape it was when the person occupied the space time. It is a different entity because of the change of essential properties, namely the loss of the psychological or embodied person that served as the nexus.

When a person dies, the value of the universe as a whole need not be the same, nor does the value of the various organic wholes have to equal, or be proportional to, the sum of their parts. That is, the value of universe at time_t need not be equivalent to the value of the universe at time_{t+n} minus the value of the person who will die. In addition, the value of the organic wholes to which the embedded, psychological, immediate, intermediate, or overall person belonged need not be the same. The individual person's value has been altered as well as the relationships and other intrinsically valuable states of affairs that existed, in part, because the person existed. Below, I will consider how to evaluate the intrinsic worth of each individual person, people in general, and the overall intrinsic value of various organic wholes.

Although the value theory in the preceding chapter claimed that immediate and longer-term mental states are intrinsically valuable, what has only been hinted at above is the value of living things and psychological subjects and types of persons. There are valuable states of affairs in which a person is pleased about something or has a mental state that is valuable in and of itself, but little has been said about the value of the person in and of herself and *qua* that person. I will begin with the general intrinsic value all living things share as living things and that all people share qua being a person, and then describe how individual persons *qua* individuals have particular intrinsic value.

The most intrinsically worthy and least controversially valuable entities, all things considered, are those instantiating the properties of being a person (Feinberg 2014; Frey 2000; Rollin 2006; Warren 1983). That is, a person is intrinsically valuable because it is a person.

The intrinsic worth of persons as members of a class and as individuals is the result of a number of features moral agents have. First, as argued above, any entity that is a person is intrinsically valuable for that reason alone. Increasing each moral agent's intrinsic worth is the possession of other intrinsically valuable characteristics. As will be discussed below, being sentient, being alive, and having the telos of having a flourishing life as a moral agent all make a person more intrinsically worthy than merely being a person. Moreover, individual moral agents can be more or less valuable based upon certain mental and physical properties they possess at particular times. A person *qua* moral agent is intrinsically valuable because he has the characteristics of moral agency, while an individual person is more or less valuable as an individual person based upon the actual experiences he has now or in the past.

Besides the particular, immediate, discrete experiences a person has, the telos of persons *qua* person is intrinsically valuable to possess and to fulfill. As has been stated before, the function of a person is to have a good life, which is a flourishing life. According to Frankena, a good life is a:

> "mixed life,"...consisting of activities and experiences of the kinds listed earlier, that is of activities and experiences that are enjoyable or both excellent in some degree and enjoyable. [That has] form, not just in the sense of pattern, but in the sense of being inspired by a certain attitude, posture, or "life-style." (Frankena 1973, 92).

In other words, happiness is the end of all human life, and I assume the same for all sentient life, if possible, which might be achieved in a variety of ways. Moreover, a happy life is always a good or ethical life. Hence, the more ethical an individual person is, the more the person flourishes and is able to fulfill her telos as a person, and the more intrinsically valuable as an individual she is.

What allows us to classify individual human entities as more or less valuable are the qualities they possess other than being a human, such as deserving punishment, having certain types of consciousness, being alive, being capable of achieving the flourishing of persons, or being a person. For example, a living human baby is more valuable than a dead human adult, not because the two are different species, but rather because one is living and the other is not. Hence, in one way, I am endorsing Regan and Taylor's claim that all human beings *qua* human being have equal intrinsic value, while rejecting what appears to be their position that there are no other moral factors justifying classifying one *Homo sapien* as being more or less valuable than another (Regan 1989, 222; Taylor 1986, 77). The axiology from the

Benefit of a Doubt principle as employed here and elsewhere is very complex and might appear to draw too fine a line in places, or just be terribly vague and too inclusive, but it seems to better reflect the actual, legitimate distinctions people draw—or should draw on practical grounds—when evaluating various entities' intrinsic value.

All moral agents' overall lives can be made intrinsically better or worse in other ways. First, the number and pattern of evil/good experiences and right/wrong actions can alter an agent's intrinsic value. As has already been stated, a life of excellence, which requires certain patterns and lifestyles, is more intrinsically valuable than one that is not (Frankena 1973, 93–4). A moral agent who has harmony and proportion in her life, for example, is more valuable than one who lives in disharmony and to extremes, all things considered. In general, those who instantiate more properties in certain patterns and ways will be more valuable than those who do not.

According to Frankena, and this seems to be true in general, the ordering of the intrinsically valuable experiences in each life will affect the overall intrinsic worth of the life. If every valuable state of affairs is grouped at the beginning of the agent's life, while the end is disvaluable, then even if the worthy discrete experiences outweigh the disvaluable discrete experiences, it does not follow that the life was good on the whole. The life has to be evaluated on the grounds of how well the person fulfilled the telos of being a good moral agent, which is determined by whether or not the life was happy overall. Perhaps the person has decided to commit some terrorist action that injures many people in different ways. Being a good moral agent is not only an end of life, but it also has to be maintained until life ends. To end one's life in such a terrible way is to eschew who one is as a moral agent, thereby becoming worse than an animal that can never be anything other than it is.

Given the fact that particular experiences and lives vary, we can justifiably talk about some people being more valuable than others, not as persons *qua* person which entails equality, but on the grounds of the life they lead as individual persons and their particular experiences' value. Moral saints are more worthy than those who only fulfill the basics required to make them minimally decent moral agents. Immoral devils are less worthy than either the saints or decent agents. However, saints, morally decent agents, and devils have equal intrinsic value as persons qua persons because each is equally a person. The differences between them are the values arising only from their particular life and its experiences.

When evaluating a particular person's intrinsic value, it is necessary to look at two more ways of her being valuable through her being a person. Even though each person qua person has identical worth from possessing the property of being a person and having the telos of flourishing as a person that all persons have, individual mental states of an agent can alter the intrinsic worth of the person as a whole as well as the overall value of her life as a moral agent. I will begin with the former.

In everyday life, we make distinctions between the values of particular experiences. Some experiences are clearly better than others, such as having deserved pleasure when one is recognized for doing a heroic action versus having unmerited pleasure when one is wrongly thought to have performed the same action. Since there are better or worse experiences, it reasonably follows that the individual's overall intrinsic worth can alter according to the experience the person has. Therefore, the value of the individual experiencing one or more of Frankena's intrinsically valuable things at a particular time is greater than if she were not experiencing it at that time.[48] Again, since it is impossible to have the experience without the entity experiencing it, the perceiver must be more intrinsically valuable at the time she is having an intrinsically worthy experience than she would be in different circumstances.

Instead of adopting a cardinal calculus for determining the value of an individual person, or any intrinsically valuable thing for that matter, the most practical approach is an ordinal one. The complexity of the situation and interrelations of the whole's parts can make it impossible to add up all the values the way one can in a simple system of ordinal mathematics. Hence, the method used here is to allow, within the limits of the value theory developed so far, reasonable people to compare different individuals instantiating different mental states or other intrinsically valuable or disvaluable states, followed by ordinally ranking them on a line from worse to better.[49]

The ranking of individuals becomes more complex as we increase the properties involved or focus on more and more complex organic wholes, such as embedded people. For example, we can discuss the value of a person embedded in a family versus the same person embedded in a community. Depending on the actual circumstances, the former's value can equal, be greater, or be less than that of the embedded person in a community. If the person is in a very good family situation and a very good social structure, for instance, then it is likely that the intrinsic value of the latter will be greater, in part, because it includes more relationships and non-nexus nodes. However, the family might be a very good one, and the embedded person might have a high moral value because of it, but if the society is one that is antagonistic to that family and the nexus, then the overall value could be low or negative. The opposite could apply if the family is a bad one, but the society is good overall for the organic whole comprised of the nexus and all other nodes and their relationships to each other.

---

[48]Since many of the states of affairs Frankena lists can be experienced by higher order animals or all sentient creatures, then their experiencing of them at particular times will make them more intrinsically valuable at that time.

[49]What will also change the value of an individual is if an evil person has positive experiences, such as pleasure, when he does not deserve it. An evil person having a positive experience might make it much lower in intrinsic value than if a good person had the same experience. Furthermore, if a good person has negative experiences when she deserves good ones, then the value of her having the experience is much lower than it would have been if she had deserved it. An evil person receiving a disvaluable experience might be less intrinsically disvaluable than a good person having the same experience (Ross 1988, 136–7).

3.7 The Concept of Death

Given that there are many concepts of life, personal identity, and personhood, then it follows that there will many ways in which we can die. There are as many types of death as there are types of lives and personhoods. Besides those mentioned above, there are social deaths. A social death is when a group or society has accepted the death of the individual and begins to act in ways appropriate to the person's death. In health care facilities, for example, there comes a time when the health care providers determine that therapy is futile, and the providers' time is better spent helping those who can be helped as the patient finishes his death process in palliative care or elsewhere. This does not entail that the patient is treated unethically; it only means that the relationships that would be present between providers and patient have been severed by the providers because of their perception about the patient.

The state of death is merely a classification of kinds. "X is dead" is an affirmative that an object belongs to a certain class: namely, dead things. When something dies, then it has ceased to instantiate one of the characteristics that can be thought of as giving life or an existence that is capable of death, or that it no longer belongs in the class of things that are living person of type X. Death is more than a cessation to exist. An entity must be alive in order to die, not merely in existence. For example, if a marble statue is destroyed by grinding it into stone dust, then it has ceased to exist, but it has not died.

An interesting feature of death is that it is an odd classification given that some entities, such as many of our person types, cease to exist when they die. Therefore, there is no real object to classify. For example, there is no such thing as a dead organism or a psychological person. In the first case, life is an essential part of the entity, so when it dies, it cannot be an organism. For a psychological person, having certain psychological faculties and features is necessary, and again, when they cease to exist, so too does the psychological person.

With all of this in mind, what is going on when people try to classify the new entity in terms of the previous one? I think that it is an attempt by people to classify the world so that they can control it in some way, make sense of it, or otherwise have enough information so that they can make decisions. Therefore, if a definition of death is useful because it can serve different pragmatic uses or functions, then it will not be discarded merely because it does not capture everyone's intuitions or fails to correctly classify in every situation.

References

Atkins, K. 2008. *Narrative identity and moral identity: A practical perspective*. New York: Routledge.

Bamberg, M. 2011. Who am I? Narration and its contribution to self and identity. *Theory & Psychology* 21(1): 1–22.

Belshaw, C. 2009. *Annihilation*. Montreal: McGill-Queen's University Press.

Care, N.S. 1996. *Living with one's past*. Lanham: Rowman & Littlefield Publishers Inc.
Dainton, B. 2008. *The phenomenal self*. Oxford: Oxford University Press Inc.
Damasio, A. 2011. *Self comes to mind*. New York: Pantheon Books.
Engelhardt, H.T. 1986. *The foundations of bioethics*. New York: Oxford University Press.
Evnine, S.J. 2008. *Epistemic dimensions of personhood*. Oxford: Oxford University Press Inc.
Feinberg, J. 2014. The paradoxes of personality. In *Potentiality*, ed. J.P. Lizza, 69–71. Baltimore: Johns Hopkins University Press.
Feldman, F. 1992. *Confrontations with the reaper*. New York: Oxford University Press.
Fletcher, J. 1997. *Humanhood: Essays in biomedical ethics*. New York: Prometheus Books.
Flowers, B.S. 1998. Death, the bald scenario. In *Death and philosophy*, ed. J. Malpas, and R.C. Solomon, 50–56. New York: Rutledge.
Francis, M. 2012. Quantum entanglement shows that reality can't be local. *Ars Technica*, 30. http://arstechnica.com/science/2012/10/quantum-entanglement-shows-that-reality-cant-be-local/. Accessed 24 March 2014.
Frankena, W. 1973. *Ethics*. Engelwood Cliffs: Prentice Hall Inc.
Frey, R.G. 2000. Rights, interests, desires, and beliefs. In *Living well*, ed. S. Luper, 436–442. Fort Worth: Harcourt Brace College Publishers.
Garrett, B. 1998. *Personal identity and self-consciousness*. London: Routledge.
Gell, A. 1998. *Art and agency: An anthropological theory*. Oxford: Oxford University Press Inc.
Glahn, J.A. 2009. Dignity of the dead? In *Re-imaging death and dying: Interdisciplinary global perspectives*, ed. D. Cooley, and S. Steffen, 33–42. Oxford: Inter-disciplinary Press.
Gould, S.J. 2003. What is a species? In *The environmental ethics and policy book*, ed. D. VanDeVeer, and C. Pierce, 465–469. Belmont: Thomson Wadsworth.
Helm, B. 2010. *Love, friendship and the self*. Oxford: Oxford University Press.
Hofstadter, D.R. 1979. *Godel, escher, bach: An eternal golden braid*. New York: Vintage Books.
Human Microbiome Project Consortium (HMPC). 2012. Structure, function and diversity of the healthy human microbiome. *Nature* 486: 207–214.
Kagan, S. 2012. *Death*. New Haven: Yale University Press.
Kampis, D., E. Somogyi, S. Itakura, and I. Király. 2013. Do infants bind mental states to agents? *Cognition* 129(2): 232–240.
Kastenbaum, R., and R. Aisenberg. 1976. Death as a thought. In *Death*, ed. E.S. Shneidman, 369–422. Palo Alto: Mayfield Publishing.
Kelly, E., E.W. Kelley, et al. 2009. *The irreducible mind*. Lanham: Rowman & Littlefield Publishers Inc.
Kretzmann, N. 1999. *The metaphysics of creation: aquinas's natural theology in summa contra Gentiles*, vol. 2. Oxford: Clarendon Press.
LeDoux, J. 2002. *Synaptic self: How our brains become who we are*. New York: Penguin Books.
Lee, P., R.P. George. 2006. Human-embryo liberation: A reply to Peter Singer. *National Review Online* (25 Jan). http://www.nationalreview.com/articles/216588/human-embryo-liberation/patrick-lee. Accessed 27 Aug 2014.
Lindemann, H. 2014. *Holding and letting go: The social practice of personal identities*. New York: Oxford University Press Inc.
Little, M.O. 2014. Abortions and the margins of personhood. In *Potentiality*, ed. J.P. Lizza, 174–190. Baltimore: Johns Hopkins University Press.
Lloyd, G. 2007. *Cognitive variations: Reflections on the unity and diversity of the human mind*. New York: Oxford University Press.
Locke, J. 1992. An essay concerning human understanding (abridged). In *Central readings in the history of modern philosophy*, ed. R. Cummins, and D. Owen, 93–271. Belmont: Wadsworth Publishing.
Mason, O.J., and F. Brady. 2009. The psychotomimetic effects of short-term sensory deprivation. *Journal of Nervous and Mental Disease*. 197(10): 783–5.
McAdams, D.P., and K.C. McLean. 2013. Narrative identities. *Current Directions in Psychological Science* 22(3): 233–238.

McCormick, R.A. 1974. To save or let die. *Journal of the American Medical Association* 10: 253–266.

McMahan, J. 2002. *The ethics of killing: Problems at the margins of life*. Oxford: Oxford University Press Inc.

McMahan, J. 2014. Potential. In *Potentiality*, ed. J.P. Lizza, 140–173. Baltimore: Johns Hopkins University Press.

Mendola, J. 2008. *Anti-externalism*. Oxford: Oxford University Press Inc.

Milojev, P., and C.G. Sibley. 2014. The stability of adult personality varies across age: Evidence from a two-year longitudinal sample of adults New Zealanders. *Journal of Research in Personality* 51: 29–37.

Moor, James H. 1982. Split brains and atomic persons. *Philosophy of Science* 49(March): 91–106.

Murphy, G.F. 1993. Rationality and the fear of death. In *The metaphysics of death*, ed. J.M. Fischer, 41–58. Stanford: Stanford University Press.

National Institutes for Health (NIH). 2012. NIH Human Microbiome Project defines normal bacterial makeup of the body. http://www.genome.gov/27549144. Accessed 27 Aug 2014.

Noonan, H. 2003. *Personal identity*, 2nd ed. London: Routledge.

Ouellette, J. 2014. *Me, myself, and why*. New York: Penguin Books.

Parfit, D. 1984. *Reasons and persons*. Oxford: Oxford University Press.

Paulhus, D.L., and O.P. John. 1998. Egoistic and moralistic biases in self-perception: The interplay of self-deceptive styles with basic traits and motives. *Journal of Personality* 66(6): 125–160.

Regan, T. 1989. The case for animal rights. In *The right thing to do*, ed. J. Rachels, 211–225. New York: Random House.

Rollin, B.E. 2006. *Science and ethics*. New York: Cambridge University Press.

Ross, W.D. 1988. *The right and the good*. Indianapolis: Hackett Publishing Company.

Sagan, A., and P. Singer. 2014. The moral status of stem cells. In *Potentiality*, ed. J.P. Lizza, 140–173. Baltimore: Johns Hopkins University Press.

Searle, J.R. 1981. Minds, brains, and programs. In *The mind's I: Fantasies and reflections on self and soul*, ed. D.R. Hofstadter. Cambridge: MIT Press.

Shoemaker, S. 1984. Personal identity: A materialist's account. In *Personal identity*, ed. S. Shoemaker, and R. Swinburne. Oxford: Blackwell.

Smith, C. 2010. *What is a person?* Chicago: The University of Chicago Press.

Smith, E.R., and E.C. Collins. 2009. Contextualizing person perceptions: Distributed social cognition. *Psychological Review* 116(2): 343–364.

Steffen, L. 2012. Moral death. In *New perspectives on the end of life: Essays on care and intimacy of dying*, ed. L. Steffen, and N. Hinerman, 131–143. Oxford: Inter-Disciplinary Press.

Stevenson, C.L. 1944. *Ethics and language*. New Haven: Yale University Press.

Taylor, P. 1986. *Respect for nature: A theory of environmental ethics*. Princeton: Princeton University Press.

Taylor, C. 1989. *Sources of the self: The making of the modern identity*. Cambridge: Harvard University Press.

Tooley, M. 1983. *Abortion and infanticide*. New York: Oxford University Press.

Vellman, J.D. 1991. Well-being and time. *Pacific Philosophical Quarterly* 72: 48–77.

Walters, J.M. 1997. *What is a person: An ethical exploration*. Urbana: University of Illinois Press.

Warren, M.A. 1983. The rights of the nonhuman world. In *Environmental philosophy*, ed. R. Elliott, and A. Gare, 109–33. University Park: Pennsylvania State University Press.

Witkin, S.S., I.M. Linhares, and P. Giraldo. 2007. Bacterial flora of the female genital tract: Function and immune regulation. *Best Practice & Research Clinical Obstetrics & Gynaecology* 21(3): 347–354.

Woodger, J.H. 1948. *Biological principles*. London: Routledge and Kegan Paul Limited.

Chapter 4
What Harm Does Death Do to the Decedent?

Abstract The vast majority of people believe that death harms the person who dies. After all, it seems to be common sense that death is a bad thing, and bad things happening to people are generally injurious in some direct way to those individuals; therefore, death must harm the people who undergo it. The trouble with the common sense view is that it fails to recognize that in order to be injured a person must exist. Since death annihilates the person by destroying an essential characteristic the person has, then no person can ever experience her own death, and if she cannot experience it, then it cannot harm her. In what follows, I examine various ways that philosophers and others have argued that death harms the person who has the misfortune of dying (Calling it a misfortune assumes the common sense view is correct without critical evaluation of it). Firstly, options incorporating an afterlife are found wanting. Secondly, Cambridge changes, Priorism, and other attempts to show that a person is injured when she dies are considered. These attempts are shown to be either based on fundamentally flawed reasoning that creates the illusion of existence when all that actually exists are ideas and reflections in the survivors' minds, or the entity that is harmed is not that which common sense says is injured.

4.1 Introduction

The death as harm argument generally breaks down in one of two ways. The first is firmly grounded in the Epicurean tradition that death cannot injure a person because there is no person to be harmed by the death or what happens after that event or process. That is, death itself eliminates a necessary condition for personhood or personal identity; hence, the person cannot experience injury because of the simple fact that the person has ceased to exist, and the ability to experience requires existence.[1] One formulation of the Epicurean argument is:

[1]See Mothersill (1999) and Suits (2004).

D.R. Cooley, *Death's Values and Obligations: A Pragmatic Framework*, International Library of Ethics, Law, and the New Medicine 62,
DOI 10.1007/978-94-017-7264-8_4

(A) A state of affairs is bad for person *P* only if *P* can experience it at some time.
Therefore, (B) *P*'s being dead is bad for *P* only if it is a state of affairs that *P* can experience at some time.
(C) *P* can experience a state of affairs at some time only if it begins before *P*'s death.
(D) *P*'s being dead is not a state of affairs that begins before *P*'s death.
Therefore, (E) *P*'s being dead is not a state of affairs that *P* can experience at some time.
To summarize, *P*'s being dead is not bad for *P*. (Rosenbaum 1993b, 121–2)

Basically, since death causes us not to be, then we cannot be harmed by our death and postmortem events. Moreover, given that we cannot be injured by our non-existence, we should not be afraid of it. After all, the man who is not on the stairs should not frighten us because he is not there to be able to endanger us.[2]

Despite this, people mourn those who have passed not only as a result of the survivors' loss, but also because they believe the deceased suffered a tremendous harm by dying. If asked, survivors would say that they really mean it hurts each and every person who dies to be dead. They will also claim the frustrations of the deceased's plans by his existence's cessation are bad things, not solely for others' sakes, but for the originator and holder of those goals. "The bearing of this on the Epicurean argument is that if there can be misfortunes that involve no painful experiences, then death may be one of these." (Scarre 2007, 89). Perhaps the way we should understand the conundrum is through Scarre's example of Fred ridiculing Freda without her knowing about Fred's actions. Is Freda harmed or injured by the betrayal of someone she thought was her friend? It depends on how we define how something experiences an event.

An Epicurean way of defining experience does not allow the dead to be harmed or wronged by any consequences resulting from their demise, nor can the death itself do wrong or harm.

<u>O experiences E, where O is an object and E is an event</u> = df. O's mental states are altered at a conscious level.

In other words, a person must be aware of an experience in order to be affected by it. Freda never finds out about Fred's action, and hence, she does not have a negative psychological experience from the betrayal. Therefore, she is not injured. In the same way, there is no consciousness after death; hence, there is no experience of harm from or after death.

However, this type of experience is too limiting to those who believe that an individual can be affected by his death and relevant events occurring post-mortem.

Intuitively, death seems bad because the dead miss out on the good things of life… However, it is plausible to suppose that there can be goods and evils that do not consist in positive or negative experiences and of which the subject may never be aware…Instead, we realize for the first time that it was always true that many of Smith's ambitions, including that of living a normal lifespan, would come to nothing. (Scarre 2007, 86, 88, 94–5)

---

[2]See "Antigonish" by Hughes Mearns. The poem's impact comes from Mearns' ability to create fear in the reader using fiction about a non-existing person, who can never pose an actual threat.

Initially, Scarre's claim, along with those of others like him, seems rather an odd one. How can an injured person not know immediately that she is harmed? After all, if someone kicks another person in the shin, then the latter's pain is felt instantaneously and can only exist because the person is feeling it when she is feeling it.[3] In order to injure someone, the person must know that she has been harmed.

We need to broaden experience's definition to make the death-as-an-injury claim work. Freda would be able to experience injury from Fred's betrayal and deception if experience meant that there was change to the person as an embedded entity.

> O experiences E, where O is an embedded object in the organic whole of the universe and E is an event = df. A property of O's changes as a result of E.

If true, then all events will affect all objects because a change in one will entail that all relationships to that object, no matter how trivial those relationships are, will alter. Since everything that exists is interconnected through the relationships, an object's relational properties have changed, which alters the embedded object, although this change might be trivial. Freda, therefore, remains significantly the same as individual psychological and embodied persons, but Freda *qua* embedded person is significantly altered from the entity she was before Fred's action.

Although it has its uses as a metaphorical way of thinking about experiencing, this definition's problem is that it does not do the work most want it to do. This definition is unable to make distinctions between people or intrinsically valuable entities that are affected and things which are not. A stone, for example, can experience its destruction over time from the wind, rain, and cold. That experience, however, is not the same as that Freda must have in order to be wronged or harmed by Fred's action.

A more pragmatic way of understanding experience in the Fred and Freda case is by combining the two definitions above in a plausible way.

> O experiences E, where O is an intrinsically valuable object and E is an event = df. O's intrinsic value or life-existence as O is altered by E.

If we consider the various organic wholes composing the embedded person called Freda in the universe that she inhabits, then we can see that she experiences a number of harms, although she is not consciously aware of them (Nagel 1993, 64).[4,5] Freda believes she has a friend in Fred, but Fred is not a friend and betrays O's confidences

[3]Once again, to have a mental state, such as pain or pleasure, something must be experiencing that state.

[4]Nagel adds the unnecessary condition that the person, if she is made aware of the experience, would be injured by it. I am arguing that the person is injured even if she does not know it because one of her relevant selves' intrinsic value has been decreased.

[5]Ben Bradley makes a distinction between *prima-facie* harms and all-things-considered harms. The former are harms that might not be bad overall because they can cause much greater good to happen to a person (Bradley 2009, 66).

as well as ridiculing her. This harms Freda as a psychological person in that she has a false belief that makes her belief set less valuable because it contains a falsehood. In addition, the false belief makes Freda less able to make informed decisions. She might spend time thinking well of Fred when that is a waste of her resources; her flourishing would be far better served if she focused on a far more deserving individual. Moreover, as an embedded person, Freda's relationship with Fred is defective. She does not have an actual friendship but rather an abusive relationship. Hence, her value as a socially embedded or just plain embedded person is reduced. Finally, Fred has made Freda's social groupings and other collectives to which she belongs far less valuable than they needed to be. Fred did not have to be the moral wreck of a person he is because that not only damages Freda's flourishing but also his as well. He is harming himself by choosing to be the type of person who would do such an intimate betrayal of someone who considers him to be a friend, performing the betrayal and deception required, and involving others in his insidious actions. Hence, Freda is injured and wronged by Fred's illicit actions, even though her mental states are not altered in any way.

This raises a new question: if death is like Freda's experience in that the person is unaware of being injured, can she be injured or wronged in some way because of the cessation of her life plans, or is Rosenbaum correct in saying that an inability to experience entails an inability to be harmed or wronged?

Consider three main approaches to incorporate intuitions about death injuring the deceased into initially plausible arguments.[6] The first is that the person as the individual psychological person survives his body's death; the second is Priorism; and the third relies upon Cambridge changes. I will address individual psychological person survival first.[7]

---

[6]Feldman (1992, 114) states that "Being dead is an unusual property in this respect: it appears that a thing can be dead at a time even though it does not exist at that time." In other words, one need not exist in order to have the property of being dead. "So, in a way, our statement does not entail that Adam has any 'solid' properties today, just that there were some times in the past when he had some." (Ibid.) Feldman's position of eternalism claims that death is eternally bad for the deceased.

Two other positions are subsequentialism and concurrentism. Luper's concurrentism, maintains that since death is a process, then the person can experience her death as she undergoes it. Bradley's subsequentialism, on the other hand, claims that death is bad for the person after the person's demise (Bradley 2009, 84, 89). The arguments against the effectiveness of Cambridge changes and Priorism in establishing the death as a harm claim also show these positions must be mistaken.

[7]The *Oxford Handbook of Philosophy of Death* (2013) provides a useful introduction to various positions on this subject.

4.2 Individual Psychological Person Survival

Both Eastern and Western cultures traditionally believe that some part of an individual psychological person or moral agent survives the death of the embodied person's body.[8] In Plato's *Meno*, it is the moral agent's pure reason component that leaves the world of Becoming to enter the world of Being to contemplate the Forms until that "time" when the soul is once more burdened with a body in Becoming's realm (Plato 1961). Unfortunately, the immortal Platonic soul does not have the identity of any one of its incarnations as an individual psychological person. The irrelevant identity founded on personal history, name, language, and every mental and physical state that makes us particular individuals is solely the result of being in the less perfect realm. When the soul enters the perfect universe of the Forms, such useless information is lost. Hence, Meno, the slave, cannot remember his previous lives, but with careful prompting, he can recall complex mathematics. As Plato concludes, the memory must show that the slave's reason had been able to contemplate the Forms before being re-installed in the slave's body.

Although Plato's argument is a rather poor one because it employs leading questions designed to elicit the responses desired, it also has the flaw of not keeping an individual psychological person alive in any meaningful way. The entity that survives bodily death is, much like Rawls' rational self-interested creature under the veil of ignorance, something without personal identity other than not being one of the other purely rational souls. That is, when John dies, it is not John who survives, but the purely rational soul that used to be part of the embodied John. When Mary dies, her purely rational soul also lives on without it being Mary. If we encountered them in the realm of Being, the only difference we could use to identify them would be that one is not the same as the other—that is, they lack numerical identity—but they would otherwise be identical. Hence, Epicurus could still claim under Plato's account that no one is injured by dying because the individual person who existed before death no longer exists after death to be harmed. Moreover, given that the life experienced after death is a far better life than lived in the realm of Becoming, especially for the poor slave boy, Meno, improving one's lot in life to perfection cannot be an injury to the entity perfected, although it ceased to exist in the other realm.

Philosophers argue that individual psychological persons can survive death, much along the lines of how various religions believe that they do.[9] David Lund's work is a representative example of this type of philosophical reasoning that does

---

[8] Resurrection of the body or mind could be a way to recover from death. Georg (2010) has an interesting anthology with articles trying to figure out how a person can be resurrected. Since death appears to be a permanent state given our scientific technology, I will focus on the permanent death.

[9] John Martin Fischer argues that the only form of immortality that might work is an atomistic non-serial approach, but even that would not guarantee personal identity survival (Fischer and Curl 2009, 100).

not include religious dogma. For Lund, the individual psychological person is an immaterial individual, while the body is a material object (Lund 2009, 62). What is interesting about this argument is how much it owes to Descartes' argument found in the 6th *Meditation* for the body and mind's separation. Descartes defines the mind as a thinking thing that is not extended in space unlike physical bodies. Since minds and bodies do not have identical characteristics in all things, if they exist, then they must be two different things. If they are different, then we can see that it is metaphysically possible that they can be separated and exist apart, which is also the language Lund adopts (Ibid). Metaphysical possibility is defined as a state one can imagine happening, which entails that anything one can picture in one's mind could occur, although it might not, in fact, happen in any actual world. Descartes, unlike Lund, goes on to conclude that because he can clearly and distinctly perceive or understand the mind and body existing apart, then the infinitely powerful and good God Descartes thinks he proved existed in both the 2nd and 5th *Meditations* will guarantee that his perception is veridical. Lund, on the other hand, uses adductive reasoning to establish a case for an afterlife's existence.

Lund recognizes that the mere fact that something is metaphysically possible has no real bearing on it being causally possible. Here is where it gets a bit confusing. Lund argues that any of the following hypotheses are metaphysically possible and, although he does not explicitly use the term, he implies they are epistemologically possible:

1. The self is causally dependent on its brain's activity.
2. Intentional consciousness depends on its brain's activity, but the self and its non-intentional consciousness do not.
3. The brain has a dissociative effect, which allows the self to survive with its intentional consciousness intact or enhanced.
4. The brain is a consciousness enhancer, which when it dies causes the self to have diminished intentional consciousness.
5. Other possibilities (Lund 2009, 102).

The first is the Epicurean position, as well as that of 20th century Interactionists, who believe that the mind supervenes on a properly working brain. When the brain ceases to function in any significant way, then it cannot act as a platform supporting the emergent mind that is the psychological person. Lund's second hypothesis is a bit odd because there is survival, but the new entity is not a rational agent, nor does it seem to have the same personal identity. The third position is clearly that of Descartes, Plato, and anyone else who believes that the body is an entity that needs to be shed so that the mind can act according to its higher nature. The fourth hypothesis is similar to the second in that the intrinsic value of the person is diminished by retarding certain abilities in considerable ways. It is unclear whether the entity remains a person with the same personal identity, a person, or even a moral subject. From these metaphysical possibilities, he then believes that we can start examining what he considers to be all relevant evidence, including "well-documented reports of paranormal occurrences" (Ibid.), to determine which hypothesis is better supported.[10]

We should take a moment to note something worrisome about the dualist survival position—even if any of Lund's alternative hypotheses is true, there is no reason to believe that the individual psychological person survives death. The damage to reasoning abilities and consciousness in hypotheses 2 and 4 looks too great to maintain adequate psychological continuity for numerical identity. On the other hand, the entity that survives in hypothesis 3 might be so greatly different from us as individual persons that it is merely a thinking thing without a unique personality. That is, it is not us in any meaningful way.

A more significant problem faces Lund's argument and those like it. The mere fact that something is metaphysically possible has no evidentiary weight in abductive reasoning. Given that abductive reasoning tries to find the best explanation or hypothesis using the evidence available, any hypothesis should be falsifiable; otherwise, it is not a scientific hypothesis. For example, suppose a repressed memory syndrome researcher claims that every person as a child has been sexually abused, and that every person will either admit to being abused or repress it. If a person states she was abused, that admission is used as evidence that she was abused, thereby supporting the hypothesis. If she states she was not, then that is evidence that she was abused, thereby supporting the hypothesis. Since this claim is not falsifiable, then it is not a scientific claim.

Lund's hypotheses are not scientific hypotheses, which means that abductive reasoning will not be able to justify them. The 5th hypothesis is unfalsifiable because it could mean just about anything. Hypotheses 2–4 might be untestable given that there is no way to really know if an experience happened to the person or if she merely hallucinated it as a result of some physiological or psychological condition of her own mind or brain.

In order for the empirical evidence to fall firmly in favor of any of hypotheses 2–4 over that of hypothesis 1, then there would be surprising observations if hypothesis 1 was true, but not surprising if the competitor was true. Paranormal activity would be adequate evidence against 1 if it is reliable information. That is, if near-death experiences (in which the person expires physically but not mentally), apparitions, or reincarnation exist, then that is excellent evidence that there is some type of soul or psychological personhood survival after death. The surprising element is that these experiences are relatively rare, given how many people have existed and then died. After all, it is extremely unusual for any person to go through the day without encountering another person or information showing that others exist, such as seeing them in the flesh. Our everyday experiences with others are excellent evidence to believe that other people exist. The same abundance of data should be true if people survive death. If we assume that there have been at least one billion people who are no longer alive, then with a world population of seven billion, paranormal or

---

[10]Lund argues that there is no evidence that any of the metaphysical or epistemic possibilities are also causally possible.

supernatural experiences should be commonplace and, therefore, deemed normal and natural. However, such experiences are not, which means that something more has to be added to the hypotheses to explain this phenomenon. That is, there would have to be an explanation beyond what hypotheses 2–4 offer for why the data is not plentiful when it seems as if it should be.

The danger from adding exemptions to explain away the lack of evidence or the poor quality of what some people use for their supernatural or paranormal claims' foundation is that the hypothesis becomes less likely to be falsified. If the phenomena are declared to be supernatural, then by definition, natural measurements such as electrical currents, photography, and other devices that record natural phenomena can no longer lend support to the claim. If a special type of person is needed to transmit information, such as a medium or psychic, then it has to be explained why those who died have to resort to such an unreliable method of communication as determined by the number of false statements, lack of specifics, cost, and the impossibility of falsifying the vague, ambiguous claims made by paranormalists. If the fuzziness of images and recordings are explained through the difficulty the dead have in communicating with the living, then the communication problems have to be explained as to why the dead can communicate naturally while simultaneously being supernatural.

If hypothesis 1 is the best explanation for observations, then we would expect a dearth of evidence of an afterlife. Encounters with those who have died would be rare relative to people as a whole. There would be better natural explanations for any perceived phenomenon, and those that people claim exist would not be verifiable beyond what the person having the experience can relay to others.

Even when we put aside the seemingly odd lack of evidence and consider that which has been cited, the best supported hypothesis, not surprisingly, remains hypothesis 1. To try to establish hypotheses 2–4, one has to rely on evidence that is dubious to anyone who takes science seriously. Jeffery Long argues that there is sufficient empirical evidence from near-death experiences to show that individuals can survive the death of their bodies, but we should be wary of it (Long 2010, 46–50). Lucid death, out of body experiences, blind people having sight, impossible consciousness, reunions with the dead, and worldwide and across chronological age consistencies of experience would be convincing had it not been the case that all the phenomena is better explained by obvious natural causes, such as having sufficiently similar bodies that would react to sufficiently similar stimulation in similar ways. In addition, some of the evidence seems to have little to do with an afterlife, as in the case of changed lives and perfect playback.

Lund cites studies by Karlis Osis, Erlendur Haraldsson and others that show near-death experiences (NDE) in which the person "perceives" dead loved ones and others waiting to usher the soon-to-be dead to their next destination. One natural explanation for why people believe that this event happens is a lack of blood to the brain that causes non-veridical perceptions from oxygen deprivation. Lund replies that even if this is a legitimate explanation for some aspect of NDEs, "it could provide no understanding of how someone would have in an NDE an apparent encounter with a deceased person whose (often recent) death was unknown to her."

(Lund 2009, 114). The interesting fact is that Lund provides no empirical data to back this claim. What is surprising is that Lund has already stated what he considers to be the paradigmatic studies of Osis and Haraldsson showed that though "apparitions of the living do occasionally appear, the frequency is low." (Ibid, 106). Thus, it does not matter whether the entity encountered is actually alive or dead for her to have an encounter with a person undergoing a near-death experience. Therefore, the actual empirical question should focus on how these interactions work, which need not include any information about one entity in the encounter being dead.

More worrisome is the fact that Lund attacks one natural explanation of alleged paranormal phenomenon at a time, as if each one is the one and only story of what is happening in the world. If he finds that it cannot be used in every case, then it is found deficient. Much the same approach is taken by Kelly et al. in their comprehensive evaluation of NDEs (Kelly et al. 2009, Chap. 6). Besides the blood gases explanation, they reject both psychological and cultural theories, such as expectation and depersonalization, and physiological theories, such as neurochemical theories and neuroanatomical models, on the grounds that none of these on its own explains away every NDE (Ibid.).

However, that is not how science is done. A drug might work for one person, but not for another. The fact it fails in one case does not entail that it is never efficacious. In addition, a different drug might work for the person whom the first drug failed. In these cases, the same result can be achieved through different courses of natural treatment.

The same approach can be useful in explaining near death experiences. Some can be explained by lack of blood to the brain or by the medication's side effects. Others stem from a strong belief in an afterlife. It could be a combination of them. In fact, that is what the science says happens. A perfectly plausible explanation involving the temporal-parietal cortex does not involve dubious claims to dualism in which the mind can be separated from the body (Thonnard et al. 2008, 438).[11,12] Given these and other studies that point to natural physiological and psychological explanations rather than paranormal conjecture, a reasonable person would have to accept the former rather than the latter.

In addition, the sleeping mind's power should never be underestimated. Although consciousness is often linked with wakefulness or the wakeful mind, confusion often results when someone refers to a person being unconscious. There is absolute unconsciousness when the psychological person has ceased to exist because the brain cannot support any of the higher reasoning processes associated with being a person. On the other hand, there are sleeping people who remain psychological persons with minds that are not awake. Their minds are busily creating dreams that engage their thought processes and mental states. In dreams, in

[11]See also Britton and Bootzin's (2004) work on the temporal lobe and near death experiences.

[12]Also see Woerlee (2005) for an in-depth explanation for what happens to people in near death experiences.

times of stress and for other reasons, we can spend awhile with "apparitions" of people who are not there and are not interacting with us other than in ways that our individual, unaware minds are causing to happen. Although the situations are self-created, they still engage our other mental faculties. Basically, it is an interaction of one part of the mind of which the higher thinking processes are unaware, but the mind is still engaged in its central task of thinking.

Given the power of the mind to create realistic images in dreams, there is a powerful reason to believe that those undergoing near-death experiences are actually experiencing one part of their mind—of which they are unaware—acting on a part of their mind of which they, as individual psychological persons, are aware. That is, as in dreams, when someone sees an apparition, what they are perceiving is an image conjured by part of their mind operating in a manner of which they are unaware. For instance, recurrent isolated sleep paralysis, which occurs during the transition from sleep to wakefulness or wakefulness to sleep, is associated with out-of-body experiences, which are very similar to NDEs (Terrillon and Marques-Bonham 2001). Hypnagogic and hypnopompic hallucinations, which occur upon going to sleep and waking up, respectively, of intruders in the room, alien abduction and paralysis can happen when the brain is unable to process information in its normal way (McNally and Clancy 2005). Using Locke's distinction of sensations and reflections, where a sensation is caused by external world objects acting upon our senses and reflections are operations of the mind, the apparition feels as if there is a sensation, but it is merely a reflection. Hence, no actual other mind or external world object is acting upon our sense organs, but merely memory and other internal mental faculties are at work presenting ideas to our active minds.

To overcome this plausible hypothesis that does not multiply entities without necessity, there would have to be scientifically legitimate data supporting the experience's veridicality. However, the possibly insurmountable problem to obtaining such evidence is the private mind, which makes it difficult at best for others to know what is going on in another person's mind. Since paranormal experiences are clearly not shared, according to those who believe in near-death experiences providing evidence of an after-life, then abductive reasoning is not going to be very useful in establishing life after death.[13]

Survival of death by psychological persons, moral agents, or whatever makes us who we are as individuals might be a comforting belief, but we have a duty to take evidence seriously when it is available to us. Hence, until there is adequate data collected for an afterlife in which we survive with numerical identity to at least one

[13]Testing whether or not people perceive ghosts and other apparitions would have to involve some form of deception. First, we would have to create test groups of people with the same credulity level, i.e., high credulity to low credulity. The groups could be run through similar situations in which a guide they trust tells them that they are seeing ghosts, when there is nothing and when there are images from hidden projectors there. If there is no difference between the two experiences, then there is less reason to believe that people are actually seeing an apparition because it would be clear that they are not seeing the same thing, if anything at all.

of the persons we were prior to death, the best available explanation is that the self ceases to exist when the brain no longer is able to provide a platform on which it emerges. Therefore, postmortem events and death cannot harm or wrong the person because the person no longer exists to be injured or wronged.

4.3 Immortality

There has been considerable interest in immortality, meaning life after death or other ways of eternal existence, such as transferring our minds to computer matrices. Some have claimed that the desire is essential to our largest accomplishments. Stephen Cave argues that "civilization itself has been driven onward by our quest to triumph over death, indeed the founding raison d'etre of many civilizations is the promise of immortality." (Cave 2012, 258). Samuel Scheffler, on the other hand, states that our ability to lead value-laden lives depends primarily on the continued existence of the human race and our continued existence as individuals, although the former greatly outweighs the latter (Scheffler 2012, 73). The desire for immortality, hence, appears to drive us to individually and collectively flourish, which would make the desire useful to us. However, the desire for immortality does not prove that immortality is possible or even desirable in the worthy-of-being-desired sense.

To be honest, I find the writings on immortality to not only be vastly premature in regard to our scientific abilities but also unrealistic in their view of human nature and privilege bias. Those against immortality, such as Jonathon Swift, argue that people will have a miserable existence once they run through everything that is of interest to them. In fact, it is alleged that such lives are neither attractive nor human (Williams 1973, 82). Those in favor of immortality trumpet the benefit of it as people have time to experience all that is worth experiencing, which can take an infinite amount of time because there are an infinite number of goods, or that the person grows infinitely as a person as she matures through the unimaginable number of experiences she will have.[14] Of course, the question of whether numerical identity is the same from the very start of the entity being a person to some infinite distant future poses serious conceptual tensions as to whether the entity survives as the individual person, but there is a practical problem that should concern us far more than these more abstract reasonings.

When people have the best motivation and evidence to know that they ought to live a life of great value—that is, they know that they will die—we see them squandering their opportunity without a second thought. Many people overindulge in experiences and actions that will actually harm them, such as wasting their time and resources with technology that distracts from life rather than improving it,

[14]Both sides of the issue seem to base their reasoning primarily on whether they think that they will enjoy themselves, which says a lot about them but little about what actually might happen.

eating large quantities of food that is of low quality and high risk to health, and otherwise engaging in activities that do not enhance their own or anyone else's flourishing. They are shortening their lives and making it harder to flourish than it should be, keeping in mind that we are not even talking about flourishing at the very top of their abilities, but merely improving their lives in ways that make them significantly better than they are. For example, how many of those who should know better about wasting time have long lists of the classics to read, but never find the time to do so because some trashy novel or other distraction grabs their attention? Quite a few are discouraged by the facts that those things we should appreciate, such as great literature and art, are not to their taste. As a result, their lives tend not to be the best they can be, nor do they even rise to the standard of better lives when that is easily within our grasp with a little effort on our parts.

Since we squander opportunities that we should take when the stakes are so high in improving our lives, there is little reason to believe that an immortal individual will do anything but lead the same sort of existence, if not worse. Since there is no end, there is never an incentive to take stock in life or to say that life is short, so we should live well. Even if other things would be adequate incentive to make such a life worth living—Fischer, for example, argues that negative experiences such as depression, anxiety, and boredom would give meaning to the virtues—there is even less guarantee that this would have any positive effect on the immortals' behavior than it does on that of mortals (Fischer 2013, 122). What we are likely to get is merely a bunch of people who can figure out a lot of new ways to muddle through the way that many do now.

Immortality does not make practical and moral sense for other reasons. First, immortality would be a giant waste of resources to achieve worse lives than we can achieve now through normal practices of life creation, experience, and death. Second, making matters worse is that the resources used to give a few, very wealthy people immortality would be far better spent on those without lives worth living or lives barely worth living.[15] Those goods could help improve the subpar and marginal lives to a much better level of well-being in the moment and overall. To divert resources to someone who is already having a very prosperous life so that the person can continue that life at the cost of opportunity to others who need help to achieve minimal or secure flourishing is morally indecent. It would, therefore, be better to worry more about helping people now than spending scarce goods on something that might eventually prove to be causally impossible and is not worth the doing if it could be done.

[15]For some psychological reason, those who believe that immortality is a good idea to pursue assume that they or people like them are going to be the people benefitting from it. These folks have very good lives of privilege, but imagine immortality for those who have terrible lives. Their agony would never end.

4.4 Priorism

According to Joel Feinberg, Martha Nussbaum, Barbara Baum Levenbook, and others, a person who died before his interests were fulfilled is harmed at the moment he formulated, held, and invested in those interests just as long as the interests were reasonable (Feinberg 1993, 186).[16] For example, suppose that a person chooses to be head of a Fortune 500 company, but there is no casually possible world in which that happens. Hence, it would be unreasonable for him to have that impossible goal, and it is no harm to him when he dies and does not fulfill his desire because he never could in the first place.

Other unreasonable goals can be highly unlikely ones. A person might aspire to be the Fortune 500 CEO, and it is possible that could happen because there are a few worlds open to the agent with that as a state of affairs. However, for that world to be authenic, many others people would have to act in ways in which they are unlikely to behave. Given that there are a vast majority of worlds in which the person never becomes a CEO of any corporation, and the ones in which he does are remote, then it is not reasonable for him to have the desire he has. In this instance, it is an unreasonable option and, once again, the person is not harmed by the frustration of his deeply impossible self-selected goals (Scarre 2007, 92–3).

In Priorism's most plausible versions, the only time that people are injured is when their desires and goals are reasonable and they die before being able to live in a world containing the states of affairs that satisfy their ends[17] (Luper-Foy 1993, 270–1). George Pitcher writes, "An event or state of affairs is a misfortune for someone (or harms someone) when it is contrary to one or more of his more important desires or interests." (Pitcher 1993, 162). Death injures the person dying because "The continuance of our lives for most of us, at most moments in our lives, is something so manifestly in our interests, and that being so, the sudden extinction of life would, as thwarting that interest, be a harm." (Ibid., p. 173). In Freda's case, her choices of which world to live in, namely one in which Fred is her friend, are eliminated by Fred's malicious actions. She can no longer choose to live in worlds in which she is not the object of abuse, which has harmed her ability to choose her own path in life. For Priorists, the dead are like Freda in that worlds in which their preferences are satisfied are no longer open to them because they are dead.

What is rather strange is when Materialists, who believe death is the end of the person, try to argue that the person is harmed by her death. Basically, they have rejected any rich version of dualism beyond, perhaps, that of Hofstadter's Interactionism, but they want to cling to an afterlife that allows harm, only now placed into the life prior to the cessation of existence. Feinberg's Priorism avoids

[16]Other preferentialists argue that the harm occurs when the desire or interest (preferences) are stymied, such as the person holding them dies or events after the death make it impossible for the deceased's desires and interests to be fulfilled.

[17]The world in which the person is not harmed is one in which the states of affairs satisfying his choices exist but also include his death.

the problem of a non-existent, non-experiencing subject being injured by his own death by placing the harm firmly in the person's life. Since the person is alive when the goals are set, then he can experience the injury because the goals cannot be fulfilled, even though they are reasonable. Hence, it seems possible to account for the death-harms-the-decedent intuition by making it consistent with reality.

Many Priorists justify their positions on death harming the deceased and against the Epicurean objection by appealing to the use of intuitions. At times, linking a person's death (and harm the person suffers because of his death) is a "natural" or common-sense thought about dating the harm to the person's demise. Many intuitively maintain that death is a bad thing, so it must be injurious for the person.

The Priorist approach has a number of problems. First, although we can agree that Freda is harmed by the false friendship in various ways, Priorists and others do not quite understand how she has actually been injured because the analogies used for her case and that of the deceased are inadequate. Basically, the degree of similarity between the target and analogue is insufficiently small, given that one is a living being and the other does not exist. If Freda had not existed, then she would not have been harmed. Feldman's Existence Condition states that "Nothing either good or bad can happen to a subject s at time t unless s exists at t." (Feldman 1992, 205). Had Freda not existed, then the intrinsic value from her being a psychological person in relationships could not have been altered since it no longer exists.

A second problem for Priorism is that it is better for us to die sooner rather than later if we have many or significant desires that will not be fulfilled by our deaths. "[W]e have seen that although [Priorism] provides a reasonably convincing explanation of some of the evils associated with death, it has the snag that it represents the harm of death as being over as soon as we die." (Scarre 2007, 105). That is, the agent is harmed by the frustration of his desires, interests, and plans only as long as he is alive. As soon as he dies, the continuing injury that devalued his life will immediately cease. On these grounds, the rather odd conclusion follows that it might be better for someone to die sooner rather than later depending on the injury being caused him by his unsatisfied desires, goals, and interests.

If we follow this line of reasoning further, we should become even more squeamish with what it entails for those who die before they can achieve their goals. For the sake of argument, assume that we have a person at 25 who has a number of plans that will take a considerable amount of time to achieve. Let us call her Sara. Her plans are to attend graduate school to become a great jurist who will sit upon the supreme court of the land; she fully expects to be an active member of her various communities to make them better; she wants to have a family, raise children, and then see her grandchildren make their way in the world; and she plans on having a rich and fruitful retirement with her partner. Tragically and unbeknown to her, none of these plans are possible in her actual world because she will die at 30, although in other worlds she will continue to exist.

According to the Priorists, Sara suffers injury from the moment she has these goals to the moment she dies with them unfulfilled. If Sara adopted them at 24, then she suffers from her death for longer than if she had adopted them at 25. Therefore, the life in which she suffers longer because of her normally morally permissible, if not laudable, goals is worse overall than one in which she had the goals for a shorter period of time, *ceteris paribus*. In fact, she would be injured least if she set the goals, and then died immediately. In Sara's best interests, she should adopt her goals when they do the least damage to her, even though more reasonable people would see that as a rather strange position to take.[18] Her life with the great goals adopted earlier would have to be a better thing for her than one in which she delays.[19]

Perhaps the evil suffered is balanced through some means. The good she generates from having the goals earlier in life will more than offset the injuries she has. She leads a far better life than she otherwise would. However, we should not forget that any long-term plans that include her, including those of her children, partner and so on, are also injuries to those who have them. Sara's death, in fact, starts harming the survivors at the very moment that Sara and her loved ones make them. Priorism entails that many are extremely and unknowingly harmed at the very time when most people would say that the individuals are flourishing and their well-being is very high.

In addition, Priorism entails that in order to increase intrinsic value or limit injury, it might be best to make only small plans that the immediate person can undertake and finish in a relatively short period of time. That is, those individuals who look at the long-term self, such as those who care about the flourishing of their intermediate or overall persons, or that of the psychological or embedded person, are harmed when their plans fail. Great enterprises, such as starting a family, finding a significant other, living a long life, and other projects are detrimental to those who die before they can be fulfilled. On the other hand, those who have nothing but a life of little goals safely satisfied are not injured the moment they create and adopt their mundane goals. A very cautious person who lives day to day without making any plans for the future will be harmed much less than someone who has a life narrative that takes time and dedication to achieve.

However, this discrepancy makes little sense if we take flourishing as a goal. Although one can thrive in the moment by instantiating a number of intrinsically valuable states, overall and intermediate person flourishing requires much more than that. It demands organization of one's life to achieve large projects that make the individual a much better person, e.g., becoming more virtuous. A narrative for the overall life shows value that cannot be reduced to or explained in terms of a cardinal calculation of individual bits of that life. As argued above, a full life is a mixed life that produces more intrinsic value as an organic whole than a string of immediate person days added cardinally together.

Moreover, for the best lives and narratives, certain goals that a person plans will come to completion long after the person is alive, and the person knows this for a

[18]It would seem that those who commit suicide are ideal candidates for not being harmed by their deaths, as long as they did not have interests that survive their deaths. If this is the case, then suicides would not damage their lives as much as those who have posthumous interests.

[19]Even one moment difference in times of death could make the person dying sooner better off.

fact. If we put into place projects that will not bear fruit until long after we are gone, then we have not accomplished our goals of helping future generations at the moment we die.

This caring act in our existence's twilight would seem not to be an action that harms us, but according to Priorism, it is. First, we are injured by acting in this manner because our goals cannot be fulfilled during our lifetime. Second, as rational people, we know we will die and that our death will cause injury because our goals are unfulfilled; therefore, we must rationally say that the long-term projects we set in motion but cannot see to the end during our lifetimes bring injury to us the moment we formulate the plans to act in such a manner. Such an utterance contradicts what we think overall flourishing to be. Antoni Gaudí, architect of the *Basílica i Temple Expiatori de la Sagrada Família* in Barcelona, knew that his creation would not be completed in his lifetime, in part because cutting the stone would take at least 100 years. In fact, the Sagrada Família is expected to be completed in 2026, 100 years after Gaudí's death. It is rather difficult to see how Gaudí's death caused him injury from the moment his unfulfilled plan was created in 1883 until he died in 1926. Although there is no decisive evidence for the following claim, most would say that if the person had a flourishing life, and had goals that would benefit others but could not be accomplished during the person's mortal span of existence, then no injury occurred. None of the organic wholes in which he is or is part of loses intrinsic value. In fact, planning for such monumental gifts to others is a symbol of greatness rather than harm.[20]

To salvage Priorism, several replies could be made. First, it might be said that Gaudi's and other's after-life goals are unobtainable, and thus, are not rational, but no reasonable person would draw such a conclusion. Gaudí never expected to live to see what he wanted to accomplish, and his goal is obtainable in a way given that the Sagrada Família will be completed.

In addition, it would be even more unreasonable to state that only when the goal is fulfilled that the harm to us ends or that the fulfillment of the goal by others after we are dead entails that we are not injured in the first place. Many, many people have to help fulfill Gaudi's designs. For anyone with unmet goals, our survivors and others have to work to realize our plans. If all these others do not do their parts, then we are harmed. Therefore, whether we are harmed or unharmed while we are alive can be mediated by what people do after we die. Existentially, that is a very odd thing to say. We have no power at all, whereas others, even people who will be born after we die, have total power over our well-being and flourishing. One case in point is the damage done to Nietzsche's reputation by his sister and the Nazis.

[20]Martha Nussbaum argues "events that happen after a person's death enrich the life retrospectively and alter its causal significance." (Nussbaum 2013, 33). Events can also harm the person's life, according to Nussbaum.

Although personal identity and lives can be construed as narratives that continue after the psychological person has died, narratives without that person being the central narrator are not the people/lives we are talking about when discussing if death harms the deceased.

A third problem for Priorism is that, again, we see intuitions used in a manner that distorts reality rather than being merely in accord with what is actual. The intuition is the driver in the argument; it is reality being forced to fit theory rather than theory reflecting reality. Since almost by definition the detection of harm by and after death is impossible, then it is much like the physical and mental substances Hume labeled as meaningless. The case of Freda and Fred is unlike this.[21] In the case of those who die, most of them can do nothing to improve their intrinsic value and flourishing because their deaths are in the future, undetected. If they knew of their deaths, then the rational thing for them to do is to change their plans to accommodate their actual future. That is, with the additional information, they would have to change their desires and interests because their goals can no longer be rational for them or for any rational person evaluating the situation.

Unfortunately, those who will die cannot be allowed to mitigate the damage to their intrinsic value and flourishing; otherwise, the injury done to them by their death is not all that great, and the intuition that death causes harm goes unfulfilled again.[22] That is, suppose that Sara knows that she will die at 30, and being a perfectly reasonable Priorist, adjusts her goals appropriately so that she has no desires, interests, or plans that will remain unfulfilled at the time of her death. She even changes her goals to reflect her fate, so one of her plans is to die at 30, which will be fulfilled by her dying at 30. Her injuries from her death cease to exist the moment she revises her goals, and her death will no longer be an injury to her. If people still intuitively feel that not only is Sara's dying an injury to her, but also that her early death compounds that harm, then they would be wrong. On Priorist grounds, Sara cannot be harmed unless we force generic desires and interests on her, such as for a long life that fits within the current life expectancy for women in Sara's society. Of course, this means that we are taking away Sara's power to mitigate or eliminate injury to her merely to satisfy the survivors' need that Sara's death will injure her, which seems callous at best.

Another very odd result of Priorism is what it says about the rationality and reasonableness of the dying individual's beliefs. If someone thinks Priorism is true, then it is only those who irrationally continue to have desires, interests, and goals that are likely to exceed their actual, reasonably known fate that are harmed. That is, since they know that their death will harm them, and they have the power to at least mitigate that harm to enhance their well-being and flourishing, then they act

---

[21]There is a reasonable probability of Freda finding out about the injury done to her, which would allow her to make decisions to distance herself from Fred, call him to account for his action, and so on.

[22]Scarre states that there would be "something peculiarly tragic about the death of a person who had recently taken up fresh projects or commenced new relationships to which she was deeply committed." (Scarre 2007, 95). We should ask, "For whom is the situation tragic?" There is little evidence to suggest it is tragic for the person who is no more. It could not be. It is tragic for the perceiver who has a very negative experience from seeing such a promising life end at an early age. This negative experience does not provide any evidence that the dead person has been harmed.

irrationally by clinging to desires they know cannot be fulfilled. The only reasonable thing would be to take Sara's course of action and give up. If the dying do not, then they will be guilty of wrongdoing by foolishly pursuing their impossible goals, desires, and interests rather than trying to make their lives better. Should we take this uncaring line for those who are so vulnerable? Our flourishing as decent people would be against such an uncaring, un-empathetic position; hence, it might be more practical that the reality of each psychological person's demise, not an injury to him or her, should be the evidence used to evaluate the intuition's legitimacy rather than the other way around.

A fourth problem for Priorism is that it is always assumed that the psychological person's plans are good ones, but what of evil plans for a rather bad person? "[I]t seems plain that it is not generally correct that what thwarts a person's desires is bad for the person." (Rosenbaum 1993a, 300). Suppose that the person has plans to kill a number of her co-workers and then herself, or she could be a serial killer who wants to have more murders to her dubious credit than the most proficient serial killer known.[23] On the way to her nefarious deed, the killer dies in a convenient car accident. If Priorism is correct, then is the agent injured by her accidental death? In one way, she is harmed every moment of her existence when she has the plan until her death because her goals are not fulfilled, and her desires were thwarted from being brought into fruition.

There seems to be a better case to be made that her death benefited the would-be killer (that is, if we believe that people can have posthumous benefits and harms). Given that she cannot commit the atrocity, her death in the car accident prevented her from enacting those things that would harm, without adequate moral justification, her or others' flourishing. If we consider those worlds in which she survives to kill others and herself versus the worlds in which she is killed first, then in general, the latter will have greater intrinsic value than the former. Moreover, the embedded person she was would have greater intrinsic worth because she did not alter herself into a mass murderer or serial killer. Therefore, we cannot automatically say that the death was an injury for an individual merely because her plans were stymied.

The intuitive thwartation-of-desire-satisfaction-is-bad principle is unreasonable to hold for any situation unless we add another condition to it.[24] Given the Practical Principle and the end of flourishing espoused throughout this work, the condition would be something along the lines that a person is harmed if one or more of his desires or interests that are consistent or required for his flourishing as a general type of person or as an individual person are thwarted. The person is illicitly harmed if the thwarting is done in a manner that does not respect her intrinsic worth. If developed in this way, then it is impossible to thwart a non-existent person's desires and interests relative to her flourishing. Since the person does not

[23] These violent, bloody scenarios, although rare, happen too many times to be ignored in the discussion of harm.

[24] See Heathwood's (2006) "Desire Satisfactionism and Hedonism".

exist, then it is nonsensical to think about her flourishing because flourishing is impossible for something that does not exist. Hence, the wartation-of-desire-satisfaction-is-bad principle has no application in death cases, although it does have some value for the living.

The fifth problem is an existential difficulty that Priorism and the multi-verse theory adopted in this work have in common. The future universe need not exist at all, so it might be impossible to do the comparisons required by those who know that death as harm can be shown only by comparing two sufficiently similar worlds. That is, some people believe that the universe is being created at each moment rather than existing in its entirety. Therefore, there is no set of existing worlds from which to pick one that will become actual for the person. There is just the actual world and those fictional worlds that are dreamed up by persons to make sense of morality and the idea that we could have acted or been other than we are. The other worlds are possible only in the sense that they could have come into existence, but they have no actual reality other than being objects in the mind of a perceiver.

Assuming that time is linear,[25] for the Priorists, there can be no harm for the individual because there is no fact yet in existence that makes it false that the individual will be able to fulfill her goals. Perhaps all we have are wave functions that collapse as the universe keeps existing. If John intends to marry Mary, then John's future death cannot harm him at the moment he begins to intend to marry her, or even later. John's future death does not exist in the future in any meaningful way to be able to harm him other than as a possibility or an object in one's mind. When we speak of something that will happen, then we are making claims that have an undetermined truth value to them, at best, since there is no fact in existence that corresponds to the proposition of our utterance.

A somewhat standard line of comparisons shows that people can not only be harmed by their deaths, but by postmortem events as well, through the thwarting of the interests or desires the person had prior to his death. Ben Bradley's Difference-Making Principle is one case in point[26]:

> Difference-Making Principle (DMP): The value of event E, for person S, at world *w*, relative to similarity relation R = the intrinsic value of *w* for S, minus the intrinsic value for S of the most R-similar world to *w* where E does not occur (Bradley 2009, 50).[27]

---

[25]Four-dimensionalists would argue against this position, but their view of reality should worry us on a number of grounds. First, it might be that they have the physics wrong about space and time, but that is beyond the scope of this work. Second, according to Adam Taylor, if they are right, then it follows that a person who dies in horrible pain after leading a life not worth living will spend eternity in that state because if time and space are identical, then the suffering entity is merely in a space different from this one and continuing to exist in horrible circumstances.

[26]Bradley is not a Priorist.

[27]Steven Luper also uses the comparativist approach to argue that there is post mortem injury or wrongs that can be done to the deceased; unlike Bradley, Luper examines both harm and benefit as well as the fulfillment of interests (Luper 2009, 86–7). However, both Luper and Bradley resort to a cardinal mathematics when doing their calculations, which leads them to a conclusion based on an illusion of existence (Ibid., pp. 84–5).

According to Bradley, DMP explains how things can go badly or well for a person, and is an intuitively plausible, simple, elegant, and flexible principle for the work we want it to do (Ibid., 50–1). Suppose that S gets a well-deserved promotion in one world, and that leads to a much better life for S. In the nearest R-similar world in which she does not get the promotion, her life is worse off. The promotion's value for S would be positive given that the intrinsic value for S in the world in which she receives the promotion is greater than the intrinsic value for her in the world in which she is denied it.

For comparativists, S's death could also have positive or negative value for her depending on how it works out. Assume that a world in which S dies entails an intrinsic value for S of zero, since she no longer exists from the moment she dies until the moment that S dies in the other world. If the world in which S survives offers her positive intrinsic value from the time of death for the other S in the other world, then survival is good for S in the world in which she survives. On the other hand, if the world in which S survives offers negative intrinsic value for her after the death of S in the other world, then survival is bad for S. So, S's death harms her if it is bad for S in the other world and benefits her if it is good for her in the other world (Luper 2009, 100).[28]

The main problem here seems to be an existential difficulty in what should be compared. In the case of John, the value of World A in which John survives can be compared to the value of World B where John dies. That comparison makes sense. We can ask ourselves the sensible question of whether World A is better than B without John's existence in it, and that is plausible and reasonable to do. We can also make sense of comparing the two lives to determine which had greater overall value. However, one universe having more or less value than the other is not a harm or benefit for John if John does not exist to be benefitted or be harmed in that universe. It merely means that one universe is better than the other as a result of John dying or being alive. Furthermore, one life having more value than another also does not entail that the life has more value *for* John.

What does not make sense is comparing John in World A to John in World B after John has ceased to exist in World A, and then drawing conclusions about whether John's death or life benefits or injures him. This can be seen easily with Luper's and Bradley's equations if we reject looking at the values of the entire lives and focus on the actual crux of the matter, which is the difference between the life of John in World A from the moment World B's John dies to the death of World A's John. By definition, the lives of John in the first and second worlds will have identical values until World B's John dies; therefore, those segments of life and value can be ignored in the cardinal calculus.[29] What is now interesting is that we

[28]Luper uses preferentialism based on his version of the thwartation principle and versions of pluralisms to argue for post-mortem injuries and wrongs (Luper 2009, 100).

[29]However, if we use a narrative or organic whole and ordinal calculus approach instead, then we would be evaluating each full life against the other.

can have a value of the remaining life for World A's John, but there is no value possible for that of World B's John because John no longer exists in that universe.

Since John in World B does not exist, talking about any intrinsic value for him makes as much sense as discussing the nonexistent man who was on the stairs,[30] but an odd thing happens when we begin thinking or talking about nonexistent objects—they begin to appear to have some form of existence.[31] Bradley's DMP creates an illusion of existence when performing cardinal math calculations using the values of entire lives.[32] If Value is determined by subtracting X from Y, then we must use values for X and Y. The problem arises when the value for one of the terms is merely a placeholder, as in the case of World B's John. Let us say that both worlds' Johns have an overall life value of +10 for the time both Johns identically exist. Let us also stipulate that the remaining part of John's life in World A is +5. According to DMP, the value is

Value $= +15 - +10$, or Value $= +5$, if we make continued life the event that does not occur in World B.

The conclusion is supposed to be that death for World B's John is bad for him because World A's John's life has +5 more value to it.

The difficulty is that we are not really interested in the value of the lives that are identical because they will cancel each other out in any calculus like this. The actual DMP should be focused on comparing likes to likes rather than to radically dissimilar things of the living versus the non-existent. That is, we should be considering the same timespans in both worlds as determined by which John lives longer or just what happens after the first dies until the second John ceases to exist. If that was done, then it would become clear that a nonexistent object is being compared to an existing object. For example, DMP for the time period between deaths would be:

$$\text{Value} = +5 - \text{Nonexistence}.$$

For the same timespans, it would be:

$$\text{Value} = +15 - (+10 + \text{Nonexistence})$$

Neither of those is going to give us Value in a meaningful way because nonexistence is not a value.

Perhaps the clearest formulation can be had by focusing on the fact that death causes something to cease to exist; therefore, it cannot experience anything after it is no more. Instead of making this all about John, let us focus on the worlds themselves. Rather than having S be a person, let us make S be the World *w* in each case, and then apply DMP to the destruction of World B as the event. Worlds A and B are identical up to the point at which World B ceases to exist. DMP would give us

---

[30]See footnote 2.

[31]This issue will be discussed in more depth later.

[32]Belshaw (2009) also uses the comparisons of lives in different possible worlds this way.

a value of [X − (X + Y)] or Y where Y is the value of World A after the destruction of World B. Hence, is World B's value lower than that of World A? Yes, because World A has more time in existence than does World B. Is World B's lower value worse for World B or anything in World B? No, because it is exactly the same as for World A as long as World B exists, and then nothing exists in World B, and World B does not exist. In fact, nothing can make World B better or worse after its destruction because there is nothing to be affected by anything that happens; nothing can happen in a world that does not exist. The same applies for John in World B. His life is less valuable than that of World A's John, but the former John's life is never worse for him, and it is never better for World A's John. Hence, for these reasons, the calculations and comparisons of the living versus the dead cannot be performed legitimately.

Since the issue of non-existence has been raised, then preferentialism Priorists of all varieties have difficulty explaining why the interests based on the desires of the individual do not cease to exist at the very moment that the individual ceases to be. Although there is a claim that "we have a prima facie duty to respect, hence fulfill, the desires of self-determining beings, even ones who have died, in at least some matters concerning themselves, and to avoid acting against their interests, expect perhaps when that is what they wish." (Luper 2013, 332). Notice the slippage here between an existing thing and one that no longer exists. It makes perfect sense to claim we have a *prima facie* or an actual duty to respect self-determining beings because they are intrinsically valuable, but it does not follow from that plausible and probable conclusion that we have a *prima facie* duty to a non-existent object.[33] The reason is obvious: "Not only does [death] prevent the fulfillment of the desire, but also removes the desire by removing the person." (Rosenbaum 1993a, 298).[34] Once again, death destroys the person, which means that the desires and interests are terminated, and there is no continuation of that person after the event or process of death has been completed. Hence, although our language and thinking give the illusion of existence because we can have the concepts of the deceased in our minds, there is no surviving entity that is numerically identical to the living being.

For a moment, suppose that desires can outlive the person having them. What would that mean practically? If there are desires and interests that can survive the death of the being that held them, then there would be something to which such things attach now that the person is dead. That is, they cannot be independent things

[33]It is also problematic to assume that fulfilling someone's desire respects the person. I would suggest a number of caveats for such a principle, including whether the action of fulfilling helps the various individuals involved to flourish.

I would argue that there is a duty to respect promises and such, but that duty is based on our own flourishing and being a good person who has the integrity to fulfill his or her promises even when the person promised no longer exists.

[34]Assuming that death is a process, then someone might be alive and dead simultaneously in a way that would allow for the person to still have interests and desires that can be thwarted illicitly yet still be dead. Luper suggests such a possibility when he writes, "death unfolds over a period of time, and it obviously can affect us while it occurs." (Luper 2009, 70).

because they require a mind to adopt them. We could argue that the dead body of the physical person is capable of having these things, but then we would have to accept the odd result that dead bodies have desires and interests.

If we granted this existential claim, merely for the sake of argument and to fulfill the intuitions of death as an injury to the deceased, then, among other conundrums, we would be required to evaluate such interests against those of the living. For example, would the most trivial of interests of the living outweigh the most important interest of the dead body? Assume that Tom has left strict orders that his body be preserved by mummification, which can be done without much cost to anyone. However, Suzzanne, his apathetic widow, slightly prefers cremation over mummification. In this case, should Tom's dead body's interests override the insignificant interests of his apathetic widow? Without resorting to unreliable intuitions, it is difficult to see how to answer the query. Perhaps we should merely take the expedient and practical step to say that a person's interests and desires end when he does, which would prevent such questions from arising.[35]

What is more worrisome is that if Priorism is true, then the desire and interest status quo must remain in force for the body even if circumstances were altered in such a way that the living person would no longer have the interests and desires he did when he died. Suppose the costs of mummification rose precipitously because the chemicals used were suddenly no longer readily available on the market. If Tom had been alive, he would not have wanted anyone to bear the costly burden of his desire for mummification, but no one knows that, nor does anyone have sufficient evidence to guess that would have been his desire. However, the desires and interests that Tom had before his death become eternal and unchanging the moment he ceased to exist as a psychological person. Tom is harmed when his pre-mortem desire is thwarted by circumstances beyond his control which would have changed his mind had he been alive. This is an odd result if desires and interests survive the death of the person. Normally, for living beings, interests and desires can be changing things. Perhaps the very core significant ones shared in common with all people, such as the goal of flourishing, do not change, but there can be changes in how such a goal is achieved, or in trivial matters. We make adjustments to our trivial-to-important desires and interests as circumstances warrant. Nonetheless, in death, the decedent cannot do this. He becomes a helpless victim or beneficiary to his own destiny according to the whims and dictates of others as well as whatever frame of mind he was in when he died.

Finally, perhaps the best way to evaluate the debate between Priorists and Epicureans is to ask what the point is in thinking that death has to harm the individual who died. If we understand death as Epicurus did, then we can maintain the value of the plans without incurring the costs from Priorism, and we can be consistent with our stated goal of flourishing.

---

[35]What actually happens is a legal fiction of the person's estate is created. The desires and wishes are attached to this entity.

4.5 Cambridge Changes

One way to try to account for harm to the deceased, yet not encounter Priorism's problems of a non-existent being having to experience something, is to base the injury on Cambridge changes. Cambridge changes are one way that we can "sustain the intuition that death is bad for a person during the period that he or she might have lived." (Scarre 2007, 109). There is no need for people to be injured by their unfulfilled goals, nor do non-existent people have to experience something they cannot experience. "Cambridge changes are purely relational changes that happen to things in consequence of real changes in other things." (Ibid. 106). Thus, the deceased does not alter when a Cambridge change occurs.

Cambridge changes do not alter a thing's intrinsic properties, even when the intrinsic values of another object in the relationship change its properties. If Karen becomes a better chess player than Christine, then Christine remains the same person she was, but a Cambridge change has occurred (Scarre 2007, 106). Christine's ability to play the game does not alter by Karen making Christine worse than she was, but the relationship between Karen and Christine in regard to this ability does change for the worse. If Christine cares even one iota about the relationship and remaining the superior player, then she had been harmed. As with Freda, she need not know that the relationship has altered to her disfavor, but she is still injured because she is no longer in the relationship she desired.

The question then becomes about how Christine is adversely affected. She cannot be directly affected because that would alter some internal property she has. The best way to explain how this works might be to return to the notion of organic wholes. Christine and Karen are in a relationship, which creates an organic whole of Christine being a better chess player than Karen, and Christine desiring to keep this relationship as it is. When Karen becomes a better chess player, the relationship alters, as does the organic whole, into a significantly new one. That is, the organic whole's identity has undergone a transformation because one of its necessary features has been altered. The organic whole's intrinsic value might now be different based on the new entity. The values of the two wholes can be compared. If the new whole has less intrinsic value, then Christine has been harmed because she exists in a situation that is less valuable than it otherwise would have been. Karen, as a psychological person, might be more valuable, but it is possible that she, as an embedded person in this relationship whole, is worse off than she would have been.

Christine can have a self-induced injury when she undergoes the Cambridge change with regard to her chess playing ability. Kim, being the better player, might harm Christine because it makes the latter feel bad about herself, or her dislike of Kim makes it impossible for Christine to feel anything other than degraded if Kim bests her at anything. In this situation, both parties of the relationship exist. Christine suffers an injury as the indirect result of the Cambridge change. However, Tolstoy, since he no longer exists, cannot harm himself when Turgenev becomes more popular.

One of the puzzles about Cambridge changes is whether they actually do the work their proponents claim they do. We are told that individuals can be harmed by their death, but are they injured if only the relationship changes? Can Cambridge changes happen to the dead as Scarre and others claim? It is not as clear as one would hope because what is missing is often a clear understanding of how time affects the situation. Assume that Tolstoy is the most popular Russian author. If the public starts reading Turgenev more, then both Tolstoy and Turgenev undergo Cambridge changes while remaining essentially the same individuals. Since Tolstoy disliked Turgenev, it is a bad change for Tolstoy.[36] In this case, the alteration in popularity occurs after Tolstoy's death. Therefore, at the moment of the change, Tolstoy is no longer any type of person. He was a person, but persons, by definition, are living and existing objects. Unlike Christine and Freda, Tolstoy becoming less famous than Turgenev does not create a less valuable organic whole because his non-existence cannot be in an organic whole in a relationship with Turgenev. In fact, there cannot be a relationship at the current time because two elements of the relationship are missing; hence, there is no relationship.

Non-existence is what makes the Freda and Fred example inapplicable to Cambridge changes and harm after death. Even though Freda is unware of what has happened, her intrinsic value is illicitly reduced. For Tolstoy himself, there is no intrinsic value to be affected since he does not exist to experience the value reduction.

There are a variety of other puzzles facing how well Cambridge changes answer the question of whether an individual can be injured by his own death or post-mortem events. First, there can be an infinite number of Cambridge changes to any person after his demise, but we must assume that the person who died is frozen in time as the psychological or embedded person he was at the very moment, or period,[37] of his death. According to the Cambridge change view, whatever his desires, ambitions, goals, and other future oriented plans are, those are the only ones capable of giving him harm if they are affected by his death and later Cambridge changes.[38]

However, assuming a multi-universe theory, why not examine what else could have happened to him, or what alterations could have occurred to his goals, desires, and who he was as a person? That is, why should we think that only certain counterfactuals matter while others do not? Given that we are expected to continue saying that, "If S had lived, then S would have cared about E in the relevant way," looking at other alternative worlds for these other types of counterfactuals has to be legitimate as well.

---

[36]Of course, Cambridge changes need not be caused by persons or other living things. A fire started by a lightning strike might destroy a glen that had been cared about by someone now deceased.

[37]The possibility that psychological death does not occur in an instant needs to be considered.

[38]This argument also poses difficulties for Priorism.

We know that people change their minds. What looks like a good idea now might not be such in a year's time with new information and alterations in one's circumstances. If the deceased had lived longer, he would likely have changed some of the trivial and significant goals he had when he died. Instead of being a ploughboy who would be married to Mary, John, had he lived, would have discovered his aptitude for poetry. Being a poet would have required him to give up his earlier choices to marry Mary and do many of the things that would have been appropriate if he had remained a ploughboy, but now merely interfere with his revised interests.

Now, why would this matter? When someone states that the decedent is harmed because, at this age he would have married his girlfriend, (Scarre 2007, 108) they are assuming knowledge of harm that they, at best, can only reasonably guess at merely from a static impression of the individual. This privileging of one individual psychological person over another is a rather strange thing to do. In the actual world, each embedded and psychological person, whether it is an immediate, intermediate, or overall, had certain interests, so worlds consistent to the one set of goals should matter no more than those associated with a different set of goals. In some worlds, the relevant person could change his goals, plans, and interests. Hence, he would not have been harmed if a now irrelevant Cambridge change occurs. In the counterfactual worlds, his ends and interests would have come to naught through some actions of his or others, or states of affairs that close off those worlds in which his interests are met while he is alive. Perhaps he is left at the altar instead of marrying his fiancé. It might turn out to be the case that it is better overall that he died before his interests were so ill-served than to be alive to see it happen. With claims of harm after death, it is not clear that anyone is harmed even if they do undergo a Cambridge change, unless it can be shown that only those worlds in which he would have met his goals matter if he had survived.

There are some attempts to alter the deceased's interests and desires to account for new developments that arose after their death. Adam Yauch, a Beastie Boys member, did not want the group's songs used for advertising, and left written instructions in his will to this effect. However, one song, "Girls", was parodied with new lyrics in an attempt to attract girls to science and science careers. Would Mr. Yauch have allowed this use for an excellent cause even though it is an advertisement? If we use any of Mr. Yauch's immediate, intermediate, or overall individual identities, then the answer would be a refusal.

Privileging the static Mr. Yauch and others is problematic because it is inconsistent with how desires and interests work. They change as we change—perhaps we have acquired new information or insight—which should be reflected in how we think about them. Unfortunately, the actual desires, interests, and goals of the deceased cannot be changed in the normal way as a result of the simple fact that the deceased does not exist to have the power to alter them.

The problem with people other than the deceased modifying his interest and desires is the possibility that the deceased's personal identity will be lost. First, survivors assume the power to make such alterations when they were not permitted to have it when the person was alive. If someone can be posthumously harmed, then a loss of autonomy and narrative power would count as an injury as it would if the person were alive. Second, many people would make the changes without really being able to capture what the person would have done because no individual is able to fully understand the private mind of another being. We could attempt to change some desire or interests of another, but we would have to be careful. We cannot alter essential desires lest we change the person to a new entity. Important interests and desires are not essential, but they matter significantly. These would not seem legitimately alterable because they do matter so much. Mr. Yauch's request in his will would be one such. Finally, trivial desires and interest would not appear to be of great concern, but these could add up over time to change the identity of the person.

A rather unsettling question is also suggested by the previous paragraph and Cambridge changes: do significantly different later versions of me harm previous versions when the former prevent the latter's plans from coming to maturation?[39] That is, a succeeding immediate person could have different interests from those immediate and intermediate individuals of the same overall person who came before. For example, as a youngster, the person had the goal of eating whatever sweets she wanted when she wanted them. A later immediate person of the same overall person could prevent this goal from being fulfilled. If the Cambridge change position is correct, then it seems to follow that the former person is injured by what the later person does. The succeeding person has stymied the fulfillment of a goal the former person had, thereby adversely affecting the relationship between the two while not altering any of the former's psychological or embedded person characteristics. Hence, later immediate persons can harm, through Cambridge changes only, previous incarnations of themselves as immediate people within the overall person. The same would hold true for intermediate-immediate people and intermediate-intermediate person relationships, as long as the people in the relationships were sufficiently different.

Although it might appear a bit silly to worry about a sweets acquisition relationship being altered between different people within the construct of an overall person, it does matter a great deal in cases of mentally debilitating illnesses. Those afflicted with Alzheimer's and similar diseases undergo a gradual process in which they change from being a particular psychological and embedded person before the disease's effects are noticeable to being a different immediate or intermediate person and then losing that status altogether in the disease's end stages. The final stage, if the individual lives through it, is being a moral subject, which is not the same type of entity as being a person. If we are considering harm from death and Cambridge changes, then which entity is harmed by the relationship change? If it is the immediate or intermediate person pre-the-effects of the disease, then that entity was long gone before the moral subject and physical body died. Is it the final immediate person, who has lost the ability to make major decisions in her life and significant portions of her identity as the previous immediate and intermediate

[39]The same problem arises for the Priorists.

person, yet she remains a person in her own rights? That entity is not the person used in most ordinary language. In fact, it is the entity with fully operational individual and general species faculties upon whom many focus when, in the attempt to comfort, making such statements as, "He would not have wanted to live this way" at a dementia patient's funeral. There does not seem to be a mistake about privileging one immediate person over another within the overall person, but care should be taken not to assume that Cambridge changes of the same type *must* be harmful to the overall person.

The mistake here seems to be caused at least in part by equivocation. What a person is can have many different definitions, including being a person in general and having various individual personal identities, as has been argued previously. The ones I have been using generally are psychological and embedded persons in their various forms of immediate and intermediate persons because these notions seem to be the ones that can do the most work and fit best with what people identify as their referent. In fact, Cambridge changes proponents, such as Scarre, often seem to be talking about different entities existing (Scarre 2007, 112–3). Obviously, there is at least one psychological person, but we need to be careful about the context being used when talking about a person and harm that is done to him or her by the death.

4.6 Is There a Psychological Need/Desire for Death to Harm the Decedent?

There are many philosophers who use intuitions and the "common-sense view" that death harms the entity that dies as the basis of their argument (Silverstein 1993, 95). In *The Ethics of Killing*, Jeff McMahan spends 93 pages trying to evaluate the harm suffered by the individual who dies as a result of that death. "Since the badness of death cannot be intrinsic, it must be comparative. Death must be bad by comparison with what it excludes." (McMahan 2002, 98). Now, this is true in one way explained above, but from this truth it does not follow that death is bad or a loss for the person who has ceased to exist. We are comparing what might have been with what is—a possible world that interests us because it is meaningful in what we want to think or talk about and the actual world in which the person is dead, or one life against another. Instead of addressing what makes one death worse than another because we need not have it be a loss or injury to the deceased to account for that, McMahan and others appear to use common intuitions that death harms the deceased as the central grounding to their arguments. For example, McMahan writes that:

> Our intuitive judgments about the comparative badness of death for different people are responses to [how bad is it for a person to die this death now rather than a later death from a different cause]. And in exploring this question, I will be respectful of those intuitions. Just as in moral theory, I take intuitions that are deeply and pervasively held to be presumptively reliable, so in thinking about the badness of death I will take strongly held intuitions…to be starting points for inquiry that are not to be lightly abandoned. (McMahan 2002, 104)

I have argued earlier that intuitions should be considered to be suspect until they are shown to be reliable beyond merely being deeply and widely held.

There appears to be a plausible and probable explanation to why many people believe that death is a harm to the individual. My hypothesis explains the belief's force and widespread nature in human history and across cultures, while at the same time indicating that there is no evidence supporting the view that death harms the deceased or that the person who ceases to exist can be harmed or wronged by things that happen to him after his death.

Freud was right to say that we fear our own death. An odd thing is that we have no experience of our own death, so it seems odd that we are afraid of something of which we have no experience. I think that the reason lies in how our brains work in trying to make sense of the unknown. First, as I stated previously, the brain on which our minds depend is at least a partial result of evolutionary forces acting upon it. Our brains and minds have become very apt at being able to classify things to be afraid of, things to fight, and things that are safe or not of immediate concern. Since this mental processing has worked very well for us so far, and our brains and emergent minds are partially the result of it, then we tend to classify the unknown the same way even when we are without adequate evidence.

Making these classifications, we must be using some evidence to make decisions in this matter. Without any evidence, we would be like Buridan's ass, unable to choose into which class—to be feared or not—our death should go. Either there must be some data that we use as adequate evidence to make the classification into things to be afraid of, or there is an automatic brain or mental reaction to fear anything that is unknown so that we can take flight from it and thereby save ourselves. Of course, if it is the latter, then our fear of death is non-rational at best and should not count as adequate information since we have Epicurus' plausible and probable argument appealing to reason as to why that fear is unjustified.

In the case of our deaths, although we have no evidence of what it will mean to us, we make do with information that seems to be close enough to be able to help us in our mental processing of the issue. After all, by definition, each moment of our lives is different from the moments we have lived, which means that each new moment is something we have not strictly experienced before. We are generally able to function well enough by remembering what worked before and then using that to make decisions in the present about the future. Basically, we adopt induction to classify general kinds of things and make predictions about the future.

One reason we fear our own deaths is that we fear and are injured by others' deaths, in part, because of what those events and processes do to us. We are emotionally or otherwise severely harmed or hurt to lose those who have close, positive relationships to us. We are likewise injured, only to a lesser degree, by the deaths of those who are not as central to who we are as psychological, embodied, and embedded persons. Even if we do not know someone, if she is sufficiently like us or has other features able to engage our empathy, we are negatively affected by her death by feeling a loss.[40]

Perhaps the intuition's power is that our empathy or sympathy, which allows us to have morality and, possibly more importantly, gives us common cause with all other persons, makes us think that death has to be a harm to the deceased because their deaths harm us. First, we place high value on individual selves, which gives us motivation for our fight or flight actions. If we did not have this, then we would merely accept whatever happens to us no matter how harmful it might be. Second, we also place high worth on our relationships with other people. When they die, it harms us in significant ways. Hence, the greater the empathy or emotional connection, then the greater the harm to those who survive the death of another. We have more than adequate evidence that death of others can injure us in very important and deep ways.

Now that we have the data, we naturally look for patterns in order to understand how to behave. Most of the time, our minds process data from external and internal sources without our noticing it the way that we do for anything that is a fundamental habit in our daily lives. However, there are instances in which the normal does not work, as in near-death experiences. There have been a number of reports of a person's life "flashing before their eyes." Although this phenomenon has been interpreted in religious ways, e.g., allowing the person to put life in perspective before death, there is actually a better hypothesis that does not involve the supernatural. Basically, what is happening is that the person's mind is doing a rapid memory dump of life events, patterns and processes for a purpose the mind has always fulfilled, which is decision making. Near death experiences, such as car crashes, are generally unique, stressful situations,[41] which means that our minds have inadequate patterns of behavior on which to rely.[42] Therefore, what the mind is actually doing is trying to find a pattern, process, or experience sufficiently similar to what is happening in the present so that it can take action to understand the situation and preserve the individual's life.

Given the unique experience of an individual's own death as something beyond what is normally stressful and the lack of evidence for what it means and how to behave, each person's mind automatically uses the sufficiently similar pattern in which others died and how it affected her. The mind subconsciously makes the pattern fit the situation. Since the person's death hurts us, then it must be the case that the death injured the person who died. In addition, since we as living beings

---

[40]This engagement can be seen in the unfortunate racial bias many have when informed about the disappearance of children of color versus white children. There appears to be less empathy in the former case, which explains why, even though they are morally identical cases of children in danger, that the news media and others pay far more attention to the white child's case than to that of the child of color's.

[41]There is one theory that the mid-brain region is releasing noradrenaline which is associated with brain regions that mediate emotion and memory (Mobbs and Watt 2000, 449).

[42]Appelby argues that the mind is acting as it normally acts in stressful situations (Appelby 1989, 976).

would not want our greatest projects thwarted or our most central desires left unfulfilled, then that must be true of the deceased as well.

Continuing this armchair psychology, which seems to explain things in at least a plausible sort of way, the transference on others of the negative emotions, such as pain, that we feel as a result of the impact of their death upon us is performed by the mind-reading powers we have (Liebermann 2013, 131). Of course we cannot actually do the mind reading given the brute fact that the deceased is not really there, but we have at least one thing that can perform as a stand-in. Our minds try to fit the situation to how the mind normally operates. First, we have to find a substitute for the mind reading. One possibility is that we use ourselves as the other person so when we are doing the mind reading we are actually merely seeing ourselves in a mirror. Our reflections utilizing our own experiences and emotions, therefore, produce a result that reinforces what we already believe and feel because they are them. We fear death and we think it will harm us; therefore, death must harm those who are dead.

A second possibility is again based on a mirror but considers our reactions to our own demise rather than trying to figure out what happens to others. "My only resources for reacting to [my existence ending] seems to involve turning back on myself a set of attitudes—such as sadness, grief, rage, anxiety—that are tailored to circumstances in which the self endures and undergoes a loss." (Scheffler 2012, 86). In this situation, the mind tries to make sense of death but can only do so by assuming that the individual doing the introspection remains in existence before and after her death. Although it is hard to understand why there would be fear or a sense of loss when the essential being remains the same, the mind reading being done here is for the same person only at different periods in the person's history. Without merely the assumption that the future self will sufficiently resemble the present self—and the present self is having negative attitudes toward its demise—then the belief that there is a loss or injury to the deceased by the deceased's death is reinforced in a very small echo chamber.

A third possibility for how the mind reading is performed in the absence of one of the two parties is that we subconsciously bring to mind the person who has died. That is, our conception of the person is used in the process of determining if death is an injury to the deceased, as well as whether the person is wrong or harmed after the individual has ceased to exist. Given our conception of the living person (because we have no real understanding of a dead one since the person no longer exists), we make do with how we now perceive the person's identity, including his desires, emotions, and such.

It is here that we begin to see again an illusion that is created when thinking about death; to talk about the deceased person, even when we know that the entity cannot exist, gives the object in our mind a form of reality. We cannot think about nothing; when we think, we think about concepts and other things, which then have existence in the mind, and we cannot perceive the non-instantiation of a concept when we are thinking about the concept. Hence, when thinking about a dead person, the perception-conception of the person has a form of being along the lines of Meinong's sosein, even though the actual person does not exist.[43] By subconsciously reading the perception-conception's "mind", we think that death harms the person because we illicitly believe that the person is still alive or existing in a relevant way. Mind reading, which Liebermann claims is genetically programmed into us so that we will have the same beliefs and values, as well as being able to harmonize our thoughts and actions with those around us, then processes the information (Liebermann 2013, 8). The results are that many intuitively believe that a deceased person can be harmed because the person has existence that allows it to be injured, and the person is sufficiently known to the perceiver or assumed to be like the perceiver so that she believes that she would not want to be treated in such a manner. Given these beliefs, it therefore follows that the "living" concept of the person that represents the actual person who is no more would likewise be harmed against her will.

Intuitions that death harms the deceased also rely upon the fact that we do not realize the suspect sources of these intuitions and common sense beliefs. First, if these intuitions reflect reality, then we should see them across age groups within a society that believes that death is an injury. However, we do not find that. Starting with the very young and death experiences, it is clear that children do not measure danger and harm the same way that adults do. They have no fear because they do not have the mental patterns to be afraid. That is why they do not perceive death as a harm to the person who died. As B.F. Skinner showed, those fears have to be learned. Middle-aged people do not have enough experiences to find a pattern of acceptance for death but do have sufficient data to have fear of dying. It might be the case that most philosophers and others who believe that death is a harm are at this stage and are using their experiences as the norm by which to judge beliefs, conclusions, positions, and arguments. The very elderly and terminally ill have lots of experiences with death, so it has lost a great deal of its power over them. Their minds have developed a pattern of coping and understanding, which allows them to accept the inevitable in ways much more difficult for anyone other than the very young. Death is not an enemy, but something that will give them release from their existence in this world.

We should not forget the impact of social conditioning and genetics on how we react to death. To accept death or write it off as not an injury to us is to go against our flight or fight instincts created through a very long line of natural selection. In addition, it is to acknowledge that if we fight, even if we put everything we can into the battle, then eventually, we will still lose. It is to say that no matter how clever we are, no matter what paths are open to us, no matter what resources we have, each person ends in the ultimate result of death and the death of his loved ones. This is an attack on our self-esteem, worth, and will to survive. It shows that we cannot control the situation, although we need that power to guide our lives and flourish.

---

[43]Although these individuals do not have sein, they can be objects of our psychological attitudes (Rosenbaum 1993b, 130).

Those who fear death and claim that it injures the deceased are also operating on information that is based on conventions taught to them by society and their own personal experiences. On social conditioning and genetic grounds, we think that there is something bad for the person in her dying. To seek death or to claim that death is not an injury to the entity that ceases to exist is to call into question cultural beliefs in a way that challenges the very basic natures we share with other people and all animals capable of fight or flight choices. This, in addition to other factors, is what makes the death-is-not-an-injury position so difficult to accept for anyone not in the same circumstances as the individual seeking death or those who support him in his efforts to accomplish this end. It is even more difficult for insufficiently involved people to accept it as a legitimate, rational position. To accept that death is not a harm to the deceased is a rejection of common belief that almost all share, and it denies the very function of what makes us animals and people with our shared commonality. We talk about being a member of the community or society and how difficult it is to define these terms, but one thing we clearly share in common is our attempt to remain alive by either fighting or running from incipient destruction. Hence, it is understandable that seeking death and not believing death is an injury occasions so much vehement social condemnation.

In conclusion, there seems to be a deep psychological need or desire for death to harm the injured. This need or desire is the result of genetics, socialization, and the way our mind thinks about death and deceased persons. However, no matter how deeply held the belief is, if it is false as Epicurus shows it to be, then we need a great deal more evidence to show us why we should continue to believe that death can injure the dead. In what follows, I will try to chart a middle course that accepts the Epicurean argument while still allowing this psychological need to be satisfied, although not in the manner that some folks might want.

4.7 A Pragmatic Solution: Reified Distributed Personas

Fred Feldman argues that people can survive death at least for a little while after their biological body has died and their psychological personhood has permanently ceased (Feldman 1992, 112). In his example of Adam, Feldman claims that Adam exists up until the time that his body is cremated, and "in order to live, a thing must exist." (Feldman 1992, 112, 114.).[44] These are plausible claims to make. Although existence does not require life, it is impossible to understand how something can live without existing.

The history of an entity, which is "the smallest event that contains as parts all the events that happen to that entity during the time it exists", can continue after the psychological entity has perished (Feldman 1992, 116). For Feldman, the history of

---

[44]Feldman believes that the before and after death entities must be numerically identical. I argue that they are not because they are similar but essentially different organic wholes.

Adam, or any entity for that matter, persists until that overall entity ceases to exist. Adam's history, therefore, continued until his body was cremated, whereas Adam as a psychological person died before that time. As a consequence, the good news is that if someone is wealthy enough and the technology is available to her, she can keep a departed loved one in existence by carefully preserving the body. Lenin resting in his museum, hence, is alive in a biological personhood way.[45] The bad news, of course, is that those without the financial resources must settle with a shorter time period for their loved ones to linger after their physical body has died.

Feldman's view suggests a middle course that will allow us to answer the Priorists and others' intuitions about harm after death. His or her death cannot harm the person because he or she no longer exists to be injured, but there is now something that, in a way, has the former person's identity, and that entity can be damaged.

William James argues that the self is not located solely in the psychological person, but is distributed over space and time in ways that go far beyond the socially embedded person discussed in the preceding chapter:

> The sum total of all that a man can call his, not only his body and his psychic powers, but his clothes, and his wife and children, his ancestors and friends, his reputation and works, his lands and horses, and yacht and bank account. (James 2007, 291)

The accumulation of the ideas, works, memories, and other artifacts the person leaves behind have to be part of the reified distributed persona because they were part of the embedded person.

More specifically, a ***reified distributed persona*** is an organic whole that comes into being the moment that a person ceases to exist (dies). The reified distributed persona is a new organic whole made up of all the parts of the embedded person that has ceased to exist as an embedded person which includes the physical artifacts the embedded person created; any relationship in which the psychological person or embedded person was a member that is transferred to the reified distributed persona without the relationship being significantly altered; any perception, thought memory, or other mental states about the psychological or embedded person that is not significantly altered; and all the new relationships and mental states about the reified distributed persona that comprise essential elements and relations of the new organic whole that is the reified distributed persona.[46] For example, after the psychological entity is gone with its intrinsic value as a person, living entity, instantiation of mental states, and all the other components to the entity's intrinsic value, then there is a new organic whole that is made from parts of the former organic whole, such as relics,[47] as well as any new feature that falls within the

[45]An art historian, Kris Groberg, contends that the Lenin on display in Moscow is not really the preserved body of the individual psychological person known as Lenin. Her visual perception of it and knowledge of conditions at the time Lenin died leads her to believe that it is a wax sculpture on display.

[46]Of course, there will be many legitimate different conceptions of the reified distributed persona.

[47]See Strong's "Relics." (1989).

organizing organic whole of the particular reified distributed persona.[48] However, many essential and significant components have been altered including, but not limited to, the relationships the entity has to other entities and internal psychological conditions.[49]

Often, the dead body serves as the nexus of the reified distributed persona's organic whole and will continue to do so until that time that the body ceases to exist. Given its long history in law and relative understandability and usefulness, we can think of this entity as the personal identity estate that remains after the person is no more. Immediately after the death, it is the physical body, which is in whatever state the body is. It might be a living body that is the result of the physical body still being a living thing but without the complex mental structures in place that made up the psychological entity. In this case, there could still be an embedded moral subject, but the organic whole would no longer be a person. It might be a dead body in which the biological functions that kept the body alive so that a psychological entity could survive no longer serve that purpose but rather have begun to decompose the body in a different life process.

Even when the body is no more, the reified distributed persona can still exist. In fact, reified distributed personas need not have a nexus the way that a person does. Such personas require that a person exists, dies, and leaves enough relevant physical or mental artifacts that an organic unity, such as a body of work or a concept of the reified distributed persona, can be formed.

Instead, there is merely an informal collection of mental states, relationships, and artifacts. The reified distributed persona can be held together in part by her estate, will and planning as occurred in London's Sloane Museum.[50] That is, it can be created and maintained as a legal fiction that represents the final desires of the person in regard to how she wanted her possessions to be treated. In addition to the legal fiction and other binding things, the reified distributed persona is comprised of the memorials she has made of herself in the thoughts of others about her, and most importantly, by the works she has created that carry the essential elements of her as an individual psychological person, moral agent, and embedded person. The latter artifacts are especially valuable if they record at least part of the personality of the person who created them. If she was a teacher who helped a student become a better critical reasoner, for instance, then that work stays alive as long as that skill exists

[48] I think that Eric Olson would find this claim odd for much the same reasons that he argues that corpse concurrentism—the corpse existed before the entity's death as the same matter as the living entity—and corpse creationism—the entity's death brings the corpse into being as a new object—are untenable (Olson 2013, 88–9).

[49] Ivan Soll claims that we make a serious mistake thinking that death is a change when it is actually a destruction (Soll 1998, 37–8). With a change, there is numerical identity before and after the event, but with destruction, there are two different entities. The one prior in time is destroyed when it ceases to exist.

[50] Luper and George Pitcher argue that we can be wronged by what others do after we die, but the "we" they use is better explained by these legal fictions, and as I will argue, reified distributed personas, than it is by retroactivity reaching back in time to harm the psychological or embodied person (Luper 2013, 318; Pitcher 1993, 161, respectively).

and is passed on to others. The things the teacher has published, written or created that tells something about her, including but not limited to journals, books, plays, crafts, recipes, or anything else, will affect others in some way whether that effect be positive or negative. In fact, getting to who the psychological person was might be best served through the person's work rather than direct interactions with the embedded person. What she found important, her views on them, and whether her opinions were well considered can be discerned through the effort she took to write. Those must have been important to her if she spent her time and resources on them. Therefore, writers and other creators who have essential works still exist as reified distributed personas, but those who lived as individual persons who left little trace while alive and none after they are gone never are reified distributed personas and cease in their entirety as organisms, biomes, or people once their bodies are destroyed.

Moreover, the more that a person puts herself into her work, then the more intimately tied that artifact is to who she is as a person. It is a revelation of innermost thoughts and properties of the person as she consciously, subconsciously, and unconsciously sees herself. Hence, when that work is under attack by critics or folks who do not understand it, then it is understandable why the creator becomes upset. The rejection of the work is an attack on her because the work is what she is.

Given that we help create the reality around us by giving value to things that without us would not have such value, it is reasonable to believe that reified distributed personas exist. We place great worth on the objects that were owned or came into contact with those people we find laudable, such as Shakespeare, Gandhi, and King, as well as those possessed by the infamous, such as Hitler and the Borgias. We have great wonder at the property of the boy pharaoh, Tutankhamen. To touch it is to touch what the other has handled, but is it rational to feel as if these objects have greater worth than the sheer monetary or aesthetic ones that they possess? It seems that on purely rational grounds, they do not. They are merely physical objects with no intrinsic value at all. If all human persons should cease to exist, then the value of such things would also cease to be.

Let us explore the valuing of these objects, and decide if such judgments and the beliefs on which they are based are permissible even though they might not be rational.[51] There is Donald Davidson's thought experiment on identity over time that involves swamp gases, lightning, and an unfortunate hiker—Davidson himself (Davidson 1987). Suppose that Davidson is studying swamp flowers during a lightning storm. A stray bolt strikes him, disintegrating him so that no trace remains of his ever existing in that spot. A different lightning bolt interacts with swamp gas nearby to create an entity that is identical in all respects to the dead Davidson save for the fact that the new person did not exist until created by lightning at the same moment the other ceased to exist through the same general cause. Both before and

[51] By providing some external justification for the judgments and beliefs, I am trying to avoid the unexamined intuitions as evidence problem I raised earlier for those who think that death injures the deceased.

after entities have identical functional mental data, which allows them to act in exactly the same way. They have the exact same physical appearance so that none could tell that they are different. The question is whether the two are the same or two different people.

Davidson argues that the two entities are different on the grounds that they actually do not have a contiguous set of mental states that gives them a causal history. The new Davidson, for example, has never actually met the old Davidson's friends and family, although he thinks he has. As a result, under Davidson's theory of semantic externalism, the new entity cannot have new thoughts or a mind until such time as it sufficiently engages its language with real world objects to give the words actual meanings that are determined by the new Davidson's causal history and internal states.

Putting aside the issue of meaning, the ontological puzzle of identity posed by Davidson's example is interesting here because it says something about how we create identity and value based on the objects' causal history and how we view what the objects are. If we buy a silver ingot we believe contains silver from the Great Seal of England, then we value that object in a way determined in part by what we think is its causal history. If, in reality, the ingot has not one silver atom from any Great Seal, and we learn that fact, then rationally speaking, we should not automatically feel as if we have lost something of value. It still looks the same as it did before we learned that we had a false belief. In addition, let us assume that the value we could get on the market is exactly the same for ingots of this type even without the link between it and a Great Seal. The objects are so expensive that the additional property that cannot be shown without a provenance has little financial value. Since we are not out anything by the object not having the history we thought it did, then rationally speaking, it should not affect our personal valuing of the ingot.

However, the causal history and the value it confers matters to a great number of reasonable people, especially those in the museum trade. Some might not want the ingot anymore for the pure reason that it does not have this historical causal connection with England's power.

The alteration in how the ingot is valued helps support, in small part, the notion that reified distributed personas exist. An individual's personal objects are given value by those who care about such connections, a worth that is caused, in part, by the fact that the person possessed or had some other relevant connection to them. These objects are not essential to the deceased's identity, nor do they bear the deceased's essence in or on them, other than as some possible stray genetic material. They can be considered as mere mementoes of the person, much as the clothes of a deceased person are, but still can occasion great emotion in his survivors. When a person ceases to exist, she leaves behind these objects that were her or of her. The stronger the connection to her, for example, because she used them a large number of times, the more they remind survivors and others of one of her essential properties. Thus, we heavily associate the objects with the person, especially if they were intimate things revealing portions of the deceased's identity, which creates a greater mental state about the person generated in those thinking about him. The reified distributed persona is stronger or weaker in part based on these remembrances and mental states.

The reified distributed persona's power is determined by a variety of factors. First, the existence of the dead body, which is what is left of the living body that was part of the embedded person organic whole, serves as a nexus. The more intact the dead body is, the more it looks like the original person, thereby being better able to sustain those relationships, feelings, and mental states that require the existence of that body, as well as serving as a nexus for new relationships which might be very similar to the ones that ceased to exist when the person did, new relationships *qua* dead body, and mental states that the embedded person had or caused. For example, if a dead grandmother looks like the live grandmother, her body can still make her grandchildren feel the love they had for the embedded entity that was their grandmother. Therefore, the body maintains that love and causes the survivors to feel grief because of their loss. In addition, it can make people feel sadness and other strong emotions that they would feel from viewing her body.

Second, the number and strength of relationships, feelings, memories, and other mental states people have about an embedded person are part of her reified distributed persona and can strengthen or weaken it. As those who have these mental states about the reified distributed persona cease to have them without passing them on to others, the strength of the reified distributed persona fades. As we have seen before, an embedded person is not merely the psychological person or the psychological person in the person's living body. Part of the identity of a person, if we are considering a wider scope than merely the internal, is how others perceive her, her relationships to others, and her affect in the world. When the psychological person and living body die, then all relationships to her as an embedded person end. However, new ones can arise. The survivors might have duties to fulfill the estate's obligations, bury the body, and take care of unfinished business which create new relationships that will be part of the reified distributed persona. In addition, those special memories about, feelings toward, and other mental states caused by the embedded person keep memories of her alive, and form components to the reified distributed persona, much as our views of the embedded person form part of her embedded identity. Finally, the reified distributed persona can create and include new perceptions about itself and what the psychological and embedded persons were in their immediate, intermediate, and overall stages. The funeral director might never have met the embedded person, but will have an idea about her after talking to her survivors.

The intrinsic value of a reified distributed persona is much more difficult to gauge and argue than it is for the psychological or embedded person because the latter instantiated intrinsically valuable attributes, while the reified distributed persona has no organizing individual psychological or organism identity. Most reasonable people can appreciate why an organism or psychological or embedded person has such worth, but with an entity that might seem a mere useful classification along the lines of species, it is harder to come to a consensus. However, the reified distributed persona does have mental states distributed over the living entities within the persona that can have positive or negative intrinsic value.

Of course, they are the mental states other people have, such as a fond memory of a friend who is no more. Others might look at the person's writings, for example, and be changed by them intellectually or emotionally for the better or worse. That power for good and evil in the reified distributed persona is part of what gives it inherent worth, which can cause intrinsic value changes in the perceiver. Adding to the reified distributed persona, and therefore its power, are the living developmental states about the reified distributed persona's identity. That is, since the reified distributed persona is a collection of artifacts, loosely understood, of the deceased, as well as old and new relationships with the reified distributed persona, and mental states about the reified distributed persona, then any additional mental state or relationship builds the reified distributed persona's power as an organic whole.

Once again, the reified distributed persona's identity and intrinsic value is not merely determined by the sum of its parts, just as in the case for all persons. It is an organic whole, in part, because of the mental states included within it and the impact it can have on other entities. If there is no perception at all of the reified distributed persona from the list of intrinsically valuable and disvaluable mental states, then there is no intrinsic value.

If the non-perception means non-existence hypothesis is correct, the sad truth is that those for whom no one has any perception will have died in every way possible. To be forgotten is to be non-existent in every way. However, if the person has left works that still exist as physical entities, then there is a possibility for reincarnation or regeneration. With mental states about and in the reified distributed persona, the reified distributed persona will have intrinsic value and identity again. Suppose that the works are buried in the desert, and everyone who knew of the person and the reified distributed persona's work is dead. The reified distributed persona still exists through the works because they are artifacts showing identity, but there is no intrinsic value there until the works are found and mental states are created as a result of the work.

The greatest benefit to the reified distributed persona way of thinking about harm to the deceased is that it eliminates the dodgy ontological assumptions made about unfulfilled goals reaching back through time and Cambridge changes actually being able to harm the deceased, while preserving the ability to talk about harm in a way that many reasonable people can accept.

4.8 What Harm Does Death Do?

The answer to the question about the losses caused by death can be answered using the value theory outlined above and in earlier chapters, but specific enquiries require specification of a number of variables. First, which entity is being addressed when discussing the harm? If it is the psychological or embedded person, then the psychological or embedded person's termination cannot harm the person, given that the person no longer exists to be harmed. The same conclusion follows for every type of person defined.

However, the reified distributed persona could be a different matter. According to Frankena, an entity is inherently valuable if and only if the experience of contemplating it is good or rewarding in itself (Frankena 1973, 82). That is, a beautiful work of art is not intrinsically good because it is not good in and of itself. However, the art is inherently good because it causes the perceiver to have intrinsically valuable experiences from the mere contemplation of them. If this claim is true, then inherent value is intricately linked with ends in and of themselves, such as the state of affairs in which a person is contemplating the art.

Reified distributed personas can include both inherent and intrinsic value. First, they have intrinsic value if there are mental states of others that comprise the reified distributed persona. Second, they have inherent value because contemplation or awareness of them, or portions of them, might be good or rewarding in themselves. For example, Shakespeare's works are inherently worthy yet are not intrinsically valuable because they are not mental states, persons, or any other object being good in and of itself. Since the works are part of his reified distributed persona, and the works give insight into who he was as a psychological and embedded person, then the intrinsic value generated by the contemplation is evidence that the reified distributed persona has inherent worth.

The reified distributed persona is that which can be benefitted or harmed after the psychological and embedded person's death because it exists partly or wholly through mental states and relationships to other things. Harm occurs when some intrinsically valuable component of the reified distributed persona is altered so that it is less valuable, and perhaps, the overall value of the organic whole that is the reified distributed persona is reduced as a result. If it turns out that Tolstoy is less popular that Turgenev, then it is Tolstoy *qua* reified distributed persona that is harmed. When John dies, Mary is no longer married *qua* marriage relationship to him, but she is in a relation with the reified distributed persona that is John as a biological body. Mary's misery and pain become part of her relationship to the biological body entity, which significantly lowers the value of the overall reified distributed persona that is a deceased John. This, in turn, reduces the value of the various organic wholes in which the reified distributed persona of John resides.

Another way the reified distributed persona's value can alter is through new beliefs about the former person. If Mary finds out that John was stealing from her, then her view of John will change, thereby changing the reified distributed persona for the worse in comparison to the world in which Mary remains ignorant of John's thievery. If she thinks less of John than she did, then John as reified distributed persona's intrinsic value is lowered by Mary's new perception of him because they are part of the reified distributed persona. The alteration in Mary's mental states about John and any other changes to intrinsically valuable entities will negatively affect John as reified distributed persona's intrinsic worth. In addition, the reified distributed persona also loses some of its inherent value because thoughts of John no longer simultaneously create intrinsic value in Mary's mind and in the minds of anyone who now thinks less of him.

Besides harm and benefit to the reified distributed persona, death harms other entities of which the person was a part as well. Of course, what that harm is will be determined, in part, by the scope, context, or organic whole to be considered. On a world or universe view, the loss in intrinsic value might be insignificant, or even a net positive. Since the universe is a very large place, then small components might not affect it very much even as an organic whole. Of course, if the intrinsic value for the universe is small relative to the universe's size, i.e., there are few intrinsically valuable things in existence, it could be the case that the loss of one end would affect the world's worth in a significant way. Then again, it would depend on whether we are looking at a slice of the universe, the universe slices added together, or the universe as a narrative whole.

If we are considering societies, then a death might range from a very large loss to a very large gain given the removal of the component of the social organic whole. When a good person dies in a society that does not require that death, then the overall worth of the organic whole of the society is less. Because it no longer has the intrinsically valuable person with its intrinsically valuable states contained within it, then it has to be a less valuable entity than it otherwise would have been. If the person is a needed pillar of his or her community, then the loss is greater than if the person had neutral or negative intrinsic value within that whole. For other groups that make up various organic wholes in which the individual is part, the same sort of considerations would have to be taken into account to evaluate the loss' value.

Death's harm can be measured according to the change in intrinsic value from the value of the organic whole containing the psychological or embedded to the value of the resulting organic whole that now contains the reified distributed persona. Since the calculus is ordinal, the organic whole that is the person serves as the base measurement by which the new organic wholes' value is compared. If the reified distributed persona is dynamic enough to incorporate inherent or intrinsic values so that it exists in the organic whole being studied, then the organic whole containing it can be readily compared to the one that contained the person. How great the change is might not be a matter that can be proven beyond any reasonable doubt, but a reasonable person should be able to agree to a suitable range to reflect what has occurred.

Besides the loss of the intrinsically valuable individual person and that person's mental states, other areas of disvalue need to be taken into account when evaluating the harm caused by death. Both the positive and negative mental states of those who survive and are affected by the death need close consideration. Some people might be pleased that the person is dead or that the relationships the person had are significantly altered, e.g., John's rival now sees an opening to woo Mary. If the person was regarded positively, then the negative mental states of affairs that are caused by his passing will affect the various new organic wholes in different ways. Mary, being in a state of sorrow, can make her various organic wholes less valuable. Finally, we need also to consider desert and any other state that would affect the value of all the various mental states of affairs. For instance, if Mary deserves to suffer because she was unnecessarily cruel to John near the end of his life, then her state of affairs might be less intrinsically disvaluable than if Mary had neutral or positive desert state. The person who is pleased by John's death might not be the monster we think he is if it turns out that John was bullying him to the brink of self-annihilation. Although difficult to discern, the particular situation with its states of affairs and relationships will determine the actual intrinsic value gain or loss for whichever person and scope is selected to evaluate.

Generally, death is a harm because it causes a drop in intrinsic value, but that is not always the case. In some circumstances, the society might be far more intrinsically valuable than if the person had survived. If it is a society with limited resources, and the person's death freed up enough of them so that each of the remaining members can have a flourishing life, then the society is far better off than it would have been with one too many members to allow all to flourish.

It should not be thought that the loss is without intrinsic value at all. Reality is far from it, which can be seen if the lens is narrowed to focus solely upon the individual person's worth. That entity with its psychological personhood and universal and idiosyncratic traits that made it a person, as well as this particular person, was intrinsically valuable in and of itself. However, in the larger context, what is intrinsically valuable as a smaller organic whole might have a different meaning when the context is broadened to be more inclusive.

References

Appelby, L. 1989. Near-death experience: Analogous to other stress induced physiological phenomena. *British Medical Journal* 298: 976–977.

Belshaw, C. 2009. *Annihilation*. Montreal: McGill-Queen's University Press.

Bradley, B. 2009. *Well-being and death*. New York: Oxford University Press Inc.

Britton, W.B., and R.R. Bootzin. 2004. Near-death experiences and the temporal lobe. *Psychological Science* 54: 254–258.

Cave, S. 2012. *Immortality: the quest to live forever and how it drives civilization*. New York: Crown Publishers.

Davidson, D. 1987. Knowing one's own mind. *Proceedings and Addresses of the American Philosophical Association* 60(3): 441–458.

Feinberg, J. 1993. Harms to others. In *The metaphysics of death*, ed. J.M. Fischer, 169–90. Stanford: Stanford University Press.

Feldman, F. 1992. *Confrontations with the Reaper*. New York: Oxford.

Fischer, J.M. 2013. Epicureanism about death and immortality. In *Our stories: essays on life, death, and free will*, ed. J.M. Fischer, 103–128. Oxford: Oxford University Press Inc.

Fischer, J.M., and R. Curl. 2009. Appendix to Chapter 6: Philosophical models of immortality in science fiction. In *Our stories: essays on life, death, and free will*, ed. J.M. Fischer, 51–62. Oxford: Oxford University Press Inc.

Frankena, W. 1973. *Ethics*. Engelwood Cliffs: Prentice Hall Inc.

Gasser, G. 2010. *Personal identity and resurrection*. Farnham: Ashgate.

Heathwood, C. 2006. Desire satisfaction and hedonism. *Philosophical Studies* 128(3): 539–563.

James, W. 2007. *The principles of psychology*, vol. I. New York: Cosimo Inc.

Kelly, E., E.W. Kelley, et al. 2009. *The irreducible mind*. Lanham: Rowman & Littlefield Publishers Inc.

Liebermann, M.D. 2013. *Social: Why our brains are wired to connect*. New York, NY: Crown Publishers.

Long, J. with P. Perry. 2010. *Evidence of the afterlife: The science of near-death experiences*. New York: HarperOne.

Lund, D.H. 2009. *Persons, souls and death*. Jefferson: McFarland & Company Inc.

Luper, S. 2009. *The philosophy of death*. Cambridge: Cambridge University Press.

Luper, S. 2013. Retroactive Harms and Wrongs. In *The Oxford handbook of the philosophy of death*, ed. B. Bradley, F. Feldman, and J. Johansson, 317–335. Oxford: Oxford University Press.

Luper-Foy, S. 1993. Annihilation. In *The metaphysics of death*, ed. J.M. Fischer, 267–290. Stanford: Stanford University Press.

McMahan, J. 2002. *The ethics of killing: problems at the margins of life*. Oxford: Oxford University Press Inc.

McNally, R.J., and S.A. Clancy. 2005. Sleep paralysis, sexual abuse, and space alien abduction. *Transcultural Psychiatry* 42(1): 113–122.

Mobbs, D., and C. Watt. 2000. There is nothing paranormal about near-death experiences: How neuroscience can explain seeing bright lights, meeting the dead, or being convinced you are one of them. *The Lancet* 355: 447–463.

Mothersill, M. 1999. Old Age. *Proceedings and addresses of the APA* 732: 9–23.

Nagel, T. 1993. Death. In *The metaphysics of death*, ed. J.M. Fischer, 59–69. Stanford: Stanford University Press.

Nussbaum, M. 2013. The damage of death. In *The metaphysics and ethics of death*, ed. J.S. Taylor, 25–43. Oxford: Oxford University Press.

Olson, E. 2013. The person and the corpse. In *The Oxford handbook of the philosophy of death*, ed. B. Bradley, F. Feldman, and J. Johansson, 80–96. Oxford: Oxford University Press.

Pitcher, G. 1993. The misfortunes of the dead. In *The metaphysics of death*, ed. J.M. Fischer, 157–168. Stanford: Stanford University Press.

Plato. 1961. *The collected dialogues of Plato,* ed. E. Hamilton and H. Cairns. Princeton: Princeton University Press.

Rosenbaum, S.E. 1993a. Epicurus and Annihilation. In *The metaphysics of death*, ed. J.M. Fischer, 291–304. Stanford: Stanford University Press.

Rosenbaum, S.E. 1993b. How to be dead and not care: a defense of Epicurus. In *The metaphysics of death*, ed. J.M. Fischer, 117–134. Stanford: Stanford University Press.

Suits, D. 2004. Why death is not bad for the one who died. In *Life, death and meaning*, ed. D. Benatar, 265–284. Lanham: Rowman & Littlefield Publishing Inc.

Scarre, G. 2007. *Death*. Montreal: McGill-Queen's University Press.

Scheffler, S. 2012. *Death and the afterlife*, ed. N. Kolodny. Oxford: Oxford University Press, Inc.

Silverstein, H.S. 1993. The evil of death. In *The metaphysics of death*, ed. J.M. Fischer, 93–116. Stanford: Stanford University Press.

Soll, I. 1998. On the purported insignificance of death. In *Death and philosophy*, ed. J. Malpas, and R.C. Solomon, 22–38. New York: Rutledge.

Strong, J.S. 1989. Relics. In *Death, afterlife, and the soul*, ed. L.E. Sullivan, 51–61. New York: MacMillian Publishing Company.

Terrillon, J., and S. Marques-Bonham. 2001. Does recurrent isolated sleep paralysis involve more than cognitive neurosciences? *Journal of Scientific Exploration* 15: 97–123.

Thonnard, M., et al. 2008. Near-death experiences: Fact and fancy. *Revue Médicale de Liège* 635–6: 438–444.

Williams, B. 1973. *Problems of the self*. Cambridge, UK: Cambridge University Press.

Woerlee, G.M. 2005. *Mortal minds: A biology of the soul and the dying experience*. Amherst: Prometheus Books.

Chapter 5
How Should We Feel About Our Own Death?

Abstract Sigmund Freud argued that people believe in God because they fear their own death. If there is a divine entity that takes a very positive personal interest in you, and has the power to keep your personality alive after your body has died, then death need not concern you because you will never undergo it. Hence, since death is unable to harm those God preserves, then fear and other feelings and emotions should alter accordingly. Fear, for instance, should be assuaged by the belief in this god, although in reality it might be simply repressed. In this chapter, I examine the morality of some of the most significant feelings and emotions we have toward our own deaths. Depending upon which view is adopted for understanding emotions and feelings—and I advocate for a pluralistic, pragmatic approach—we can not only have morally right/bad or wrong/bad emotional reactions and emotions, we can also have obligations to feel a certain way at a certain time.

5.1 Introduction

In ethics, there is a long history, dating back to at least Aristotle's time, of requiring moral agents to have appropriate emotions and emotional reactions and feelings to each particular situation. However, this idea is not merely found in philosophy's bailiwick; in Dante's (1971) *Inferno*, for example, Dante the Traveler is sometimes chided by Virgil whenever the former feels pity, sympathy, or has any other discordant caring emotional response for the sinners he encounters.[1] The reason for the rejection of what we would now see as morally appropriate reactions to the suffering of others is the hubris it demonstrates. By feeling sorry for those God is

[1]Emotions and feelings are not treated as the same thing in the literature on emotions. If feelings are discussed, then they tend to be an involuntary emotional reaction. I will argue that emotions are actually comprised by different things. In some cases, they are involuntary feelings, and in others they are cognitive judgments. However, I will assume that there is a unique feeling to each emotional reaction which might or might not be controllable in some way by the entity experiencing the emotional reaction or feeling.

D.R. Cooley, *Death's Values and Obligations: A Pragmatic Framework*,
International Library of Ethics, Law, and the New Medicine 62,
DOI 10.1007/978-94-017-7264-8_5

punishing with acutely fitting preciseness, Dante the Pilgrim is placing himself at a higher level than the Divine Entity. If God says what is happening to these shades is deserved, then no mere mortal should question its rightness or offer mercy and compassion where God saw no need for it. In fact, Dante's emotional responses are lauded only when he finally rejects the pitiful entreaties of Hell's denizens and feels what he ought to feel for those who are getting just punishment for their moral failings.

Dante's view can be informative with regard to contemporary emotional reactions to issues involving death and dying. Given the value theory developed earlier, having an appropriate emotion at an appropriate time could be an intrinsically valuable state of affairs. Even if the emotion is negative in and of itself, it might be that, as part of an organic whole of desert or moral appropriateness, the feeling becomes good or mitigates the bad in that context. A good emotional reaction or mental state of affairs might, in the context of a larger whole of the entire psychological or embedded person, be intrinsically disvaluable on the whole.

The question now becomes the appropriate reaction to death, especially one's own demise. Montaigne, for instance, argues that people should feel nothing about the death of oneself or another. In order to achieve this Stoic emotional apex, he suggests that everyone acclimate herself to death by exposing herself to it enough times so that it is no longer a frightening unknown (Montaigne 1958, 56). Instead of familiarity breeding contempt, it will breed indifference built on reason and evidence. Part of Montaigne's view was echoed recently by Peter Caws.

> If death is not felt, and God not feared, I am free to direct my life to higher objectives than craven avoidance or propitiation. My own death, I concluded, could not really matter much to *me*. If it was sudden, quick, unexpected, and painless, it would not matter at all. (Caws 2007, 4)

Stuart Walton adds that the fear of the unknown compounds the fears of death and of the deity involved in it (Walton 2004, 9). However, once we know that death is not something that can happen to the person, since it destroys such an entity and its interests while creating a reified distributed persona, it is irrational to fear life's cessation. If one cannot be harmed, then no grounds can justify such an emotional reaction to identified danger to oneself, much as it is irrational to fear the boogieman once one knows that no such creature exists.

However, since death is a complicated, contentious subject, there are those who do not agree with Caws and Montaigne. For instance, it might be the case that the fear of death is instinctual in a way that cannot be controlled, which, if true, takes it out of the realm of reason's ability to alter. Morally speaking, whether a person fears death or not is of no moral weight and, therefore, the person cannot be legitimately praised or blamed for his feelings.

It also might be the case that the fear of death is justified. Jeffrie Murphy argues that the prudent fear of death can be rational (Murphy 1993, 56). In this way, losing one's fear of death is actually acting against one's self-interests. If we allow that people should be pursuing their self-interest to flourish, and death is clearly a detriment to one's flourishing, then on the grounds of ethical egoism, one should be afraid of death. Moreover, if a person morally ought to do something, then it must be the case that the person has sufficient ability to act; otherwise, "ought" will imply a "can't." Hence, Caws and Montaigne are wrong; each person should fear his own death.

Once again, what is permissible in the situation seems to depend on the circumstances and which ethical groundwork we put into place. To ground claims that people are responsible for their emotions we must show that:

1. They knew or should have known that their failures to control their passions will result in their becoming the sort of person who will later be unable to control that emotion;
2. They have and know they have techniques to accomplish this;
3. They were not blamelessly in an uncontrollable emotional state when they failed to do that which would have led to more properly formed emotions (Sabini and Silver 1987, 171).

Using the Practical Principle's RPU and QCI, with the ultimate measure being flourishing, then the emotional reaction one should have to death-related issues depends on three questions. Would a reasonable person reasonably believe that feeling in such a way would maximize utility? Can feeling a certain way for a particular situation respect all people affected by the feeling? Finally, how do both tie into the flourishing of the individual instantiating the feeling and to that of relevant others affected by the person having that feeling? I will show that beyond the constant requirement to respect intrinsic value in the proper way, there are only general requirements that people feel or cultivate a disposition to emotionally respond in a particular way about death issues. The main idea is to do what is pragmatic.

5.2 What Are Emotions?

First, it is important to acknowledge the vital role that emotions play in our morality for without them we would not have morality in the first place. Antonio Damasio has demonstrated that emotions at their core are "*Bioregulatory reactions* that aim at promoting, directly or indirectly, the sort of physiological states that secure not just survival but survival regulated into the range that we, conscious and thinking creatures, identify with *well-being*." (Damasio 2004, 50, 2010). Although Damasio's theory is too reliant on an evolutionary basis of morality, it does show that the very platform that enables morality in the first place functions to match emotions and emotional reactions to thriving. Jesse Prinz argues that emotions are essential to morality, in part because moral concepts are bound up with, among other things, emotional reactions (Prinz 2007).[2] Moreover, moral emotions can lead

[2]Prinz seems to define emotions as emotional reactions to a particular stimulus rather than as habits or traits (Prinz 2006).

to moral behavior, and moral behavior can lead to moral emotions (Bandura 2001; Tangney et al. 2007). Without emotions, we would be unable to make the decisions we need to flourish over time, be motivated to act ethically, or have the morality we do in the first place.

Although emotions clearly play a very large part in our activities and what it is to be human, there is no consensus on what emotions actually are or how they work. However, there seems to be general agreement that each emotion has several kinds of objects:

1. *Target*—that at which the emotion is intuitively directed,
2. *Formal object*—each emotion type has a characteristic way in which it evaluates the target—we fear something because we think it is dangerous,
3. *Focus* of emotions—the background object having import to which the target is related in such a way as to make intelligible the target's having the evaluative property defined by the formal object (Doring 2010, 320).[3]

There also seems to be strong agreement and an argument for emotions being appraisals of some sort (Nussbaum 2001).

However, there is general disagreement on most other things about emotions. Some, such as William James, claim that emotions are feelings that are not under our control. Robert Solomon and others believe that emotions have an enormous amount of rationality in them. Additionally, those, such as Prinz, attempt to strike middle ground—emotions are perceptions. I think that each of these positions has something right to say about emotions. There are emotions that are feelings beyond our control; there are more complex emotions that involve judgments; and finally, there are emotions that are perceptions. Let us begin with emotions as feelings.

James argued that emotions were the recognition of the physiological changes our bodies undergo in some cases: "we feel sorry because we cry, angry because we strike, afraid because we tremble, and [it is] not that we cry, strike, or tremble, because we are sorry, angry, or fearful, as the case may be" (James 1884, 190). Basically, we physically become aware of an alteration in our circumstances which then creates physiological changes. The emotions we feel are merely an awareness of the bodily changes.[4]

Although James's view is inadequate to explain how all emotions work, e.g., anger and euphoria have the same physiological conditions but manifest themselves differently according to the situation, (Schacter and Singer 1962) some studies have shown that there are a few emotions that have unique physiological reactions that allow us to identify them by those responses (LeDoux LeDoux 1998; Panksepp

[3]Angst might be an exception.

[4]This seems to be Damasio's position as well: emotions are "constituted by a patterned collection of chemical and neural responses that the brain produces when it detects the presence of an emotionally competent stimulus... Emotionally competent stimulus depends on evolutionary history, personal history or current context." (Damasio 2004, 50–1). Moreover, the responses are produced automatically, which entails that we would have no free will other than trying to alter them through a different stimulus.

1998). For example, the emotional reaction of being startled is both universal to all people and cultures and does not have a cognitive aspect that can be manipulated by reason. Moreover, the simpler emotions are universal and associated with evolutionary processes (Ekman 2004). It is also clear that genes can influence personality which is created by and influences emotions (Lewis and Bates 2014). P.E. Griffiths argues that feelings are not learned but are innate, and that emotional responses are complex, organized, brief, quick, and typically involuntary. They cannot be cognitive because there is insufficient time and resources for them to be so, (Griffiths 2003, 262) nor can they enhance rationality, although they are instrumentally useful (Elster 2010, 264).[5]

Cognitivism, on the other hand, holds generally that emotions are appraisals of some sort (Nussbaum 2001, 125–4). Shame, guilt, embarrassment, and pride are "self-conscious emotions" that are evoked by self-reflection and self-evaluation (Tangney et al. 2007, 2). Robert Solomon[6] goes so far as to claim that emotions are judgments:

> [Emotions] may be said to be rational in precisely the same sense in which all judgments may be said to be rational; they require an advanced degree of conceptual sophistication, including a conception of Self and at least some ability in abstraction. They require at least minimal intelligence and a sense of self-esteem, and they proceed purposefully in accordance with a sometimes extremely complex set of rules and strategies. (Solomon 1993, 181)

Moreover, "such emotions as love and anger might sometimes be better understood in terms of the choices we make rather than in terms of visceral reactions, metaphorical or neurological 'chemistry,' or as passively undergone feelings." (Solomon 2004, 12).[7] Solomon states that we can test his hypothesis by reflecting on our emotions to see their logic, much as we can whenever we make a judgment (Solomon 1993, 182). For example, anyone can ask himself why he thinks that a particular action is wrong enough to be repulsive and, through reflection, fashion an argument justifying his decision that the action is so forbidden that it violates decency's standards.

Claire Armon-Jones argues that our emotions are cognitive and stem at least in part from social construction. "According to constructionism, emotions are characterized by attitudes such as beliefs, judgements and desires, the contents of which are not natural, but are determined by the systems of cultural belief, value and oral value of particular communities" (Armon-Jones 2003, 182). If emotions, even the primary ones, are fully socially constructed, then we have the same control over them that anyone has over the culture's values and conventions. That is, as we can

---

[5]Elster (2004) also argues that, at times, emotions can subvert rationality.

[6]Solomon's Chapter 8 of *The Passions* (1993) is an excellent resource on particular emotions, including how they work.

[7]As usual, there is no one version of cognitivism that has been adopted by those working in the area. William Lyons, for example, argues that not only are emotions judgments, but also they are affect-laden judgments (Lyons 1980).

act to alter cultural values and conventions by recognizing areas we think should be changed, we can also change our emotions. We have some control because we are part of the culture and, therefore, have the power to cause the culture to change it as has happened in the case of the emotional reactions people have to death and dying.

Armon-Jones is correct in the partial construction of our emotions. Some of our emotions, such as fear and empathy, are at least partially learned from others. However, a shared learning platform among people's minds gets learned emotions off the ground in the first place. There has to be, at bare minimum, a platform in the brain or mind that allows people to experience such emotions, although the full experience of many emotions is also crafted in part from learning. Moreover, those platforms have to be similar enough so that the emotional reactions can be generated by one person teaching it to another; over time, with enough instances of that emotional reaction, the person has to be able to acquire the disposition to have those emotional reactions in situations that would stimulate them.

At least two problems with the cognitivist approach to emotions show that it cannot fully tell the story of what our emotions are. First, we know that human infants and non-human animals are capable of feeling emotion, i.e., being startled (Robinson 2003b, 244–5). The problem, of course, is that they do not have the language that would allow them to make judgments as required by Solomon and Armon-Jones. What we have here is an example of emotions as feelings that are most likely the result of evolutionary processes that have created the brains of those creatures that can be startled or have these most basic and universal emotional states (Deigh 2010, 27). Second, Jenefer Robinson describes an instance in which one has an emotion that is unwarranted and not rationally justified, but that emotional reaction cannot be controlled by reason. Suppose that someone is afraid of ladybugs, even though she knows that ladybugs are not a threat to her.[8] The person can and does dispassionately judge that ladybugs pose no danger, but she is still irrationally afraid of them and will fear them whenever she perceives them to be near her. Regardless of her best efforts to alter her emotion and emotional states, she continues to think that ladybugs are a threat (Robinson 2003a, 124). Hence, not all emotions are capable of being altered by reason and, therefore, those emotions cannot be a judgment of the cognitivist type.

Perceptualism is the view that emotions are perceptions, or cognitive states, which do not offer judgments or have propositional attitudes but still give us the ability to talk about appropriate or inappropriate emotional responses. Prinz, for example, argues that emotions are meaningful but not cognitive: "Representation does not require cognition. We can have meaningful mental states without deploying any concepts…A mental representation…represents that which it has the function of reliably detecting." (Prinz 2007, 61). Basically, emotions were developed, acquired, or evolved to detect certain states. Pain, for instance, represents problems with the body that should signify to us an alarm that something is going wrong because it is reliably caused by the corresponding physiological problems

[8]This is Robinson's example.

and was evolved for the purpose of indicating that attention should be directed to the cause of the emotion. (Ibid.)

One merely perceives an emotion much like one might perceive a color, such as red. When I am perceiving red, that does not mean that I judge that what I see is red, as in "I see red," or that I necessarily believe that what I am seeing is red, as in "I believe that I see red," since both reports imply a judgment I am expressing through language. Instead, when I perceive red, I am merely having a perceptual experience in which I am appeared to redly.

Emotions work much the same way as other perceptions. We are not making a judgment but merely having a perceptual experience when we are instantiating an emotional response.

> The cognitive or judgmental element in an emotion is of a special kind, in that (1) it is not a dispassionate judgment that, say an object A has a property P, but rather a conception or way of thinking of A as P, and (2) this way of thinking is both governed (caused) by our desires (values, interests, goals, and so on) and also 'colored' by our desires in the sense that it shares some of the formal features of desires. (Robinson 2003a, 121)

I am afraid because I perceive danger to myself without actually believing I will be harmed or judging that what I perceive is dangerous. Robinson's ladybug case is an example in point. Reason says that one should not be afraid of ladybugs because they pose no actual danger to the perceiver, while at the same time, the perceiver is afraid because she is being stimulated to perceive fear.

Perceptualists argue that emotions focus attention on certain features of situations or arguments which make those features more prominent and give them greater weight in our experience (de Sousa 1987; Roberts 2003). For example, when looking at a painting, our attention might be drawn to one particular area in the work that allows us to concentrate our awareness on that area in relation to the painting as a whole. We then evaluate the painting from the framework that the emotions have created. Thus, emotions are not judgments, but they provide the framework which will help to determine the beliefs and judgments to come.

If Perceptualism is true, then it will not allow us to say whether an emotion is warranted on the grounds of its rationality, much as we cannot claim that a person's experience of red is warranted. Perceptualism entails that:

1. Emotional reactions are not generalizable. I can have an emotional reaction to one person that I would not have to someone who has the same basic properties, e.g., love.
2. Perceptualism explains why it is that "the same judgment" can be made either emotionally or impartially.
3. Since our emotional conception of something is determined by our desires, it may well be based on insufficient grounds, or even if the grounds are sufficient, they may not have been well scrutinized (Robinson 2003a, 123).

The framework Perceptualism gives each person will not allow us to call the emotions rational or irrational, but can produce a method for evaluating them as correct or incorrect depending on how well they correspond to the actual concerns we have (Prinz 2007, 62). For example, if someone desires not to be in the presence of ladybugs, then the fear of ladybugs would be correct. If the person has a desire to be around ladybugs or is indifferent to them, then a fear of ladybugs would be incorrect since it does not capture her actual concern.

The most promising way to understand emotions might be that suggested by de Sousa. Emotional reactions and emotions seem to be on a continuum from those felt automatically as the result of a person's genetics or education and over which the person has no real control to those that can be altered through reason. de Sousa argues that we have a two-track mind which processes information in different ways[9] (de Souza 2010, 96). The intuitive track's processes "are sometimes thought to be modular; they are typically associative, holistic, relatively fast, parallel, automatic, cognitively undemanding, highly contextualized, and organized around stable, "short-leased" goal structures." (Ibid., 101). These processes are shared between humans and nonhuman animals and were developed over time through evolution. Those animals that functioned better as a result of them were far more likely to survive and reproduce than those that suffered from slower processes.[10] My concern with Frank's position is that it tries to show that evolution explains all of the emotions, but the story/narrative he is weaving is without sufficient empirical evidence. Anyone can come up with an evolution story for any biological event, but unless we can test the hypothesis, then it is no more rational to accept it than another story about origins. Startle would be a paradigm example of this type of emotion and emotional reaction.

On the other hand, de Sousa's analytic track processes are "rule based, controlled, serial, cognitively demanding, relatively slow, decontextualized, and organized around "long-leased" and frequently updated goals." (de Sousa 2010, 101). These processes appear to be uniquely human or person based because they require far more advanced cognitive capabilities. Empathy would be an example of this type.

de Sousa argues that since they are different processes with different characteristics, the two tracks can oppose each other. They can conflict because what they use as relevant data and goals are dissimilar, much as the mind can have a conflict between deontological and consequentialist thinking when making decisions or evaluating a situation.

It is this conflict between the different tracks that allows me to claim that there is a continuum or spectrum of emotions and emotional reactions running from the very clearly intuitive to the very clearly analytical. Although we notice it more when they are in conflict, it could be the case that they sometimes work together to produce a result that is not fully one or the other, just as some ethical decisions are based on both consequentialist and deontological grounds. The context of the

---

[9]de Sousa believes both are the result of evolutionary processes (de Souza 2010, 96).

[10]Robert H. Frank argues that moral emotions are at least consistent with an evolutionary explanation of their origin. It better serves someone's self-interest if the entity is cooperative with others than if the entity was not part of a cooperative arrangement (Frank 2004, 422).

situation would be a key factor in the results produced. For example, there might be more time in which the analytic part can work; in others, there would be a shorter period. We could also imagine, as happens in the case of anger and euphoria, that each is at work evaluating the situation. Which emotion is expressed is determined by the situation's factors, such as what information is deemed by the perceiver as most relevant in the circumstances. At times, it will be anger, and in different circumstances, it will be euphoria.

Obviously, it would be harder to change emotional reactions from the intuitive track than from the analytic track, but it can be done. We are taught by society to act in certain ways that run counter to how animals respond to stimulation—for example, being good when no one is watching us—so it is possible, although not easy, to modulate our behavior. For emotional reactions and emotions, the same processes employed to change the disposition to act could be used to alter or control the habit to feel a certain way when stimulated to do so. However, our success at changing ourselves depends on how much we have to do and how much it challenges our core values to achieve the results. A purely theoretical change will generally get the result of someone agreeing to do it. On the other hand, if a person actually has to alter something that functions as a virtue or vice does, then the person is far less likely to do it. "In fiction, our attitudes are safe from commitment; we can't be expected to do anything about it, and so we can allow ourselves a broader range of sympathies than we can in active life." (de Sousa 2010, 111). It might be the case that in practice, change is impossible or not worth the doing; therefore, on the grounds of pragmatism, we need to know what can be done.

5.3 How Should We Control Our Emotions?

It is clear at this point that neither of the diametrically opposed positions that we are never and we are always responsible for our emotional reactions can be promulgated. "The innate emotional basis of ethics thus need not be limited to a set of discrete emotion types, or basic emotions, but also would seem to include general mechanisms of emotional learning capable of modifying and expanding the basic set." (Greenspan 2010, 544). Sometimes, we can modify and expand the basic set, and should, while in other situations, making the emotional reaction fit the circumstances is something other than right or wrong, or good or bad.

The issue becomes when and how we ought to make our emotional responses, and emotions themselves, appropriate to a particular situation and our individual flourishing, as well as that of our community. The stakes are high for us. If we do not act as we should, even if we are creating or controlling our emotions and emotional responses and feelings, then we can lose our rationality. "It is constitutive of being an agent to comply with the requirements of rationality, so that to violate them is, at the limit, to cease to be an agent." (Doring 2010, 283). That being said, the criteria to develop has to reflect the reality of people in general, as in their basic abilities and needs, and individuals in particular, as in their individual needs, identities, and abilities.

One limitation on being responsible for an emotional reaction is that there are instances in which a person cannot make herself feel or stop feeling an emotion at a particular time. Some individuals do not have the normal ability to feel emotions, such as those with depersonalization detachment or narcissistic personality disorder. Others might be overwhelmed by other emotional responses, especially in the case of traumatic events. For example, it might be appropriate that the agent feels sadness over her mother's death, but the agent cannot do so because her mind is too occupied with other pressing concerns. It also might be the case that the relationship with her mother was fraught with recriminations and indifference, which prevent the daughter from experiencing relationship's goods. In these situations, it would be wrong to claim that the agent should feel something she cannot feel within those specific circumstances. After all, that would be a violation of Ought Implies Can, which states that a person has a moral duty only if the person can fulfill that duty.

In addition, given that not all emotions have propositional attitudes, such as judgments, they might not be susceptible to change in the same manner as the latter. If someone believes something based on the very best evidence for it, then when she receives sufficient evidence that it is false, she is epistemologically obligated to believe it is false. In fact, according to this work's hierarchical value theory, the false belief is disvaluable, and the agent's having it is likely to make him a less valuable psychological organic whole than she otherwise would have been. The rejection of reason's demands would also count against her value as a person, because she is denying a fundamental characteristic that makes her a person, and therefore intrinsically valuable, in the first place. Emotions, however, might not be capable of change through rationality because they are different things by their very nature. Reason can operate on those mental entities that contain propositional attitudes or enough of the cognitive for them to work, but some emotions and feeling emotional states of affairs are totally or significantly non-rational by their very nature.

Putting involuntary emotions and emotional responses aside, there are instances in which people are responsible for their emotional responses and emotions. If they knew or should have known that their failures to control their passions would result in their becoming the sort of person who would later be unable to control that emotion, know they have techniques to control their emotional responses and emotions, and are not blamelessly in an uncontrollable emotional state when they failed to do that which would have led to more properly formed emotions, then they are culpable for what they did (Sabini and Silver 1987, 171). Moreover, if they are able to alter their emotional responses to a more appropriate one for the situation using these techniques, then they are obligated to do so.

With cognitivism, emotions can be appropriate or not, and we have sufficient control over them to be morally responsible for them (Solomon 1993, 213). The reason why is that of the five components making up complex emotion, at least some of them are propositional attitudes that can be altered, including "appraisals, perceptions, thoughts, and reflections *about* one's emotions" (Solomon 2004, 13).

William Lyons argues that emotions are affect-laden judgments. Emotions cause physiological changes, but the changes are not necessary and universal. That is, the changes can be different for different people and cultures or the same person at different times (Lyons 1980, 162). If true, then we can control or change our emotions and emotional reaction, and we might have to do so in order to pursue our well-being. If the emotions are causing overall harmful physiological states, i.e., stress, then we need to alter them. If they are interfering with our thriving, then we need to change to something more conducive to leading a better life.

In certain situations, people can make themselves feel appropriately in the moment in one of two ways. Paul Ekman has demonstrated through psychological studies that each individual can create emotional responses by identifying the "hot triggers" and then figuring out if the triggers can be changed or weakened (Ekman 2004). To weaken or alter the triggers, we will need to habituate ourselves to respond in a certain way by doing it enough times so that the response becomes a disposition. Ekman's nine methods to create a particular emotional response are:

1. Autoappraisers—Automatic appraising mechanisms constantly scanning the world around us;
2. Extended appraisal, which then clicks on the autoappraisers;
3. Memory of past emotional experience[11];
4. Imagination;
5. Talking about a past emotional event;
6. Vicarious emotional experience—witnessing another person with whom one is not directly engaged;
7. Others instructing us about what to be emotional about;
8. Norm violation;
9. Voluntarily assuming the appearance of emotion, which then creates the emotion. (Ekman 2004, 121 and 133)[12]

If Ekman is correct, we can alter our emotional reaction at a particular time to a more appropriate one. For example, if someone dies, and we feel little or nothing

[11]Tangney et al. claim that a person's history or past allows people to infer the emotions that will result (Tangney et al. 2007, 3).

[12]Jeanne Safer believes that we can train ourselves to respond appropriately when our parents die, which would seem to work for other instances of the death of loved ones.

1. Motivate—make a conscious decision to address and learn from your parent's death, investing time and energy in the project.
2. Anticipate—give yourself permission to seek death benefits and identify your resistance to doing so.
3. Meditate—cultivate a receptive stance. Think seriously about both the positive and negative impact your parent has had on your life, and the unfinished business between you.
4. Activate—death benefits are available at any time, and evolve over time. Actions to periodically pursue them—construct a narrative of your parent's history as objectively as possible; conduct an inventory of your parent's character; seek new experiences and relationships to support the changes you desire (Safer 2008, 5–6).

about the intrinsic value loss, then we can create a fitting emotional reaction by assuming the appearance of the emotion until we finally feel as we should.

Second, using the above and other methods, we can habituate ourselves to feel appropriate ways in certain circumstances we encounter so that we automatically feel a certain way providing the right stimulus. There is adequate justification to believe that people can alter their character by becoming better or worse through their emotional reactions and emotions. If Solomon is correct, then reason should be able to play a role because the cognitive is part of most emotions (Solomon 2004, 13). Habit can control the emotions, at least those with a cognitive component. Emotions might be sudden and urgent, but they can be compatible with skill and spontaneity at the same time. The emotions that are more hard-wired are less under our control in extreme circumstances, but we can still develop them in the way we develop virtues. Aristotle wrote that emotional reactions may constitute virtues: "at the right times, about the right things, towards the right people, for the right end and in the right way." (1985, 44 (1106b17-35)). If the emotion is appropriate—not too intense or too weak—then we are well off (1985, 41 (1105b26-81)). Even if we have inadequate power in a certain situation, we should have habituated ourselves to have the appropriate response at that time.

How to acquire the propensity to feel fitting emotions is similar to how people acquire virtues and vices.

> Just by the virtue of being human – 'by nature' as the ancients put it – we are susceptible to a wide variety of feelings and tendencies to react. Our development, and especially our moral development, consists in our getting (more or less) control over these feelings and training them in some ways rather than others. This is a process that never ceases. Since our lives are not static, we continuously react to different things and experience certain feelings and emotions, and the way we do so both reflects how we handle past feelings and affects the way we will handle future ones. It is a process that begins early; children are taught how to train their feelings by a variety of examples and sanctions, and the process becomes ever more conscious and autonomous with age. (Annas 1993, 53)

Moreover, we have a great deal of power to craft our own lives as we live them, and therefore our happiness and flourishing, within the social, cultural, and other causal factors constraining us. Even with internal and external limitations, "Our propensities for happiness clearly evolve over time…And there are many ways to fulfill one's emotional nature. Indeed, one might reasonably embark on a plan to *change* one's emotional nature." (Haybron 2008, 184–5). For example, if someone is unable to sing well, but wants to be famous for her singing abilities, then she cannot be happy if she makes such a goal something that she will seriously pursue.[13] Change is not only possible; it is morally necessary. As the constraints on a moral agent alter, then the moral agent needs to change her emotional nature to fit her internal and external reality.

Like virtues and vices, the emotions for which all moral agents are morally responsible are developed through repeated activity over time. In order to have a

[13] I am assuming a world in which merit is used in the music industry, which might be too fanciful an assumption to capture how the real world functions.

certain disposition to act if certain conditions obtain, then the person must have acted in that way a sufficient number of time to acquire the disposition. To create these habits for the emotions, we should:

1. Correct the patterns of acting that are associated with having the wrong feelings.
2. Use only appropriate thoughts and reflection.
   (a) We need to reflect before trying to change our habits; we need to know how to go about changing and how to acquire the correct feelings so that we can be successful in our endeavor to improve ourselves.
   (b) At the same time, we must continue to appreciate how our behavior is changing and monitor in an intelligent way the modifications of attitude and feeling as they develop. What we should not do is to let our behavior become mechanical because that is no longer an appropriate emotional response to the situation (Annas 1993, 57–8). That is, it lacks authenticity and does not respect the intrinsic value involved as a moral agent would respect such worth.

To be courageous requires that the person acts courageously and feels courage; otherwise, it is not a courageous action. To act cowardly necessitates an instantiation of cowardly feeling as well as cowardly action. Hence, it is possible for a psychological person to train himself into having certain feelings at certain times.

If "changes in emotional states, particularly mood propensity, are tantamount to temporary changes in personality," then alterations in how we are disposed to react emotionally to a certain situation will be a far more permanent change to one's identity (Haybron 2008, 183). That is, by manipulating our individual emotional reactions, we will eventually alter our emotions. As our emotions change, then so do the mental structures we have in place to accommodate these emotional responses, which might mean that when sufficient characteristics change internally, then the person's identity is significantly or essentially altered. Moreover, by training ourselves to have the appropriate emotional reaction for a situation, then we respect ourselves as the intrinsically valuable things we are, respect others involved with us for their intrinsic values, likely do what is best overall, and in doing all of this, make ourselves better people, which enhances our flourishing and the flourishing of our communities.

5.4 A Pragmatic Guide to Evaluating an Emotion's Appropriateness

Happiness can be short-lived and subjective, but it can also be much longer term, applying to a whole life, and be more objective (Annas 1993, 45). Deciding when there is a deficit or excess to an emotional response is fraught with the fear of appearing overly callous or empathetic. There seems to be a range for each emotional response in each situation. If someone who is emotionally very close to another feels nothing when the other person undergoes some sort of setback, then we rightly question the former's morality or the strength of the relationship. On the other hand, when a person strongly reacts when the situation does not appear to warrant it, then we ask ourselves why the person is responding in such an excessive manner. There might be evidence of a strong connection or some other morally relevant detail that we did not know about, or it might simply be the case that the person is indulging himself for some benefit he thinks it will net him, such as garnering attention.

We must be careful when trying to craft appropriate emotional responses to situations. Virtues might be a result of a need to overcome negative emotions, but we should not emphasize the virtue so much that a person's psychological condition becomes worse. There are individuals in society who know what an ideal person should feel in ideal conditions, or even what a normal person would feel given the average of the species as a whole, but cannot duplicate that emotional reaction in themselves as the result of some impediment beyond the individual's control. Social pressure might reinforce the idea that there is something seriously defective in the individual. The result is a person who has a negative psychological condition about not being able to have the virtue and, at the same time, has the negative emotion. Consider, for example, fear:

> Fear may have a clear evolutionary utility, in the sense that it once taught us a proper respect for predatory animals, but when the feeling of fear expands cognitively from self-preserving caution required on the hunt, to an unreasoning and intractable dread of such situations even when they are not immediately present, then a code of conduct for facing it down becomes useful. It is when that code of conduct for facing it down becomes an ideological imperative that a sort of vicious circle closes around the mind. (Walton 2004, 41)

Part of this phenomenon might be an individual condition created through genetics, environment, or some other causal factor. Others might very well be the result of the culture in which one is raised. A distinction exists between emotions from collectivist cultures that involve a conjoint model of agency in which good actions originate in an interdependent self, and more individualist cultures that involve a disjoint model of agency in which good actions originate in an independent and autonomous self. How one copes is also influenced by these models (Mesquita and Markus 2004, 341–54). Hence, we should remember that not all people can feel the same way, nor should they.

Let us take a moment to compare how two emotions associated with morality can enhance or degrade a person and community's intrinsic value and flourishing. Guilt can be vastly beneficial to the individual and the various communities of which she is part. Guilt focuses on action or behavior rather than on the person's identity or character, so there is no automatic defense mechanism engaged against internal and external attacks of the person's self-worth as that individual. Because of the different circumstances, the person is able to reflect on her behavior, and then come to a reasonable decision about how to make amends. In fact, guilt's effects are generally positive for the individual and community flourishing. It causes reparative actions by the wrongdoer of trying to undo the wrong; it shows and enhances the individual's other oriented-empathy; it fosters constructive emotions, cognitions, and behaviors; it leads to better social and less risky behavior; and it generally exhibits itself to be a more moral and adaptive emotion that tends to be less problematic for harming the person's psyche (Tangney et al. 2007, 5–10).

Shame, on the other hand, seems to generally be a bad emotion and feeling to have because it rarely, if ever, leads to any positive results for the person feeling shamed or the community that does the shaming. First, shamed study participants had higher negative self-evaluation if they were blamed for wrongdoing by others or if they imagined themselves having done something wrong over the control groups of no blame by outsiders or themselves (Rothmund and Baumert 2013). This should not surprise us because shame is a "negative evaluation of the global self." (Tangney et al. 2007, 4).[14] An individual feels shame as an attack on her individual character and identity, and thinks that others will perceive her as inherently a worse person because her action is degrading (Ibid.). As a result of the assault on her identity, the person is far more likely to attack others in defense of her character, which, in turn, means that not only will she not make amends for what she did, but she will also try to rationalize it in a way that is destructive to her individual flourishing. By creating a false reality, she enhances her vices while reducing her virtues at the same time she makes things worse for others in her community because she sees them as a potential threat to her self-conception. More specifically, shame's effects are intrinsically bad and damage a person's and community's ability to thrive: shame causes attempts to deny, hide, or escape rather than acknowledging wrong, making amends and growing from the reparative experience; a disrupted ability to form empathetic connections with others because of the fear of attack on one's identity; destructive reactions to anger to protect oneself; externalization of blame to others as a way of defending oneself internally; and a wide variety of negative psychological conditions, including low self-esteem, depression, anxiety, and suicidal ideation based on a perceived lack of alternatives that foster the self. There are no real positives to shame because it does not lead to positive adaptive thinking and behavior and increases risky and anti-social behavior (Ibid., 8–9).

We should not accept all instances of guilt and eschew all instances of shame. Just as every other emotion has a legitimate role to play in morality and death, so to do guilt and shame. At times, an agent has transgressed the boundaries of the morally acceptable so that the society has to shame her lest the society itself appears to condone such behavior. In other instances, guilt can become paralyzing or unhelpful as will be shown below.

---

[14]Bernard Williams argues that this is also reflected in Ancient Greek culture (Williams 1993, 83).

A useful measure by which to evaluate any emotional reaction or emotion as virtue is that of the flourishing of the individual having that reaction. If the feeling does not interfere with her well-being in the short or long term, then it is permissible. If it enhances both, then it might even become a duty to feel it, all things considered. In cases in which it does neither, then the agent should not feel what she is feeling. For example, if her emotions are too extreme for the situation, then she needs, as any type of person, to reconsider what she is doing. An emotional excess is a waste of time and resources that are better spent on other obligations that surround her. Since every person is unable to accomplish all the good that she can because of limited financial, physical, and emotional resources, then to use them for something that is not required is an insult to who she is as a moral entity as well as anyone else who could have benefitted from those resources. It is something akin to a fully-fed person taking a second dessert when someone else is starving. Not only does it injure the hungry person, it also physically and ethically harms the over-indulged individual. She should not be selfishly taking care of her self-injurious interests in the face of another's suffering.

Emotional deficits also negatively affect flourishing. Becoming a neo-Stoic is not a morally permissible route to take; fully controlling our emotions or being unemotional would be a loss rather than a gain in our overall happiness and authenticity. "Emotional learning may be seen as a way of building social norms into individual practical reasoning and motivation…what is normally in question in adult life is just the availability of emotions as reinforcement for moral reasons that we otherwise might be inclined to discount." (Greenspan 2010, 554). Instead of having an ethics that allow us to live well for ourselves and in society, we would be able to use our reason to rationalize away too much of what makes morality what it is: "the upshot is that emotions function as barriers to rational discounting and are useful to us morally because they do." (Ibid.). Greenspan's concern is that our reason can eliminate the very feelings on which morality rests so that our personal aims and priorities are served even though they should not be. It is permissible to discount emotional responses at times; perhaps there are higher order considerations at work that are contrary to the emotional response, but it is not acceptable to eliminate them altogether.

One of our duties as moral agents and socially embedded people is to feel certain ways at certain times. For example, if someone has become a significant other, then she has a duty to feel as a significant other should. This relationship is built on a certain kind of care as well as other emotions. If she does not care about the other person's triumphs and tribulations, then the latter is not getting what is expected of such a relationship. She is being treated in a way that is not good for her flourishing nor respectful of who she is and the decisions, such as being in the relationship, she has made. The callous individual also injures herself by not living according to her decisions, thereby making her a hypocrite, and what might be worse, by not developing herself as a fully realized person. Her lack of feeling keeps her embedded person value rather low when it should have been much higher.

To evaluate an emotional reaction or to have a disposition to a particular emotional reaction in a situation, hence, is to ask about flourishing. One way to make the question have a practical solution is to put it into context of RPU and QCI. First, does the emotional reaction respect all people affected by the action? That is, if the reaction respects the person feeling it and those to whom she feels it toward, then it will mean that the agent is not treating anything with intrinsic value as a mere means. Second, if the reaction is likely to maximize utility, according to what a reasonable person reasonably believes, then RPU is satisfied. Since intrinsic value is being maximized by the feeling, then there is at least *prima facie* evidence showing it is a right thing to do. The combination of the two establish that flourishing is being promoted, in this case for both the individual—mostly for the individual—and for others who are affected by the emotional reaction. Let us be honest about it; most emotional reactions will cause behavior, especially if the emotions are very strong or entirely lacking. This will be the standard used throughout to evaluate reactions such as guilt, fear, hatred, grief, and regret.[15]

5.5 Why Do We Fear Death?

Jeffrie Murphy argues that the fear of death can be rational on prudential grounds, which might work well to produce a flourishing life. Of course, there are irrational fears of death that do nothing to enhance our ability to flourish through "the successful and satisfying integration and functioning of our person" which should be overcome, especially if they interfere with our integration and function (Murphy 1993, 56). However, Murphy's focus is that the prudent fear of death is rational. "By prudent fear of death I mean simply (a) one that provokes people into maintaining a reasonable…diligence with respect to living the kind of life they regard as proper or meaningful (e.g., maintaining their health…) and (b) one that is kept in its proper place (i.e., does not sour all the good things in one's life.)" (Ibid.). That is, prudent fear of death prevents us from doing foolish things that would harm us. If I am afraid of dying from a car accident because my brakes are worn out, then I will make sure that my brakes and anything else on the car that might lead to my demise

[15]Marie de Hennezel uses the narrative approach in her therapy with the dying so that they can form a perception of their life as a whole. By doing this, she believes that "Death can cause a human being to become what he or she was called to become; it can be, in the fullest sense of the word, an *accomplishment*." (Hennezel 1979, ix). I take this to mean that understanding one's life as a result of understanding death is what she is referring to here and not that being dead is an accomplishment for a person.

are kept in good running order. Anyone with a desire to live longer will rationally pursue means that will likely result in those goals being achieved.[16,17]

The problem is that the fear of death is neither prudent nor imprudent; given the Epicurean stance, to believe that death will harm the psychological person who dies is an irrational belief. Since prudence is doing what is in one's best interests, and being dead means that one no longer has any interests, then avoiding death is neither prudent nor imprudent. The fear of death that results from the irrational mental state, therefore, cannot be rationally justified fear. Hence, reason and the prospect of flourishing tell us that our rational emotional state is acceptance or some other positive state toward our own death.

If death is not the end of the person, then it still is not rational to fear death unless there is sufficient evidence to show that the agent can or will be harmed by it. For example, approximately 1.25 billion Christians on Earth have faith in an afterlife that includes heaven and hell, and perhaps Purgatory. On faith and habit, other religions posit life after physical death. Given that they believe in survival in a better state, then it is irrational to fear death. Only death can bring about the state they actually desire as a reward.

Since our demise is no actual danger to us, then the fear of it is unfounded. The question then is why we fear our end of life. The fear of one's own death is a learned reaction.[18] Environmental and genetically linked psychological conditions create this fear in us. "The formation of fear has two sources: (a) personal experience [of a painful event]; (b) social experience acquired through communication." (Heller 2009, 74). Since the fear is not an innate idea, it comes from the personal experience of pain based on how our mind processes it as well as social interaction.

The claim that social learning is a necessary element of fearing death is borne out in social science research. Children three to five deny death to be a regular and final process; those between five and nine begin to see that death is personified and will happen to them; and it is around the age of nine that children recognize death as an inevitable life process (Hoffman et al. 2010, 24). The only way to account for this alteration in belief states is through personal and social experience. Moreover, the fear of one's own death seems to be a fairly consistently held belief across societies and cultures. Generally, most people in most societies fear their own demise even if

[16]Daniel Haybron's Systematic Imprudence thesis is that "Human beings are systematically prone to make a wide range of serious errors in matters of personal welfare. These errors are weighty enough to substantially compromise the expected lifetime well-being for individuals possessing a high degree of freedom to shape their lives as they wish, even under reasonably favorable conditions." (Haybron 2008, 227). Haybron provides a plausible and probable argument for the Systematic Imprudence thesis over the Aptitude thesis, which somewhat naively believes that if people have the freedom to do so, they will choose and live prudently (Ibid., 229).

[17]Murphy's argument could possibly work for avoiding harm that would not immediately end one's life. I keep up my brakes because I do not want to die in anguish, be disfigured, experience enormous pain and suffering, or be affected in other ways that would greatly reduce the quality of my life and my ability to flourish.

[18]William Hazlett argues that "The effeminate clinging to life as such…is the effect of a highly civilized and artificial state of society." (Hazlett 1926, 311).

they believe in an afterlife (Kerrigan 2007, 18). Finally, the pain that survivors suffer from their bereavement when loved ones die has always been severe enough that death is associated with great suffering which then impacts how they view their own deaths. (Ibid.) Given all the universal experience and reinforcement that one should fear one's own death, it almost seems natural that people have this rationally unfounded emotional belief and reaction.

Although we will probably never know with any degree of certainty why people began fearing their own death, there are a variety of suggestive factors about human nature.[19] As Kerrigan has shown, many in society and cultures around the world do not understand their own death or what it entails (Kerrigan 2007, 18). Since we have no actual evidence of what our death will be for us, then we have no cognitive framework to be able to classify and understand the process or event, and then cope with the end of our existence. Hence, as argued in the Chap. 4, we develop makeshift frameworks based upon fictions that never are cognitively satisfying because they, by necessity, are lacking information required to make them complete, efficacious, and testable for their usefulness. "If something does not fit the framework then the very inability to classify can create fear in the person. The perceiver is first uncomfortable about not being able to classify, and then fear develops as the perceiver remains unable to classify, which makes the object threatening." (Carroll 1990). If true, the fear of death is caused in part by a feeling of powerlessness on our part to control the situation through classifying or creating our own narrative.

It is not merely our inability to control by grouping that is at play in the creation of fear. Rationally speaking, until we have evidence one way or another, we should stay at equipoise when it comes to deciding if something is dangerous, but we do not. "Since we do not know it is safe, then we assume it is dangerous." (Heller 2009, 74). That is, we have a tendency to classify or perceive something as dangerous in the absence of sufficient evidence to show it safe, which might very well be the result of evolutionary processes in which those entities that were more wary were able to escape, while the more trusting creatures were killed.

The fear of death becomes far greater if the entity to be classified appears to the individual to be disgusting or impure in some way. We fear death because we are afraid of it for what it will do to us or what will happen to us after we are dead, and which we cannot control.

> Monsters are symbols of the disgusting… If we think about the limping, moldering state of most imaginary monsters, we can see our own elderly selves in much exaggerated form. In this view, part of our odium for monsters can be understood as fear and loathing of our own mortality. (Asma 2009, 266)

---

[19]In many African societies, death is not feared because there is a widespread belief in a very positive afterlife (Wiredu 1989, 24). In fact, death can be a positive for both the deceased and the survivors. It "gives the community the opportunity to give concrete expression to community solidarity." (Opuku 1989, 20).

In other words, death makes us impure, therefore disgusting, and hence gains power to create even greater fear than can be accounted for with the unknown and unclassifiable. Asma's insight can be seen in the variety of ways many individuals fear their deaths.

1. Fear of premature burial and being buried alive.
2. Fear of posthumous indignity—A primitive and deep rooted fear with the assumption there is an inviolable right not to be disturbed.
3. Fear of bodily disintegration—Many who do not mind death do not like the thought of their bodies disintegrating.
4. Fear of being forgotten and the need to be remembered—People fear that they will not be remembered because being dead is like never having existed at all.
5. Fear of ignominious death—There is a desire to die in a manner that is consistent with the person's self-identity and self-esteem (Wilkins 1990, 53, 104, 153, and 194).

Most of the above fears are a matter of impurities that are beyond our control and done to us, which cause loathing. No one wants to think of his body disintegrating into a mass of rotting flesh, nor do they desire to be dug up for medical or other experimentation. They do not want their dead body to be disposed of in a way that is humiliating to them, such as their corpses exploding from poor embalming. There is also a potential impurity in having one's dignity, identity, and narrative degraded through being unworthy of being remembered or remembered primarily for dying in an undignified way, such as Lupe Velez being wrongfully remembered for expiring with her head in a toilet after an alleged suicide attempt.[20]

The problem with each of the five fears is that they assume that we will still be alive or in existence and that we can be affected by what happens after our demise. As was seen in the preceding chapter, this assumption is caused by an existential illusion required when thinking about what happens to us after our death. When we cogitate about the time after our dying or death itself, we cannot divorce ourselves as existing beings from that moment because we think about what will happen to us after we are dead. Hence, we are always in existence in this reasoning process in the sense of sosein but will not exist in the future in the sense of sein; we cannot distinguish between our sosein and sein when thinking about our future since thinking about our sein has no experiential characteristic that allows us to separate it from thinking about our sosein. That is, when we ponder Sherlock Holmes, the golden mountain, who we were when we were children, or who we will be when we are elderly or dead, those objects have as much sein and sosien in our thoughts as our thoughts about people we have never met but know exist. They all appear to be equally existent objects as far as we can perceive when we are not directly perceiving them through sensation or reflection.[21]

[20]She actually died on her bed or the floor after she took an overdose.

[21]Descartes' clear and distinct perception seems to work well enough here for distinguishing immediate sensation and reflection from imagination.

The projection of our existence after our actual deaths is what helps cause our fear of death. We think about being made impure in some way, and feel the way we would feel about it if it happened to us whilst we are alive. For instance, no person wants to think of being forgotten so that it did not matter if the person had ever existed. Compare that to the fear of being so unimportant in society that no one notices the existence of the agent while that person is alive. It is a terrible thought, which explains why people recoil from stories about people who die alone and are not found for years after because no one cared enough to notice their absence. For the most part, everyone needs and desires to have those social connections and recognition and to be valued in order to have self-esteem and nurture the person's identity. The other four fears can be shown to be based on how we would feel now if such a thing happened to us as living beings capable of experiencing the events and not being able to control them in a way consistent with our personal narratives and autonomy.

Worrying about one's death makes as much sense as fretting about the very unlikely event of a shark attack in the middle of the Great Plains, although the shark attack worry has greater justification for it. Under extremely strange circumstances, a person could experience being savaged by a shark in the middle of the plains, but such an event would be highly improbable. To experience our own death, on the other hand, is impossible. If a plausible case can be made that death is not inherently something we should or have any reason to fear, then we should be suspicious of arguments that say fearing death is rational.

5.6 How Should We Feel About Our Own Death?

For many cultures, the prospect of one's own demise is generally not associated with an increase in one's happiness. More specifically, those who think about their deaths do not instantiate states of affairs in which they are happy at the moment. If they linger too long on the thought, or even obsess about the subject, they become less likely to have a thriving life. The prospect of death for those for whom we care, and I am assuming we care a great deal about ourselves, makes us both unhappy and have lives less worth living than they would be without the maudlin aspect.

The question is whether this reaction to death as well as fear of death are necessary consequences of contemplating one's own end of life. I contend that they are not. Although many fear their own demise, accepting the inevitability of death can actually enhance happy states of mind and overall happiness and thriving rather than decreasing them. In addition, a full acceptance of this fact governing our existence entails that we no longer fear our own death; to remain afraid is to show that one still perceives death as a danger when it is not. In this section, I will sketch out a plausible position by focusing on Heidegger's position on Dasein and the acceptance of death as a necessary condition to authentic happiness. After that, I will present good reason to believe that accepting one's death is conducive to a happy, flourishing life.

Defining the concept of happiness is not a simple matter, nor are the groundwork assumptions necessary to define happiness uncontroversial. There are those who contend the definition is purely objective, others claim it is subjective, and a third group truly believes it to be a combination of the two. On top of these three broad classifications, there is disagreement about the objective, subjective, or combination definition of happiness. Each person offers her best arguments, but no one can provide a concept that is capable of withstanding all criticism and defeating each competitor. Perhaps the difficulty lies in the fact that many philosophical definitions of complex ideas are based on intuitions rather than upon more solid evidential foundations. This or that particular definition "feels right" to the proponent, which generally entails that it is consistent with the person's beliefs without being something that can be shown to be Correspondence-Theory-of-Truth true. The end result of many debates about happiness, therefore, is disagreement about which intuition pump is working better in the situation.

For the purposes of this section, I will take biology's lesson on how to define principle, complex terms. In the biological sciences, there are 22 operational definitions of what a species is. Each meaning is correct; it is the use to which a definition is put that determines which one should be employed at that time.

Happiness has its various forms. It is a very complex state that can mean different things in different contexts. We can identify whether a person is happy at a particular moment or for a longer period, up to and including a life. We can talk about happiness for groups of people such as families, societies, and nations. Given this complexity and the various uses to which we put the word, I contend that there is no one definition of happiness. My task, then, is to find one that is plausible for a particular contextual use. In this case, I will discuss happiness in regard to a person's overall life.

I will begin with Camus' question, "what is happiness except the simple harmony between a man and the life he leads?" (Camus 2014) John Kekes says something similar when he talks about enjoyment being essential to a good life. We should develop a style of life that reflects one's individuality, where a style of life is composed of an attitude to life, dominant activities that translate the attitude into practical terms, and the manner in which these activities are performed (Kekes 2008, 136). Wonderfully vague in their meaning, these two ideas draw us to them because they seem to capture some universal necessities about happiness and enjoyment.[22] I will interpret the claims this way: Given objective and subjective constraints, which have been imposed on us or chosen by us, a happy person lives a life consistent with those external and internal constraints. I will address each limitation type in turn.

First, there are objective, unalterable restrictions on our lives. Those who decide not to accept the objective constraints on their lives will merely make themselves

---

[22]Kekes' explicitly claims that the enjoyment of life varies with individuals and contexts so that morality cannot hold universally or impersonally (Kekes 2008, 137). However, there seems to be a universal rule on how to live one's life well that applies to all people, although what one can do to have an enjoyable life in particular is under a great deal of control by the individual.

unhappy. As Abraham Lincoln said, "Most people are about as happy as they make up their minds to be." Among the limitations over which we have little to no control are the laws of nature, such as gravity, general human nature, some social conditions, what other people do or have done, and one's individual nature from genetics and socialization. For example, every human being is causally unable to fly without the help of some artificial mechanism, such as an airplane. We might choose to be entities that fly under our own power, but that is not going to happen, as any rational person recognizes. The laws of nature are beyond our power to change.

The second objective limitation is more controversial, but appears to exist. If someone chooses to do something that goes against human nature, which I believe has as its end flourishing, then the person cannot be happy. For example, a person who proposes to be a cold-blooded killer might successfully fulfill her desire, but a murderer cannot be a truly happy person, although she might get a great deal of pleasure from her illicit actions. Here is where universal laws of morality constrain the person, whereas the laws of nature allow such an action to occur. If the murder is unnecessary according to objective morality, then someone who performs it, no matter how much she might desire objective morality to give way to her choice of personal morality, does what is wrong and puts a serious moral blot on her existence's overall happiness, flourishing, or well-being. Moreover, it might be that one really bad action, such as knowingly killing a child merely to know the feeling of doing so, renders the actor unable to ever have an overall happy life. The damage to the individual's particular intrinsic value as a psychological, embodied, or embedded person might be too great to rebound.

Social rules, such as laws, are also generally beyond the individual ability of most people to alter as individuals. Suppose that an agent wants the law to be that driving 60 mph in a 55 mph zone is something that will not be punished by the state. However, without a change in the current regulations, what the agent wants comes to nothing. If she drives at 60 mph, she is acting illegally. If she receives a ticket, which upsets her, then the agent has made herself unhappy when she did not accept what she could not alter. Of course, this is a morally permissible law for the state to adopt and enforce, so this speed rule and others of similar moral permissibility are ethical limitations on citizen's lives. Other just laws are those that promote the flourishing of individual citizens and the society as a whole.

It is a different case for unjust social mores, conventions, and laws, such as Jim Crow statutes implemented in various states to segregate races from each other. The three unethically limit the flourishing, well-being, or happiness of those who are oppressed or treated by the state in a way that they should not be treated. However, even for these types of rules, we must acknowledge limitations beyond our control. Someone who wants to end her life earlier than will happen as a result of her terminal disease will be unable to do so in many parts of the world. If she cannot do anything to change the situation, then she needs to accept it in an appropriate way or make herself mad with unhappiness. That is not to say that she should take on a defeatist attitude, since that is not conducive to happiness, but she should address the situation in a way that will allow her to flourish. Putting the groundwork in place to alter the law or providing positive influence that will help cause positive change are appropriate responses. At the same time, the agent should understand that the law will not change anytime soon, or even in the agent's lifetime, and then plan accordingly.

What other people do, have done, and what they control can be outside an agent's power to alter. For instance, assume that an agent might be the very best candidate for a job. If the employment system is established so that the agent cannot get the position, perhaps because of her sex or race, then that is a reality that can be protested against but often cannot be altered.

Although this is a case of unethical restriction, a number of states of affairs caused by others will affect each agent's freedom to live life as she desires. Overall, as people live their lives, their actions constrain our choices by eliminating possible states of affairs that we would otherwise have been able to bring about either acting as individuals or in a group. If people make decisions that affect our liberty to do as we wish, then in the absence of power to create meaningful alternate possibilities, we are bound by an externally generated reality, within which we must work.

Finally, each person has a set of objective individual talents and abilities that are the result of genetics or learning over which the person has no control. If a person thinks that she has the potential to sing like the angels, but has no physical ability to produce notes corresponding to her belief, then no matter how much she practices, she will fail to actualize her choice. A person might want to acquire foreign language more easily, but if her brain is not properly stimulated during childhood, then she will find this goal a difficult, if not impossible, one to achieve. The potential neuropathways that would have allowed more easy language acquisition are undeveloped. Although some researchers in personal identity theory might want to argue that identity is more fluid than I am implying, the fact of the matter is that nature and nurture have well recognized constraining impact on personal ability.

There are likely to be other limitations beyond the ability of the agent, but these five objective constraints are a good groundwork to lay for the various areas that a person cannot change. A person should work within these types of restraints when living her life. If she wants to be authentic, then she must live as a person with these limitations. There can be no harmony when the path the person chooses conflicts with the realities of the conditions in which the person exists. Hence, she will be unhappy and withering, and her well-being will be adversely affected.

If people accept their external and internal boundaries, they can still make their life their own. Each person has power to create her own "narrative" by using the internal and external constraints and by making free choices within these parameters for herself. An important instance is the choice of good work, along the lines of Marx. All people need such activity to be fulfilled as a person, which is a truth of human nature in general. What constitutes good work is up to the individual to choose from her actual opportunities, although her abilities and incapacities do limit her to certain careers. If there are many meaningful choices[23] for the agent, then she

[23]By meaningful choice, I mean that there are at least two attractive enough alternatives from which the agent can choose. Moreover, the choices carry about the same benefit and burden so that

can select whichever one she wants. Provided that she is content, satisfied, or has whatever feeling is necessary to be living in harmony with her choice, then she will be happy in the moment. If her life is comprised of such decisions filled with contentment and arranged in an appropriate way so that they fit her life narrative, then her overall life can be called happy, and she is thriving.

5.7 Is Accepting One's Death Necessary to Happiness and Flourishing?[24]

Death is an objective limitation on one's life. When and how a person will die is actually under a great deal of control by the person, but the fact that each person ***must*** die is something beyond dispute even for those who believe in an afterlife. I contend that this reality does not necessarily lead to unhappiness in the moment or for one's overall life. In fact, accepting the inevitable end to all human beings and human persons might be necessary for a happy life.

Unlike other areas of the world, for many people in the United States of America, death is an inherently evil thing that should be fought with all resources. Even thinking about death seems to violate some sort of social more, and reflects badly upon an agent who not only prepares for it but welcomes the end of her existence. For example, many individuals put off planning for their deaths, which would require them to make wills, buy life insurance, and create advanced directives. Given death's inevitability, and the practical nature of each of the aforementioned undertakings, what can explain this phenomenon of rejecting reality? I contend that many Americans believe that they will never die, they never take their death seriously, or the mere thought of their own death is too destructive of an experience for them to perform the activities required to prepare for their death or the death of loved ones. My best guess is the latter is the standard in many cases. A certain squeamishness is exhibited when adult children talk to their elderly parents about end-stage care and final decisions, which shows how uncomfortable people are with the inevitable unknown. Although a fine distinction can be made between people fearing their own death and the process of dying with all that is entailed with it, the fact is that for many, thinking about death is antithetical to happiness.

However, as Milton stated, "The mind is its own place, and in itself, can make heaven of Hell, and a hell of Heaven." (Milton 2010) The mental states people have

(Footnote 23 continued)

none of them could be considered to be more coercive than the other. That is, if one alternative leads to a billion dollar reward, while the other had no more benefit than the feeling of a job well done, then the first would have an element of duress to it for the agent that the latter lacks.

[24]Portions of the section on Heidegger appeared in modified form in Becky DeGreef et al.'s (2010) "Communicating and Philosophizing about Authenticity or Inauthenticity in a Fast-Paced World".

about death are significantly under their control, or at the very least, can be altered so that they are no longer negative factors in their lives:

> In order for death to be with us in the way that it is, there has to be more than an awareness of one's demise in the face of a threat. There has to be some higher level awareness. In particular, there needs to be a recognition of one's life as having a trajectory. There is a beginning, and end and the period in between. There also has to be…a recognition that death could happen at any time. Without the awareness of the trajectory of one's life and its ongoing vulnerability to death, death would have a less significant role to play in one's life. (May 2009, 6)

In fact, mental states about death can become a positive aspect or force in a person's life. "The knowledge of our death is the ultimate boundary. A limit is set allowing us to make the most of our time here and to value whom we have in our lives and what we do with our lives." (Garfield 2010, 133).[25] That is, accepting one's death can lead to a much better life.

In *Being and Time,* Martin Heidegger addresses people leading inauthentic lives and how they can have an authentic existence (Heidegger 1962). Inauthenticity stems from one aspect of Dasein, or human Being, Forfeiture. When people are distracted by the irrelevancies of everyday life, they focus on mere illusion (Ibid.). Instead of concentrating on the real—the recognition of Being in its "active, dynamic, verbal sense," (Demske 1970, 12) they try to be successful by acquiring things that make them admirable in the eyes of other people living hectic, inauthentic lives. Those leading an inauthentic life have their attention shifted from what is real to that which is too abstract and theoretical, as in the case of philosophers who characterize human beings by using interesting, fanciful worlds that do not actually exist or are too focused on the external world of others, such as parents ordering their lives around work and intricate schedules to take their children from one event to the next. The result is that instead of having wonder and amazement from each recognizing her true Being, the "I" of the individual is sacrificed to the "they" mentality in which the focus is on objects outside of who we really are (Heidegger 1999).

Angst allows us to break free from an inauthentic existence to recognize what our world and being actually are (Heidegger 1962). Angst is the only emotion capable of doing so, in part, because by its unique lack of object, it makes us realize that something has been forgotten (Seidel 1964). We dread something, but we do not know what it is that we dread. Dread gives us a "weird and uncanny feeling" in which we act normally yet observe ourselves performing the actions from what seems to be an alienating distance (Gelven 1970, 115). We are aware an essential has been forgotten; we do not know that it is our Dasein and needs to be found (Seidel 1964). For many of us, angst can be caused by thinking about death in the usual way. We are afraid to die, which then makes us aware that something

[25]Ruth Garfield argues that our natural death anxiety can be dealt with appropriately by recognizing and acting upon the need to take responsibility for one's life (Garfield 2010, 129).

necessary is missing from our consciousness. Hence, thinking about death can be an extrinsic good.

Most individuals do not randomly have thoughts about their own death. Generally, some sort of cathartic event forces a person's mind to overcome the natural and socially constructed reluctance to consider the inevitable fate that awaits each of us, as well as those for whom we care.

> We are familiar with the frequently beneficial consequences of involuntary askesis… Suddenly instead of mindlessly and compulsively pursuing an abstraction – success, or money, or happiness – the person is reduced to what is actually there, to the immediately personal – family, geography, body – and begins to live freshly in love and appreciation. The change is a direct consequence of a forced realization of human limits. Pulled out of the fantasy of a god condition and confined to the reality of the human condition, the person is surprised to be living not a diminished life but a deepened life, not a crippled life but a zestful life. (Peterson 1994)

This event and the rethinking of one's life and its meaning would be a form of Facticity. The person undergoing the process is forced to realize that she is not an immortal creature, but instead of being emotionally overwhelmed by the realization, is liberated. Within her actual ontological parameters, the person can give her life a depth of meaning and appreciate things as they truly are rather than in some deluded, imaginary way.

After Facticity, Conscience enables us to grasp our Dasein (Heidegger 1962). Conscience is "The voice by which human being, speaking in secrecy and silence, calls itself out of the distraction of self-forgetfulness to the lonely avowal of its own responsibility for being itself." (Grene 1972, 461). Basically, Conscience is the final step to an authentic life. This is where Existentiality, the third aspect of Dasein, comes into play. Existentiality is the act of freely making the world, as it is, one's own, even though the world itself cannot be altered (Heidegger 1962). "Only by taking death into ourselves…is an authentic existence possible for us." (Barry 2007, 63). "Only by the anticipation of death is every accidental and 'provisional' possibility driven out. Only Being-free *for* death gives Dasein its goal outright and pushes existence into its finitude." (Heidegger 2002, 384). Hence, accepting one's death is necessary for leading an authentic life. The destitute confront their deaths on a sometimes daily basis; they are able to contemplate their reality to a degree that others do not. As a result, they understand the world as it truly is and their roles in it, which then makes the world their own; they can gain mastery over the "everyday," although they can never extinguish it (Heidegger 2002, 371). Accepting our death and reality means knowing our own possibility and that of the world in which we are; it is understanding our world for what it is and who we are. Calvin Colarusso claims that contemplative pessimism that replaces the optimization and idealization of youth is necessary for development and becoming more mature in midlife (Colarusso 2010, 110–1).

An authentic life is a happy existence. It is "living honestly and courageously in the moment, refusing to make excuses, and not relying on groups or institutions for meaning and purpose" (Soccio 2007, 522). Although it is not easy to accept the fact that we will die and actualize our Dasein, doing both liberates us because it frees us from the inauthentic (Barry 2007, 63).

Heidegger's world is a more realistic realm in which to live than many people experience. Gone are the illusions of existence and forfeiture that make us inauthentic and lead to actual unhappiness as we attempt to realize ourselves using false goals and desires. With authentic lives, we reject the neediness of our current society filled with materialistic and technological entities that distract us from our Dasein (Heidegger 1962, 1999). Knowing that we die allows us to focus on those projects that are actually important to us and drop the ones that do not matter to our authenticity or objective needs and desires (Todd, 112).

The world is as it is, we are in it, and we give ourselves our own meaning and purpose rather than having it passively thrust upon us, thereby instantiating our Being-in-the-world (Versenyi 1965). Resources are not needlessly wasted in pursuit of the irrelevant. Instead, everything is put into its proper prospective and we can be happy as any fully realized being in the world can be. Authentic people flourish better within the parameters they know will affect them than if they left things to chance or suffered a prolonged, painful dying process. This immediate limitation on life, its acceptance, and living of the life of this type would satisfy Camus' "simple harmony between a man and the life he leads". Since an authentic life recognizes that death is a central component to our existence, then accepting one's death is necessary to one's happiness. Moreover, our fear of death will be eliminated because a full acceptance of our own demise entails that we fully accept that. To continue to fear death is to not be living authentically; we still perceive death as something to be frightened of rather than embraced.

Nietzsche had a similar view in his attack on Western secular and religious thought (Nietzsche 1968). Both positions emphasize that the actual world in which we live is not Real or as Real as alternate realms. In Plato's *Republic*, for example, we, as a combination of mind and body, inhabit the world of Becoming, which is, at best, a poor reflection of the world of Being (Plato 1961). The world of Being is where the perfect, unchanging Platonic Forms reside with the form of Good being supreme over all. When human beings die, according to Plato, they will travel to the realm of being so that they may contemplate the Forms. Truth, and all else that is noble and good, lie in Being, while this existence is merely a burden for many. Death, hence, frees the soul to be in a state of perfection away from the odiousness of the world of Becoming, i.e., this world.

Christianity, for Nietzsche, degrades life in the actual world to an even greater degree than does pure Platonism. Making lives less valuable is the idea that actually liking this world—over a supernatural one for which we have no evidence other than blind faith that rejects reason and evidence—somehow makes one a very bad person. In order to gain converts who keep the faith, Christianity must convince people that there is a better world awaiting them. "The Christian resolve to find the world ugly and bad has made the world ugly and bad." (Nietzsche 1974, 130). Christianity requires belief that there is a better place that we will go once we have done what is required for salvation and our existence in this vale of tears has ended. Although salvation is an undeserved gift given from divine Grace, people still contradictorily treat it as if it can be earned. By not doing certain things or performing other actions, we think that we can make ourselves worthy in some way for the vast benefits we will receive in the afterlife. For example, we deny the body, because it is corrupt by being of this world, so that our soul can be purified and strengthened. Having sex, drinking alcohol, dancing, and other activities are to be rejected because they glorify the body over the soul, thereby corrupting the soul. Hence, blind faith also blinds us to our real possibilities, and makes us act in ways that are neither reasonable nor rational.

The puzzle is why anyone would believe such propositions. They are not based on evidence other than habit, blind faith, or faith in what a book written by unknown men says. If Nietzsche is correct, then the basic tenets of Western religion actually trace back to dreams of primitive humans:

> The man of the ages of barbarous primordial culture believed that in the dream he was getting to know a *second real world*; here is the origin of all metaphysics. Without the dream one would have had no occasion to divide the world into two. The dissection into soul and body is also connected with the oldest idea of the dream, likewise the postulation of a life of the soul, thus the origin of all belief in spirits. (Nietzsche 1986, 1:5)

Thus, the supernatural beliefs that many still take as a given have no other origin than an illusion. The Western view that pleasures of the body are somehow necessarily inferior to those of the mind, as exemplified in J.S. Mill's *Utilitarianism*, is merely the product of habit and custom based on a dream. We have been told that physical pleasures are defective in some way, but why should we believe such a strong claim without adequate evidence? To do so would be to reject what we know to be happening to us in our existence in favor of a possible existence later on, which might or might not be in our best interests, depending on the will of a supernatural entity which might or might not exist.[26]

In fact, the rational thing for us to do is to act as if this world is the only world that is actual because it is the single world for which we have existential evidence. Although I might be dreaming or be deceived, the sensory perceptions I currently have appear veridical. Using the Benefit of the Doubt argument, we should assume and act as if they are veridical until that moment in which reasonable doubt arises that will require us to question the legitimacy and evidentiary force of our evidence. I might actually be a plant in a jar in some palace in the distant past. I could be, in reality, a butterfly dreaming I am a man. I might be a whole lot of different things that I am unable to indubitably know or disprove. Whatever those are is of no actual importance to what I believe or should believe or how I should act. Given the information available to me that no reasonable person would doubt, what appears to be happening to me should be treated as if it was actually happening to me. Since I have no reasonable evidence of any alternative reality, then it would be irrational for me to reject the evidence I have now in favor of believing something for which I

---

[26] I believe that an argument can be made that they are inferior in certain situations, but are not always or generally so. The standard by which to measure a person's pleasure is related to the person's flourishing as that person exists in this world.

have no justification. Therefore, adopting Nietzsche's view that this is the world that is, and rejecting other possible worlds as existing things, seems the most practical selection we can make to understand our reality.

Given the reasonableness of accepting the experiential world as veridical, then fully adopting that one's death is necessary because it is part of the actual system to which we have experiential access is also reasonable. That is, people die in this world, and although we cannot know our own death, we can know the dying process that began the moment we were conceived. Since we are trying to flourish in the real world, our thriving is limited by what the system actually allows in the way of happiness and increasing intrinsic value in a way that makes our lives worth living. For Nietzsche, species survival is the primary foundation of morality. That which helped the species to survive, even if it was a false belief, is justified. For example, if it serves our purpose to believe in God to keep people in check so that the species survives, then it can be a legitimate false belief to hold. Perhaps we are in a society without adequate law enforcement, and people are unlikely to act in socially useful ways on their own. Hence, believing in a divine being that will punish anti-social behavior is not only permissible, but it can also be required in these circumstances.

Nietzsche's and Plato's noble lies might be permissible in certain situations, but they should be very rare exceptions to the rule. What will help survival and flourishing in general is being able to interact as effectively as possible with the environment in which we live. Instead of trying to force a false narrative onto what is real, the actual should be at least part of the story. Of course, the Naturalistic Fallacy will have to be avoided because the mere fact that something is does not entail that it ought to be, but creating hypotheses that reflect nothing more than how some think morality should be is relatively useless overall. In the case of belief in God keeping people in line with what morality requires of them, the fact is that such a society is actually counterproductive to species' survival and flourishing. First, in which God are we supposed to believe? Once we adopt the correct belief, there will be those who cling to a markedly similar one but with enough differences to cause disagreement. Since both are built on lies, it merely comes down to who is strong enough to win an altercation between them. This anti-flourishing result is unlikely to make the society better, and it might even destroy it.

On the other hand, a morality built on the natural has, one can say, an advantage to those built on what is illusion. In claims about which is more pragmatic, the natural one has the ability to use evidence from the natural world and empirical data which shows whether flourishing is enhanced or harmed. The evidence produced should be recognized by every reasonable person acting reasonably. They can evaluate it for themselves to see its quality and quantity. On the other hand, lies are relatively impossible to evaluate in the same way unless one has the exact same mindset as the liar, which defeats the lie's purpose, no matter how noble it is. Therefore, reality will generally be more efficient to species' survival than some lie that is easy to believe, but in the end proves its own impracticality.

Although it is counterintuitive, accepting the reality of a limited existence does not entail that existence will be unhappy. The fact of the matter is those whose lives revolve around technology and running from one non-essential event to the next often do not have sufficient Angst to ever start the process of understanding who they truly are and taking ownership of their lives. They are overly concerned about inventiveness to solve discrete manageable problems rather than using any theoretical insight or having understanding. Those with inauthentic lives are the world's unhappy people because they have the wrong view of what happiness is. We know this is the case simply because they do not understand the reality of the human condition; hence, they strive to fulfill an illusion's needs rather than addressing the requirements of an authentic life. They are squandering their resources in a fruitless attempt to satisfy the requisites of an inaccurate definition of happiness instead of being authentic, which entails that one is happy and flourishing.

In the real world, happiness, in part, stems from one's harmony with death. Consider when terminally ill people have fought the good fight against death, but they are informed that little more can be done to prevent death from occurring soon. People can make better end-of-life decisions if they knew how and approximately when their life will end. In situations that occur in present circumstances in which death is soon and inevitable, most people want honest communication and the opportunity to wrap up their lives, for example, by saying good-bye to their loved ones (Falls 2008). If an individual could better manage his life, then estates could be set in good order for those remaining, funeral arrangements could be completed by the decedent prior to death, and a host of other issues could be properly addressed. Also, being aware that an individual will undergo death soon can give people time to come to terms with it, although there will most likely still be a mourning process for those who care for the decedent. Every fatally ill individual can prepare for her end, which eliminates some of the stress on her and others' lives caused by fearing when death will come or lingering through a long-term debilitating illness.

Fearing death is a waste of resources and goes against a person's thriving. Since physical death cannot be avoided, and arguments for making human beings immortal are premature and would illicitly squander goods that could be better spent on dealing with problems that we can actually do something about, death should not be feared, especially if one truly believes that death is the end of the person. Even if death does not destroy the person, fearing death makes no rational sense because there is no evidence on which to base the fear. Of course, fear is an emotion or desire not to be harmed by something that is an unknown to the person in a state of fear, but emotions should be guided by reason, if such a thing is possible in the circumstances.

5.7.1 *What Happens if We Do not Accept Heidegger's View?*

Many will reject the claim that accepting one's death is necessary to one's happiness. First, some might point out that Heidegger's metaphysical theory is wrong on a variety of grounds; hence, there is no argument that acceptance of death improves the happiness of a person's life. Second, at best, happiness in the real world would be limited to those terminally ill individuals who are in the final stages of their lives. It makes sense for them to accept their impending end because it is imminent, but for hearty, hale individuals, the same conclusion does not apply. Third, even those who are terminally ill might not find happiness in their overall lives in accepting their death. Since their acceptance is likely to cover a rather short time period, their contentment in their inevitable doom in comparison to the overall length of the person's life might not entail that the life is overall happy. That is, the happiness, if it exists at all, will be happiness in a short span of life but not happiness for a life. In fact, such an act might be in discord with the agent's life narrative. Finally, those agents who are at a greater distance from their deaths will not use acceptance of death to improve the happiness quotient of their lives because there is no reason to do so. In fact, they would be better off not thinking about death until it is something they must take up; otherwise, they might become depressed or unhappy.

Accepting reality and feeling accordingly is not always necessary for morality. If the goal of morality is flourishing in its various forms through the use of acceptable moral principles such as the Practical Principle, then it is possible for a person to believe something false and yet flourish. At times, the truth might be too much for him to accept as in the case of some elderly finding out that their loved ones have died. If the former is weak and already near depression, then telling him of the demise would be cruel and not pragmatic. Provided that it does not matter to the survivor's decision-making process whether the other person is alive, then there is little justification for enlightening him. Although the state of affairs of his ignorance is bad in and of itself, when considered as part of the overall psychological or socially embedded person in that particular circumstance, then the ignorance might actually enhance the value of the person as a whole. It depends on its effects on the survivor. Is the false belief working for the person? Is he happier at the end of life being kept in ignorance? Is he flourishing? If the answer is yes, that is the path to take.

Unfortunately, there are times in which a life built on false beliefs can be far more valuable than one that captures the real external world. First, suppose that a woman is living in a morally repulsive family situation in a morally repulsive society. She is constantly being abused physically, emotionally, and mentally. She is taught that she is property, valuable only to the extent that those who own her place worth upon her head. Those decent souls viewing her sad state of existence might even claim that her life is not worth living on a number of grounds including, but not limited to, a negative balance of good and evil experience, degradation, and injustice, and the highly unlikely chance that anything will happen to change this dreadful situation.

Now, suppose one of those fanciful mental-experiment moments occurs of which philosophers are so fond.[27] Unbeknown to her, a scientist has a decision to

[27]There is a thought experiment posed by Robert Nozick that asks people whether it is better to lead a real life rather than one that is imaginary (Nozick 1974, 42–5).

make. He can kidnap her and place her in one of his life simulators, or he can leave her to live out the remainder of her days in her family and society. If the first alternative is selected, then his simulator will provide her with positive experiences. People appear to treat her as they should treat any intrinsically valuable person. She perceives herself to be part of organic wholes in which people care and support her. She learns to respect herself for who she is in general and as an individual. In other words, she flourishes by instantiating a great number of the intrinsically valuable mental states as well as realizing her full potential as a species being and individual.

The alternate world woman does not fare as well. Her experiences are real, but they never allow her to thrive. Her soul is not tempered and enhanced by her experiences, but mauled and crushed by her existence in an abusive family and society. Given these two universes, what would a reasonable person say is the better or more valuable life, if the scope of the woman's overall psychological life is selected for evaluation?

Although some might intuitively cling to the notion that lives based on true belief and knowledge are always better than those based on false beliefs, it is plausible and reasonable to claim that the woman who has simulated experiences leads a much better life than the other. It is true that the degrading life is more authentic in that it captures reality, but the positive value of the abused woman's veridical experiences is insufficient to overcome the disvalue of her negative experiences and their consequences. The oppressive life robs her of her ability to flourish psychologically as a person. Overall, the woman living in the actual world has a disvaluable life. On the other hand, the life lived through illusion has greater overall worth, and is a flourishing life. Not only does she have intrinsically valuable experiences of feeling valuable and loved, but she also realizes her potential as a psychological person and moral agent. Hence, it is the better world.

It is more controversial to claim that when the perspective is broadened to include the woman as an embedded person in her family or society, the world in which she lives in illusion rather than reality is still superior to that of the world in which she is constantly victimized by others. After all, there are more people and relationships to be considered in the actual world, while the disvalue of the scientist's theft of the woman's autonomy and how he treats her exist in the machine world. However, the sheer dreadfulness of the woman's life in the world in which she is abused must make the family or social organic whole have a negative value overall, as well as rendering it impossible for her to reach the universal end of flourishing. Not only is her life miserable, which casts its weight against positive value, but the relationships she has with others are bad as well. The value of the people who would treat another person in this manner has to be considered as overall disvaluable because they are being terrible moral agents and hence not thriving. In the machine world, the woman's solitary life built on illusions is positive except for the false beliefs she has and the disvalue caused by the scientist. Although it might cause some moral squeamishness, her good life can make the illusionary world more valuable than disvaluable, while the actual world still has a negative balance. Hence, by having an intrinsic value theory with many components, all of which are justified by and part of a flourishing life, then signaling out one component, i.e., knowledge, true belief, and false beliefs states will not establish that a person having negative value states in that area necessarily leads a worse life or is more disvaluable than an entity that instantiates the positive members of that type of end in itself.

It is useful to turn to how false beliefs can improve an overall vicious person's intrinsic value. Once again, it will be shown that a life can be more worthy if it has a number of false beliefs than it would have been with knowledge or true belief. Suppose that, for this thought experiment, the scientist selects the woman's principal abuser, her husband, for the illusion machine. Instead of being able to affect an actual person, the man will falsely perceive that he is interacting with real persons and things but is merely being appeared to in an appropriate way for the situation. If he wants to strike his wife, then his mental experiences will be identical to the world in which he actually strikes the person who has the misfortune to be his spouse. If the scientist does not place the husband in the machine, then the latter will have the same perceptions as if he had been in the machine but with real people being affected in the interactions.

Clearly, the psychological person who is the husband in each world would, qua psychological person, have the same intrinsic value. Since each one has the same mental states and perceptual experiences, they are identical as psychological persons and hence must have equal value. Both of them are vicious persons who do not have flourishing lives, if being an abusive spouse provides sufficient evidence for making a claim about who he is as a psychological person.

It is when the perspective scope is widened to include the psychological individual as socially embedded person that the values change greatly. In the world in which the man actually interacts with real persons, the embedded life he leads is much worse than the alternate world of illusion. For example, if he is an embedded person in his society, then abusing his wife counts in the evaluation of the intrinsic value of the organic wholes of which he is part. Moreover, the evil relationships and effects of his actions will negatively change the internal intrinsic worth of those with whom he interacts. Their actual pain and suffering mitigates the flourishing of individuals or the organic wholes of which the negative states of affairs are integral parts. When considering the destruction that one truly evil person can do to the value of "his" organic wholes, then the organic wholes in which he exists as an embedded person must be disvaluable to a great degree because of him.

On the other hand, if the evil person merely appears to negatively affect others, then the only real disvalue seems to be centered on him as a psychological person. Given that he is not a significant part of the organic whole that is his society, the overall value of the society must rise, *ceteris paribus*. Provided that no one else steps into do what he would have done had he been present, the abuse and other evil actions he would have inflicted on real people do not happen to real people. Those states of affairs' values do not go into calculating the organic whole's intrinsic value nor do the effects of his interactions create greater disvalue because they do not exist; therefore, they can have no value. Moreover, his overall life as an embedded person in the societal organic whole becomes better. Since he does not actually harm people, he does not owe them reparations, nor is he as evil a person.

For instance, if a person intends to murder another, but only dreams he has killed the person, then the former has not committed as great of an evil as if he had actually performed an unethical taking of life. That is not to say that what he has done is morally right. It is wrong, and he is wicked for doing what he did and being someone who could act in this way. However, he is not as evil as he would have been if he had destroyed the intrinsic value of a psychological person or embedded person. Hence, although his overall life is still intrinsically disvaluable because he is vicious, it is not as bad as it would have been if he is a vicious person who had actual impact on intrinsically valuable entities. In this latter situation, the intrinsic value ripples he causes in the reality are large and far reaching. In the illusionary world, they affect no one and, therefore, stop within the scope of his psychological person.

Ignorance of the truth can be pragmatically justifiable in other ways. Some people might be ethical only if they believe that there is some post-life reward in it for them. That is, they need to have some incentive to be good, so a strongly held belief in a beneficial afterlife would give them motivation to be good and do what is right. Of course, it would be better overall if they would use the reason human persons have in order to be persons in the first place and choose to be good moral agents acting to do what is right because it is right, but sometimes second best is the only end that can be practically obtained. In certain circumstances, therefore, what is second best for some is best for others.

Besides the desire for reward the fear of punishment is also a motivator. Some might need the stimulus that a powerful entity is watching them. If individuals think that police cameras are monitoring an intersection, they become more likely to stop and obey the rules of driving out of fear of tickets and other liabilities. Both "stick and carrot" incentives can be part of a flourishing life, although they might not be part of the best lives.

Although false belief is bad in and of itself, there are a number of instances in which it is better to believe what is not than to know the truth. As long as it enhances thriving, then there is *prima facie* reason to believe that it is permissible. In addition, if the false belief does not treat anyone as a mere means, and it satisfies RPU as well, then it is morally right not to disabuse another or to hold the false belief ourselves.

If we want to have the best life, however, then the truth is required in the vast majority of cases. We want to improve our intrinsic value as psychological and other types of people, and this can only be accomplished by accepting what is real. The truth might not be useful to us, but it will make the individual's state of affairs good in and of itself. More importantly, by being a person who can accept reality as it is without too much negative value being generated, we have developed into some of the most valuable individuals because of the enhanced moral maturity.

Accepting one's demise is not necessary for happiness, but can play an important role in a good, happy life. The acceptance of death, even if early in life, can improve a person's happiness in significant ways. First, it makes each person recognize the fact that he or she is not immortal. For the young and others, we often put off until tomorrow what should be done today because we, in the vast majority of cases, think that tomorrow will come for us. For example, we might have had an argument with a loved one over some relatively trivial matter, which has caused us to exchange words that were meant to be hurtful. The words did their work well, and now we are not speaking to each other. The problem with the antagonistic stalemate is that one of the parties might not be alive to make up with the other person should an opportunity arise. There are a number of accidents each year, which are precisely that—accidents. People cannot plan to be in one so that they can arrange their lives accordingly. In addition, serious crimes cause people's deaths and unknown or under-appreciated health conditions can end a person's life unexpectedly.

The frequency or probability that one of the antagonists will die is not the relevant point here. Rather, it is the fact that it is not only conceivable that one of them will die, but also reasonable to believe that there is sufficient risk to each person who operates within the modern world that the danger should have evidentiary weight in reasonable and rational decision making. Hence, if one accepts that she might die through misadventure of some sort, she would not allow a petty argument to prevent reconciliation with her estranged loved one. Many times, we keep personal strife going because we think that there is time to change our mind and then alter the situation accordingly. If we sufficiently understand that there might not be such an opportunity, then petty disagreements and other annoyances can be put into context as what they really are. No one in his right mind will waste time on activities that hinder flourishing if she thinks that she will die, in part, because she cares for those who will remain, and she wants to die well.

Second, the combination of knowing one will die and the likelihood of such a death will affect the actions, mental states, and plans of each individual living her life. An agent accepting that she will cease to exist requires a realistic assessment of when and how that might be the case. When one is young, the conditions in which the person lives help determine the probability of her death. Those existing in a dangerous environment will realize that their death is more likely than someone who has safer surroundings. Health conditions, resources, and a variety of other factors will influence the calculations on probability of death. As one ages, the probability rises until, of course, the inevitable cessation of life occurs.

The probability of death is an important influence on one's life. Each agent knows that she will die but not generally when. Those agents with lower probability of death in the immediate and mediate futures can choose courses of actions reflecting their greater time-span. For example, a couple might decide to have children while they are young so that they see their offspring develop into adulthood, go off to college, and then start their own independent existences. After the children are launched into independence, the couple can retire to enjoy their remaining years doing things they could not do while they were raising a family. On the other hand, the couple might want to live their adventures with younger bodies, so they delay having a family until that time in which they are able to devote more resources and care to raising children properly. Either alternative is permissible and will lead to happiness as long as there is harmony between person and life lived/chosen.

What does the acceptance of one's death add to these happy alternatives? First, it gives perspective to the alternatives so that the individuals can make harmonic choices that can be fulfilled happily. Although we might want to do many things in life, there has to be a realization that time may run out. The limitation on life makes people prioritize what is most important to them instead of trying to live their lives solely in the moment without taking into account their overall course. Each person who has accepted her death knows that as she ages, the more likely it becomes that she will die. She will therefore attempt to do those things that are most important to her first or get large projects started in good time instead of delaying too long. If she procrastinates, then she has lived a life of unfulfilled dreams rather than fulfilled ones. The latter has to be better for happiness than the former, especially if she resents her life. That is, she is out of harmony with her existence.

In addition, the awareness of death allows agents to plan the care of those left behind. In the case of choosing to have a family at a relatively young age, the parents should be aware that they might not be alive to give their offspring the care they need. This realization will lead good parents to have a will, insure their lives, and otherwise prepare for their deaths in such a way that their children will have the best chance of flourishing if the parents die before the children have made independent lives of their own. Moreover, if an agent understands that she might not be alive tomorrow, then she will not waste resources on pursuits that are not conducive to achieving the major goals in her life. She will spend time with her children with the appreciation each one deserves. The best I can do in approximating the emotion of appreciation here is to point out how people would feel about their loved ones if the former accepted the fact that they will die and that their time with their loved ones is limited. If the agent begins each significant task with this in mind, then she would truly appreciate the value of what she is experiencing in the here and now, as well as the value of all entities involved in the experience, instead of resenting the interaction or having other negative emotional reactions.

If agents accept their death, then they will purposefully seek intrinsically valuable mental states over intrinsically disvaluable states for the simple reason that they want to have a good life before they die. That is, they will not waste time on states of affairs that are not conducive to happiness and flourishing. It is my contention that an acceptance of death and reasonable beliefs about the end of the agent's existence can foster the instantiation of positive mental states, while providing good reason and motivation to avoid the negative states. If you know that you will die, and you do not know when that will be, then it is rational not to waste time and effort in doing what is bad. Moreover, the motivation that drives terminally ill people to end their lives well will, in a lesser degree here, be present. No one wants to have his or her life ended in futility, strife, or insecurity. However, everyone wants to die in peace, security, or any one of the other positive mental states or in combination with other positive mental states.

Wasted lives become less probable with the acceptance of death. So many things people do are merely distractions to keep them from being bored or are done because more interesting, difficult things seem less pressing. Everyone knows people who live in large urban areas with amazing museums, parks, and other edifying places who have never been to see any of them (They might actually be us.). They always think that there is more time to do these things, or they are busy wasting resources doing things that do not matter in their lives overall, although they might give a bit of pleasure. For example, many people spend their lives at machines, such as the one at which I am now typing, doing paperwork and mindless drudgery instead of seeing the pinnacle of human genius, having human contact, or fulfilling any of those goals that make them flourish as a species being or individual. Their life might be a happy one in a hectic sort of way, but it would be very hard to see a case establishing that it is truly happy and thriving.

5.8 Conclusion

Although it is easy to say that no one wants to die and that death is a fear that many of us have, it does not follow that death or awareness of one's own demise needs to interfere with happiness. As Calvin A. Colarusso writes, "The intense, normative midlife preoccupation with, and eventual acceptance of, time limitation and personal death is developmental prerequisite for the benign acceptance of death in old age." (Colarusso 2010, 112). Given that each person is unique as a particular individual, and that her experiences will also be unique because there will only be one time in the universe when the particular circumstances are exactly the way they are at that moment, there are general considerations people can use to accept their death with a measure of equanimity and dignity.

1. The realization that life has been long and fulfilling but the quality of life can no longer be maintained;
2. That one's abilities belong more to the past than the present and future, and one hasn't the resources to contribute significantly to society;
3. The desire to avoid a loss of dignity;
4. The desire to avoid becoming a burden on loved ones and take away their freedom to pursue their own flourishing as they choose;
5. The belief in an afterlife (Colarusso 2010, 112–3).

Although any one of the five could be sufficient on its own, a combination of them can offer more solace to those who will die. In fact, for those who embrace death wholly, it can be a way to a life happier than one that is a disordered collection of sense experiences. One should not live as a child lives in the moment without considering anything else. Rather, one should live with the awareness that each moment is precious because time is limited, and we should choose only that which will make our lives have the meaning we want them to have rather than merely plodding drearily along and being assigned a life that comes through inattention and inaction.

References

Annas, J. 1993. *The morality of happiness*. Oxford: Oxford University Press.
Aristotle. 1985. *Nichomachean ethics*. Trans. T. Irwin. Indianapolis: Hackett Publishing.
Armon-Jones, C. 2003. The thesis of constructionism. *Philosophy and the emotions,* ed. S. Leighton, 181–203. Peterborough: Broadview Press.
Asma, T.S. 2009. *An unnatural history of our worst fears*. Oxford: Oxford University Press, Inc.
Bandura, A. 2001. Social cognitive theory: An agenic perspective. *Annual Review of Psychology* 52: 1–26.
Barry, V. 2007. *Philosophical thinking about death and dying*. Belmont: Thomson Wadsworth.
Camus, A. "Happiness Quotes." http://www.wisdomquotes.com/cat_happiness.html. Accessed 27 Aug 2014.
Carroll, N. 1990. *The philosophy of horror*. New York: Routledge.
Caws, P. 2007. Pragmatic immortality and the insignificance of my own death. In *Re-imaging death and dying*, ed. D.R. Cooley, and L. Steffen, 3–13. Oxford: Inter-Disciplinary Press.
Colarusso, C.A. 2010. Living to die and dying to live: Normal and pathological considerations of death anxiety. In *The wounded of mortality: Fear, denial, and acceptance of death*, ed. S. Akhtar, 107–124. Lanham: Jason Aronson.
Damasio, A. 2004. Emotions and feelings. In *Feelings and emotions*, ed. A.S.R. Manstead, N. Frijda, and A. Fischer, 39–57. Cambridge: Cambridge University Press.
Damasio, A. 2010. *Self comes to mind: Constructing the conscious brain*. New York: Vintage Books.
Dante Alighieri. 1971. *The divine comedy Vol. I: Inferno*, Trans. M. Musa. New York: Penguin Books.
de Montaigne, M. 1958. *The complete essays of Montaigne*. Trans. D. Frame. Stanford: Stanford University Press.
Deigh, J. 2010. Concepts of emotions in modern philosophy and psychology. In *The Oxford handbook of the philosophy of the emotions*, ed. P. Goldie, 17–40. Oxford: Oxford University Press.
DeGreef, B., A. Burnett, and D. Cooley. 2010. Communicating and philosophizing about authenticity or inauthenticity in a fast-paced world. *Journal of Happiness Studies* 11(4): 395–408.
De Sousa, R. 1987. *The rationality of emotion*. Cambridge: MIT Press.
De Sousa, R. 2010. The mind's Bermuda triangle: Philosophy of emotions and empirical science. In *The Oxford handbook of the philosophy of the emotions*, ed. P. Goldie, 95–117. Oxford: Oxford University Press.
Demske, J.M. 1970. *Being, man and death*. Lexington: University Press of Kentucky.
Doring, S.A. 2010. Why be emotional? In *The Oxford handbook of the philosophy of the emotions*, ed. P. Goldie, 283–301. Oxford: Oxford University Press.
Ekman, P. 2004. What we become emotional about. In *Feelings and Emotions*, ed. A.S.R. Manstead, N. Frijda, and A. Fischer, 119–35. Cambridge: Cambridge University Press.
Elster, J. 2004. Emotions and rationality. In *Feelings and emotions*, ed. A.S.R. Manstead, N. Frijda, and A. Fischer, 30–48. Cambridge: Cambridge University Press.
Elster, J. 2010. Emotional choice and rational choice. In *The Oxford handbook of the philosophy of the emotions*, ed. P. Goldie, 263–282. Oxford: Oxford University Press.
Falls, C.E. 2008. Palliative healthcare: Cost reduction and quality enhancement using end-of-life survey methodology. *Journal Gerontological Social Work* 51(1–2): 53–76.
Frank, R.H. 2004. Introducing moral emotions into models of rational choice. In *Feelings and Emotions*, ed. A.S.R. Manstead, N. Frijda, and A. Fischer, 422–440. Cambridge: Cambridge University Press.
Garfield, R. 2010. Facing death: Intrapsychic conflict and intergenerational passage. In *The wounded of mortality: Fear, denial, and acceptance of death*, ed. S. Akhtar, 125–134. Lanham: Jason Aronson.

Gelvin, M. 1970. *A commentary on Heidegger's being and time*. New York: Harper & Row, Publishers.
Grene, M. 1972. Heidegger, Martin. In *The encyclopedia of philosophy*, vol. 3 and 4, ed. Paul Edwards, 459–65. New York: Macmillan Publishing Co., Inc. & The Free Press.
Griffiths, P.E. 2003. Modularity and the psychoevolutionary theory of emotion. In *Philosophy and emotions*, ed. S. Leighton, 256–275. Orchard Park: Broadview Press Ltd.
Greenspan, P. 2010. Learning emotions and ethics. In *The Oxford handbook of philosophy of emotion*, ed. P. Goldie, 539–559. Oxford: Oxford University Press.
Haybron, M.D. 2008. *The pursuit of happiness: The elusive psychology of well-being*. Oxford: Oxford University Press.
Hazlett, W. 1926. *Essays*, ed. C.H. Gray. New York: The Macmillan Company.
Heidegger, M. 1962. *Being and time*. Trans. J. Macquarrie and E. Robinson. New York: Harper & Row.
Heidegger, M. 1999. *Contributions to philosophy (from Enowning)*. Trans. P. Emad and K. Maly. Bloomington, IN: Indiana University Press.
Heidegger, M. 2002. *Off the beaten track*. Ed and Trans J. Young and K. Haynes. Cambridge: Cambridge University Press.
Heller, A. 2009. *A theory of feelings*. Plymouth: Lexington Books.
Hennezel, M. 1979. *Intimate death*. Trans. C.B. Janeway. New York: Alfred A. Knopf.
Hoffman, L. et al. 2010. What happens when you die: Threee-to-four-year-olds chatting about death. In *The wounded of mortality: Fear, denial, and acceptance of death*, ed. S. Akhtar, 21–36. Lanham: Jason Aronson.
James, W. 1884. What is an emotion? *Mind* 9: 188–205.
Kekes, J. 2008. *Enjoyment*. Oxford: Oxford University Press.
Kerrigan, M. 2007. *The history of death: Burial customs and funeral rites, from the ancient world to modern times*. Guilford: The Lyons Press.
LeDoux, J. 1998. *The emotional brain: The mysterious underpinnings of emotional life*. New York: Simon and Schuster.
Lewis, G.J., and T.C. Bates. 2014. How genes influence personality. *Journal of Research in Personality* 51: 9–17.
Lyons, W. 1980. *Emotion*. Cambridge: Cambridge University Press.
May, T. 2009. *Death*. Montreal: McGill Queen's University Press.
Mesquita, B., and H.Z. Markus. 2004. Culture and emotion: Models of agency as sources of cultural variation in emotion. In *Feelings and emotions*, ed. A.S.R. Manstead, N. Frijda, and A. Fischer, 341–358. Cambridge: Cambridge University Press.
Milton, John. "Happiness Quotes." http://www.wisdomquotes.com/cat_happiness.html. Accessed 3 May 2010.
Murphy, G.F. 1993. Rationality and the fear of death. In *The metaphysics of death*, ed. J.M. Fischer, 41–58. Stanford: Stanford University Press.
Nietzsche, F. 1986. *Human all too human*, vol. 1. Trans. R.J. Hollingdale. New York: Cambridge University Press.
Nietzsche, F. 1974. *The gay science*. Trans. W. Kaufman. New York: Random House.
Nietzsche, F. 1968. *The will to power*. Trans. W. Kaufmann and R.J. Hollingdale. New York: Random House.
Nozick, R. 1974. *Anarchy, state, and utopia*. New York: Basic Books.
Nussbaum, M. 2001. *Upheavals of thought*. Cambridge: Cambridge University Press.
Opuku, K.A. 1989. African perspectives on death and dying. In *Perspectives on death and dying*, ed. A. Berger et al., 14–23. Philadelphia: The Charles Press, Publishers, Inc.
Panksepp, J. 1998. *Affective neuroscience: The foundations of human and animal emotions*. Oxford: Oxford University Press.
Peterson, E. 1994. *Under the unpredictable plant*. Grand Rapids: William B. Eerdmans Publishing Company.
Plato. 1961. *The collected dialogues of Plato*, ed. E. Hamilton and H. Cairns. Princeton: Princeton University Press.

Prinz, J.J. 2006. Is emotion a form of perception? *Canadian Journal of Philosophy*, 36 (Supplement [vol. 32]): 137–160.
Prinz, J.J. 2007. *The emotional construction of morals*. Oxford: Oxford University Press Inc.
Roberts, R.C. 2003. *Emotions: An essay in aid of moral psychology*. New York: Cambridge University Press.
Robinson, J. 2003a. Emotion, judgment, and desire. In *Philosophy and the emotions: A reader*, ed. S. Leighton, 117–126. Peterborough: Broadview Press Ltd.
Robinson, J. 2003b. Startle. In *Philosophy and the emotions: A reader*, ed. S. Leighton, 235–255. Peterborough: Broadview Press Ltd.
Rothmund, T., and A. Baumert. 2013. Shame on me: Implicit assessment of negative moral self-evaluation in shame-proneness. *Social Psychological and Personality Science* 00: 1–8. doi:10.1177/1948550613488950.
Sabini, J., and M. Silver. 1987. Emotions, responsibility, and character. In *Responsibility, character, and emotions: New essays in moral psychology*, ed. F. Schoeman, 165–175. Cambridge: Cambridge University Press.
Safer, J. 2008. *Death benefits: How losing a parent can change an adult's life—for the better*. New York: Basic Books.
Schacter, S., and J. Singer. 1962. Cognitive, social and physiological determinants of emotional states. *Psychological Review* 69: 379–399.
Seidel, G.J. 1964. *Martin Heidegger and the pre-socratics*. Lincoln: University of Nebraska Press.
Soccio, D.J. 2007. *Archetypes of wisdom*, 6th ed. Belmont: Thomson.
Solomon, R.C. 1993. *The passions*. Indianapolis: Hackett Publishing Company Inc.
Solomon, R. C. 2004. On the passivity of the passions. In *Feelings and emotions*, ed. A.S.R. Manstead, N. Frijda, and A. Fischer, 11–29. Cambridge: Cambridge University Press.
Tangney, J.P., J. Stuewig, and D.J. Mashek. 2007. Moral emotion as and moral behavior. *Annual Review of Psychology* 58: 345–372.
Versenyi, L. 1965. *Heidegger, being, and truth*. New Haven: Yale University Press.
Walton, S. 2004. *Humanity: An emotional history*. London: Atlantic Books.
Wilkins, R. 1990. *Death: A history of man's obsession and fears*. New York: Barnes and Noble Inc.
Williams, B. 1993. *Shame and necessity*. Berkeley: University of California Press.
Wiredu, K. 1989. Death and the afterlife in African culture. In *Perspectives on death and dying*, ed. A. Berger et al., 24–37. Philadelphia: The Charles Press, Publishers, Inc.

Chapter 6
How Should We Feel About Another's Death?

Abstract Death most clearly harms an individual when it is the demise of someone to whom the individual has some sort of emotional relationship, such as love, friendship, or respect. Not only will the loss justify mourning or grieving, but also the process of dying and the event of death can cause additional feelings and emotions. Assuming that these emotional reactions or mental states can be sufficiently controlled by the person who will and does experience them, then there can be morally right/bad and wrong/bad feelings and emotions within a given context. In what follows, I will consider some of the main emotions associated with death and dying, emphasizing guilt. Guilt can arise for a variety of reasons—some justified, whereas others are not; hence, it can serve as a model for determining when an emotional reaction is fitting when someone is dying or dies.

6.1 Introduction

Now that feelings about our own demise have been examined, I want to consider appropriate emotional reactions and emotions with regard to other people's death. The main difference between the two events is that, unlike what happens in our own case, we can have experiences that are caused by other people's deaths. For example, a number of emotional reactions are deemed appropriate to the situation when someone else dies, and that fact is made clear to the agent through a variety of mediums, including internal conscience and external social conventions and expectations. Among other emotions that are standardly attached to death and dying of others are fear, pleasure, hatred, grief, regret, and guilt. I will consider each of these in turn, but I will spend the most time on guilt. I do so because guilt is prevalent and complex in death matters. It is often questioned when someone causes another's death whether or not that can be considered a justified killing or a permissible act of euthanasia. Guilt is also an emotional reaction present when someone dies a natural death without the survivor having any morally relevant contributory role in the process. I believe that considering guilt in detail will shed light on the appropriateness and inappropriateness of other types of emotional

D.R. Cooley, *Death's Values and Obligations: A Pragmatic Framework*,
International Library of Ethics, Law, and the New Medicine 62,
DOI 10.1007/978-94-017-7264-8_6

reactions people have when experiencing the dying process and death of others. Unlike the preceding chapter's more objective grounding, the main focus here will be on the appropriateness of subjective and objective emotional reactions.

Near the end of this chapter, I will briefly consider duties people have to prepare for their death and the death of others. That section is relevant in this context because it fits with the duties to have certain emotional reactions to death. That is, we should prepare for the death of ourselves and others emotionally and in other relevant ways, and we should act in certain ways in preparing for or bringing about the death of others.

6.2 Pleasure and Hatred

Once again, the feelings and emotional responses people have are justified in some cases and inappropriate in others based on Aristotle's position that people should feel the "right emotion to the right degree at the right time." (Aristotle 1985, 1104b13). In *The Inferno's* Canto V, for example, Dante the Pilgrim's compassion is raised illicitly by the suffering of two lovers, Francesca and Paolo Malatesta, who were sent to Hell when Francesca's husband, who was also Paolo's brother, killed them while they were in an amorous embrace. Although the lovers-lusters are clearly receiving the divine punishment they have earned, the Pilgrim states that their tears caused a severe emotional and physical reaction in him: "pity blurred my senses: I swooned as though to die, and fell to Hell's floor as a body, dead, falls." (Dante 1971, Canto V, 140–2). Granted it is now thought appropriate by the contemporary world to experience sympathy or empathy for others' suffering, but the Pilgrim clearly violated some moral rule because he did not understand the sin's nature and God's punishment. His emotional responses did not recognize that the deserved, necessary penalty meted out to sinners is determined by an unimpeachable moral judge. Hence, the Pilgrim failed in his moral obligation to have an appropriate emotional reaction, even though he had the very best evidence that he should have had that feeling.

A much different situation arises in Canto VIII in which the Pilgrim's emotional responses and actions align with reality. A sinner mired in a channel's muck rises to question the Pilgrim about why a living being is in Hell. Although this condemned creature is also weeping and more pitiful than the ill-fated lovers he encountered before, the Pilgrim states, "May you weep and wail, stuck here in this place forever, you damned soul, for, filthy as you are, I recognize you." (Dante 1971, Canto VIII, 37–9). On this occasion, Virgil rewards him with a hug, a kiss, and a blessing on his mother for what the Pilgrim says, does, and feels.[1] A little later, upon seeing

[1]Other signs of approval when the Pilgrim acts as he should can be seen in different areas of the work, including the Simonists' realm when, after the Pilgrim chastises Pope Nicholas, Virgil clasps the Pilgrim to his chest to climb back to the path. (Dante, Canto XIX, 124-6).

another sinner being tortured by other damned souls, Dante thanks God and praises "Him for that sight." (Dante 1971, Canto VIII, 58–60). Gone is the overly empathetic creature who questioned what was right and good, and in his place is a more mature person who better comprehends and conforms to reality. Hence, Dante and others believe that the situation's particulars dictate what a person should feel and think, which in turn affects whether the person is a good or bad moral agent.

A plausible case can be made that we should receive pleasure when we see someone performing a right action as well as in instances in which people receive their just deserts. When someone who deserves death or pain and suffering gets it, then, at the very least, there should be some satisfaction that a debt is being repaid. Pleasure is a normal emotional reaction because the right and just should be pleasing to us so that we are motivated to be right and just. That is, for us to desire to be moral, then morality in action has to be something that produces positive feelings and emotions for us. If morality was always negative in all ways, then no one would ever be ethical. Of course, an individual may rise to the moral saint level by forgiving the person for who she is, but it does not seem as if that is ethically required of us. Mercy cannot be something that is deserved in any particular instance but is rather given as a gift that no one merits.[2]

Besides pleasure from received just punishment, there is sometimes pleasure from an evildoer's pain and suffering that cannot be the result of her actions or character. If Hitler, to take one of the strongest examples of an evil person and wrongdoer, suddenly developed cancer after the Final Solution was implemented, then most people would have been pleased that he was going to suffer at least part of the horror that he was inflicting on innocent others. That is, he deserves to experience intrinsically negative states of affairs of pain and suffering because he is unjustifiably forcing them on the guiltless. Since no one is able to try him in a court of law, sentence him to a fitting punishment, and enforce that judgment, then his suffering from cancer has to be mitigated at least a bit by his overall negative desert. Hence, there is reason to think that he is getting a portion of what he deserves. Moreover, if justice is being served, then it might be appropriate to have a positive mental state of affairs, such as pleasure, as a result of someone receiving his just desert, at least on the grounds that we should feel positively about that which is good and right, while feeling negatively about that which is bad and wrong.

Simultaneously, taking pleasure in another's death or dying might appear impossible to justify ethically; it seems to indicate a rather defective character that will have difficulty flourishing as a moral agent. Those individuals who go to prisons to celebrate a state-sanctioned death with barbeques, signs, or parties applauding the action give many citizens a sort of queasy or "yuck" feeling that might indicate that the formers' actions and feelings are unacceptable in these particular circumstances. Those who have qualms about others being pleased at a

---

[2]Of course, a person should be merciful as a person, but is not obligated to act mercifully in any specific situation. This life narrative obligation shows that a person's life can have different properties than the events and actions making up that life.

state execution or other death process seem to think that when someone is executed, even if we believe that the death penalty and its applications are morally justified, the appropriate emotional response is one of regret, if not guilt, for what the state and those carrying out the capital punishment have to do. To have pleasure at another's death is to show callousness to people's intrinsic value. No matter how degraded the individual has made herself, there is a residual of value that merits respect. Hence, those who feel pleasure in such cases are morally defective in a very significant way.

It is important to ask who is justified in their feelings and emotions and who needs to do some work on making their emotional reactions and emotions fit the situation. We have to answer whether it is morally permissible, or even obligatory, to be pleased when someone who is evil overall is executed or stricken with a painful, fatal disease and deserves what is happening to her. Should one feel guilt owing to the *schadenfreude* status of the feeling, or is it shameful joy or pleasure at all? Does the moral status of the emotion alter if the disease is known to be fatal instead of merely trying? Relatedly, if a person is being executed for especially heinous crimes that make all decent people feel extreme moral repulsion, such as child murders, then is it morally permissible to feel pleasure or other positive emotional reaction when that person is killed? The same questions arise for whether and how hatred of evil and evildoers can be justified.

In general, hatred, which is an ill-will hostility borne toward others, seems inappropriate as an emotional response to people or their actions. After all, if we try to act as saints, or even good people, focusing only on the intrinsic worth of others, then we should never harbor hatred for who they are or what they do. In fact, in situations in which we can beneficially influence others, then we should help improve their character for a number of reasons. First, it makes them better people. Second, it improves our character by making us better people. Third, we act rightly instead of wrongly, at least according to the Practical Principle. Fourth, it makes the world a better place by eliminating negative intrinsic value and replacing it with good. At the very worst, if we take the Golden Rule seriously, we should love the sinner, disapprove of their wrong actions, and try to rectify the situation if doing so is morally appropriate, but we should not bear ill-will to others.

However, even if all of this is true, then there still seems to be something enticingly appropriate about hating certain people with a mighty hatred if the person is morally vile. Suppose there is an unrepentant neo-Nazi who has implemented her plans to harm those who have morally irrelevant characteristics such as belonging to the "wrong" race, religion, ethnicity, or group. When confronted in a rational manner along the lines espoused by Gandhi and King, the person willfully rejects the caring admonishment. She continues to inflict her morally repulsive crimes on others. If we think that people should better themselves without having some outside source forcing them, and there is no moral reason for them to continue being that type of person, then we would blame the unrepentant neo-Nazi for, at the very least, being negligent by not altering her character to a more ethical one. In this situation, the neo-Nazi is evil through her own responsibility and knows that she has this status, yet she continues to willfully act in such a way that she is a danger to others' flourishing as well as her own. It seems that for these admittedly rare but actual entities, one should hate them because they deserve such an emotional reaction; one should also make an effort to contain or thwart her activities. When they are driven by hatred toward others to commit atrocities, then that hatred merits our hatred in return for the evil instantiated. As the person has turned her hatred against innocents, we should turn our hatred against her moral repulsiveness.

Besides desert, hating evil people for the evil they do and the evil beings they have become as individuals also seems appropriate because it fits with what we should be as good moral persons. As mentioned earlier, evil and wrongdoing have a tendency to repel people from them. Hence, an aversion to extreme evil as exhibited by hatred or similar responses is fitting. Moreover, if wrong and evil are hated, then there is strong motivation for people to avoid instantiating them. A person who hates snakes will not go to places in which snakes are likely to be. A person who hates evil and wrongdoing is far less likely to do them than someone who does not have such a visceral reaction to them. With these highly emotional reactions, there is an internal force that can motivate people to make amends, try to rectify the situation, teach others not to act or be this way, and perform a myriad of other positive actions that only the strength of hatred could sustain (Radzik 2009). If these highly negative emotional reactions did not exist, then there might be reason as understood by the rational processes of our mind to pursue these actions but not sufficient will to do anything about them, especially if there are competing alternatives with stronger motivating forces behind them. For example, we might see someone being bullied and realize that we have a duty to stop it, but fail to live up to our obligations because our comfort from not being put into a confrontational situation has greater power over us than the mere recognition of what we should do. If we hated bullies, then it would be far less likely that we would sit on the sidelines while a vulnerable person is bullied.

As for pleasure in another's death or death process, the problem, of course, is that encouraging hatred and other such negative emotional reactions would appear to go against the requirement to pursue or nurture one's own and others' flourishing. Instead of following the prescriptions to love one's fellow person regardless of provocation and to exhibit other saintly or minimally decent good Samaritan behavior and character traits, the argument is that love or similar positive mental states might not be appropriate in some situations that need far greater force. In fact, hatred might be required in a situation because it is appropriate and useful. One worries that hatred is not only a disvaluable mental state of affairs, but it can also lead us into developing vices that would make us far worse people than we should be. Therefore, a contradiction might be forming: hatred and pleasure at other's suffering and death are morally forbidden at the same time they are morally required as fitting and needed.

6.2.1 A Possible Solution and Problems with It

Perhaps a distinction can be made so that negative emotions and emotional reactions are justified at the same time that hatred or overall pleasure caused by another's downfall is prohibited. Kant states that we must respect people for being people. That is, we must, among other things, have the feeling of respect for all persons because they are possessors of the Good Will, even though they have been acting and existing in a manner that is inconsistent with what the Good Will dictates. I take this to mean that no matter how evil the person is, we should still recognize their intrinsic worth as moral agents and people in general, which makes them something better than objects, such as rocks and trees, which do not possess the feature of dignity.[3] When we are pleased that a person as a moral agent is suffering, then we are wrongly rejecting what is intrinsically valuable in a way that should give pause to those of us who believe that hating the most odious individuals or taking pleasure in her death is fitting. That is, we should never disrespect the generic human dignity that all moral agents must possess if they are moral agents in the first place by not recognizing that value remains regardless of what the person has done to himself individually.

At the same time, we can be pleased that what the evil person has made of herself as a particular individual is suffering. Moreover, Hitler *qua* Hitler and other morally repulsive individuals can be despised because of what they have become through their own actions and character as the specific persons they are, while it would be wrong to hate them *qua* general person or moral agent. Since Hitler is evil to a degree incomprehensible to most, he deserves to suffer in his individual person because the horrific actions he perpetuated were the result of his own narrative. The Final Solution and many of the other atrocities would never have happened if he had not been who he was, and he is ultimately responsible for what he made of himself and what he did.[4] Hatred for him as an individual is a good thing because it recognizes the depth of evil he instantiated, and pleasure from his downfall and suffering is appropriate because he deserves them and more.

Adopting *person in general* and *person in particular* distinction allows us to use a Christian mantra to hate the sins but love the person in a new formulation. We are required to hate the evil individual person and the evil, wrong actions she does because that is what she deserves. At the same time, everyone is obligated to love (respect) the intrinsically valuable components that are worthy of it, such as the person's personhood in general, and fully appreciate any intrinsic good that is in the actions, although these might be difficult to discern. Although it is not as pithily formulated as the original, it seems to better capture where our complex, nuanced moral duties lie with regard to our emotions and emotional reactions.

---

[3]Kant might argue that some have lost their human dignity, and therefore, deserve no respect.

[4]There is some debate as to whether Hitler was responsible for his actions. There are individuals who claim that no mentally competent person could perpetrate such heinous activities; thus, Hitler was mentally insane and, therefore, a moral subject.

One defect to this approach is a fault we always have in morality when dealing with other people. That is, we might lack sufficient knowledge or evidence of whether a person is evil enough to hate. First, there is often inadequate information about another person's private mind's working. In addition, we are often deceived about reality when it comes to others for a variety of psychological reasons. We cannot see their motives, for instance, or if we do, then we sometimes attribute baseness to them if the person's goals and actions are significantly different from ours. On the other hand, those similar to us are often given the benefit of the doubt merely because we identify with them and their goals. Therefore, hating or liking someone might be based on false evidence that would not justify negative or positive feelings toward her in the first place.

Secondly, the pain and suffering a person endures might be out of proportion to her various crimes against morality. If the person is insensitive to other individuals, which merely makes her an unpleasant person, then the pain of cancer or other fatal diseases is too much punishment to fit the crime. Not only does she pay her debt, but she is also now a victim who is owed more compassion because of her unjustified additional suffering. Only those who have committed the most heinous moral actions would be owed such pain if anyone, in fact, deserves it. Even if the illness is not a fatal disease, it might be only for angels to know what is sufficient to remove the debt.

Thirdly, the distinction between the individual *qua* individual and individual *qua* person might not bear too much scrutiny. In order to be that specific individual, the fact is that she might have to be a person in general. To divorce the generic personhood and other positive attributes from her could be so artificial as to bend reality out of shape; she, as an individual person, cannot exist without being a generic person. She is an organic whole who cannot be reduced to the sum of her parts or any particular part; that is the unit of morality we need to work with in understanding what we should do and be. Therefore, when we have empathy for the stripped-out characteristics of generic personhood, we have empathy for a non-existent entity and not for the real organic whole who is the person. We would, hence, be feeling sorry for or having other seemingly appropriate emotions for a fictional character with as much reality as Hamlet at the very same time that we are neglecting an actual person with real needs. By trying to focus on the person's parts, we would be engaged in a self-absorbed exercise in futility that wastes our resources in a significant way and makes us callous to what should move us, which is what is real in the situation rather than what is imaginary. Thereby, we make ourselves worse than we need to be.

Finally, for retributive justice, we want people to be punished ***because*** of what they have done or who they are. Disease and other seemingly random events do not make sense as punishments when compared to a justice system in which the crime is casually connected to the wrongful behavior. Hitler being stricken by cancer, for example, makes sense with regard to Karma, divine punishment, or some other system in which the evil person is punished because of the evil he has willfully done by some Force that is able to link the person's past to the person's actual suffering. That is, the person is suffering because of his wrong or the evil person he has become. However, in the natural world, cancer and other diseases are unable to make such judgments about desert, much less form themselves into a force able to mete out just desert. Those who have many different types of diseases suffer as a result of imprudence at best or forces outside of their control, such as genetics, at worse. Therefore, there is no necessary moral link between the suffering and the person's evil or unethical actions, which entails that there is at least a *prima facie* argument that the person is not receiving what she deserves. If justice is not being served, then taking pleasure in her suffering would be akin to taking delight in seeing a bad dog suffer after being run over by a car when the car's driver is fully at fault for the tragic outcome.

The dangers to our moral selves are too great to justify pleasure in another's misery or hatred unless it can be reasonably linked to what the person is or has done as the result of her actions. We should be pleased that the individual is suffering from imprisonment, for example, while still regretting what is happening to the person. Her life is not flourishing as it could, although it might not be as bad as it could be. That is a bad thing. However, it is a good thing that just desert is happening. In the cases of disease, there is no adequate method of linking the disease as punishment to the crimes the person has committed; therefore, it is a moral imperative not to feel positive emotions about the pain the person suffers as a result of her illness.

For practical reasons, we should not take pleasure in another's suffering or have hatred for them.

Any pleasure in their misfortune or hatred would be misplaced, and we would be worse individuals for making decisions based on an incorrect choice. If this is the case, then we should refrain from judgment and instead feel empathy for their pain and suffering for who they are as well as for what they are as persons. Basically, since the consequences to others and us are so dire, until we know with adequate evidence, we should err on the side of caution and pursue those feelings and emotions that positively build a good character for us, regardless of whether we are wrong. For our own moral sake, we should have neither pleasure, hatred, nor any other emotional reaction that would endanger our moral character and, hence, our flourishing.

This conclusion of restraint in pleasure or hatred of the evil does not entail empathy for them when nature visits upon them some disease or other misery. There are many innocent people who suffer needlessly as the result of nature's forces or another's action; we should first expend our empathetic abilities and support on them. If we have nothing left over after taking care of the suffering and dying innocent to give to those who are evil, then it says little about our morality as persons. We are not callous or uncaring, only limited in what we can do in pursuit of our own and others' thriving. We merely have nothing left to give because of our more pressing duties in our relationships as concrete individuals embedded in a web of relationships that require our nurturing and development. However, if we have excess empathy resources or reserves not required to take care of our own or the innocent's flourishing in the situation, then we are obligated to direct them to the evil sufferer for the sake of their general personhood and moral agency, *ceteris paribus*. No one is obligated to be a moral saint, which is the reason that we revere those who are the moral rarities-treasures they are, but we are obligated to help others achieve their flourishing when it is reasonable to do so.

6.3 Grief, Mourning, and Regret

Let us momentarily assume that grief is "the intense emotion that floods life when a person's inner security system is shattered by an acute loss usually associated with the death of someone important in [the person's] life." (Jackson 1974, 2). It has been said that grief can arise from two sources: the loss of the deceased and what the survivors experience because of the death. The fact that grief is justified on two different grounds is interesting. Some have claimed that grief is rarely about the first loss. We grieve more for ourselves than the person who has died. Another's death does not harm that person but it does injure survivors who had positive, caring relationships with the deceased. No longer can the individual receive the benefits given to her when she cared for and loved another and was in a state of being loved or cared for. More importantly, the relationship's termination can have a deep impact on survivors' personal identity, such as not being a married person anymore. Moreover, depending on the trivial-to-important-to-essential significance of the relationship to the survivors' identity, they will be harmed to a greater or a lesser degree. For example, the death of an acquaintance injures survivors but not nearly as much as it would a spouse or close friend who helped make and sustain the survivor.

In keeping with the distinction made earlier between feelings as emotional responses and emotions, Jackson's definition of grief based on emotion can be revised to include the more fleeting emotional responses. The intense feeling of deep loss that threatens the person's inner-security can happen in three ways. First, the loss is linked to her understanding of what she needs for her flourishing, i.e., what is gone is thought vital to the person's thriving, and the person no longer believes or feels as if she can flourish as well as before. Secondly, it is something of such great intrinsic worth to the person that being in the world without that object is far less valuable for the person than a world in which that object existed, and the agent suffers from that loss. Finally, the loss threatens to alter the person's identity in an important or essential way that will make her a significantly different entity.

A person whose child dies, for example, feels all three threats to the grieving parent's inner security. First, the child was vital for the parent's thriving, if we understand thriving to be as a parent in a loving, nurturing relationship with the child, among other things. Second, the survivor finds the world she lives in far less valuable as the result of the loss of the child. The parent is still capable of thriving, and possibly doing better than before because of new opportunities open to her, but the parent subjectively or objectively values the world as far less worthy than if the child was in it. Finally, being a parent to this child is, as well as the various relationships that exist because the parent is the parent of the child, part of who the parent is. When the child dies, the person loses all the individual and relationship properties, which, in turn, alters who the person actually is. The parent is no longer a parent, but a former parent, survivor of another's death, survivor of her child's death, and so on.

Dealing with these threats to self is difficult. In 1969, Elisabeth Kubler-Ross published what would prove to be one, if not the, most influential work on how many people think mourning or grieving should happen. Kubler-Ross's model of grief has five stages that progress in a set order for a healthy person: denial, anger, bargaining, depression, and acceptance (Kubler-Ross 1969).[5] Although Kubler-Ross herself stated that her finding is merely one way that mourning could progress in an appropriate way, many therapists have taken it as a set-in-stone requirement that people react and act in certain ways at certain times. If a person ignored or repressed the stages, or had them in the wrong order, then many therapists using Kubler-Ross' finding believed that the person would have unresolved and painful emotions. Basically, these survivors were not feeling as they should when they should. Left untreated, the person could develop pathological conditions that would threaten her well-being.

Another's death often justifies grief, but how much grief one should have, how to express it, and how to deal with it are matters of debate. Ruth Davis Konigsberg, for instance, rejects Kubler-Ross's model and its underlying assumptions because they, among other things, have the actualized danger of pathologizing normal reactions to death (Konigsberg 2011, 52). Konigsberg argues that the Kubler-Ross model keeps people in grief because it is comfortable for them to remain enveloped in support group relationships in which people are lauded for feeling the right emotional stage at the right time and made to feel rejected or bad when not conforming to what people think should be the case. However, the problem Konigsberg identifies is far deeper than merely desiring ongoing emotional support: "The grief movement has [adopted the view] that no one should ever get over such a loss, although the rule seems to get more strictly applied to women than men." (Ibid., 45). Women, therefore, have been burdened more greatly by social expectations than men, even though until there is adequate evidence to show why discrimination is appropriate, women should be treated the same as men.

Also of great concern is the standardization of the eternal grief model. Kubler-Ross' sometimes useful but oversubscribed model actually hindered rather than helped individuals from flourishing as they should. Before Kubler-Ross's model became the obligatory standard of how one should feel to be mentally healthy, even in cases of very close relationships being destroyed, about one month was normal to get back to a functional state (Jackson 1974, 7). In addition, it took about ten months after the death for two-thirds of widows and widowers to become interested in dating (Konigsberg 2011, 47). For most survivors, the worst grief is over within six months, but small groups will mourn from 18 months up to four years (Ibid., 48). Thus, survivors need time and support to grieve, but are ready to

[5]See also Kubler-Ross (1972, 1975).

move on with their lives rather than remaining encased in mourning for their foreseeable future. Kubler-Ross's one-size-fits-all model has made the grieving process longer than it needs to be and created a standard that does not work for some, but is applied to all people who mourn.

Another problem has arisen for Kubler-Ross's model, which might show both a mistaken assumption and consequence of it. Although there seem to be universals about appropriate grief, the fact of the matter is that how people grieve can change over time, which poses a problem for any absolute, universal model. Princess Diana's death, for instance, showed a change from the British tradition of not making one's grief public to one requiring very public, socially appropriate sentimental displays. That is, if one was not in "extravagant mourning" for the dead royal, then others in society believed there was something wrong with the person that hinted at severe character flaws of callousness or inhumanity (Kerrigan 2007, 186). Without any personal connection to the living person or deceased that would have made the survivor's grief fitting, this demand for grief was unrealistic.[6] In the end, extravagant mourning became the rule, and grievers expect their losses to be registered by society as a whole (Ibid.). This expectation of recognition that one is in mourning would have to be added to Kubler-Ross' view. Once again, there is a comfort in conforming and incentive not to eschew the grief community, which might be an indirect result of being told there is one proper way to mourn. Hence, there should be a more flexible understanding of when grief is justified and the degree to which it is obligatory or permitted.[7]

Several variables determine the existence, depth and length of time for grief's appropriateness. First, how intrinsically valuable were the losses? If the deceased was a terrible individual who made the survivor's life much less worthwhile than it should have been, then deep and long mourning is not as fitting as in cases in which the deceased had enhanced the survivor's life. That rule holds in general. The more intrinsically valuable the psychological, embedded, and other types of person the individual was, then the stronger the case that regret and grief should be felt and to a very high degree.

In addition, the deceased's role in the survivors' lives should be considered. The more central a positive component in the survivor's identity, the more the survivor should grieve or regret the loss. In fact, the deceased's role's centrality or lack of it decides whether grief or regret is the more appropriate emotion to feel. If the person and relationships were part of the survivor's fundamental identity, especially to essential characteristics the latter possessed or possesses, then grief is demanded. For example, if Mary and John are married in the deep sense thought by Kant as a union of persons, then the loss of one will be a severe alteration in identity of the

[6]The news media assisted this alteration in what people require in mourning by creating an expectation that everyone needed to be very public in their feelings. The media did this by primarily showing only those engaged in extravagant mourning and attacking anyone who did not desire to publicly demonstrate their private feelings, such as the United Kingdom's royal family.

[7]Morris Wessel argues that children should be part of the process, including going to the funeral, so that they can be part of the mourning community (Wessel 1996, 78).

other. Grief is not only permissible, but also required in most cases, even though the survivor might not be essentially the same immediate and intermediate persons. When the survivor is mourning, then she is grieving for the loss of her loved one, their union, and what she used to be and what she has become as an overall person.

Moreover, when the person has accepted the other's death, then a more plausible case for the survivor grieving the loss in relationships and the impact on self makes better sense. The presence and interactions with the person are missed owing to the large contribution the person had made in the survivor's life. When someone grieves, it is mostly owing to the fact that relationships with the deceased are necessarily severed and the grieved mental states will be necessarily affected not only from the grief, but also from not being able to share with the deceased those experiences, ideas, and other beneficial things that enhance the survivor's life. This is not to claim that the experiences will be negative, only that they will not be enhanced by the sharing as they otherwise would have been. Those interpersonal interaction intrinsic values are now recognized as forever gone or missing.

A far weaker emotional relationship between deceased and survivor would generally only justify regret. For example, if two people barely knew each other, then regret would be appropriate when the survivor learns of the other's death. We should feel regret for the person who died because that respects her value in and of herself in our world. Possibly, this recognition of intrinsic worth's loss is the first thing that happens when someone learns of another's death. People do not generally think of how it is going to affect them as much as what the change in value occasioned—not if they are good people who have empathy for others. However, the loss to self will also have to be taken into consideration in order for a survivor to fully respect the values that have been lost.

The moral limitations on feeling grief or regret are, once again, set by the individual's flourishing, as understood through the Practical Principle. "[Mourning's] function is to console and strengthen the self." (Safer 2008, 58). Safer rejects the time limited, linear progression of universal stages in favor of an idiosyncratic process that is ongoing, episodic, and fluid (Ibid.). If, as in the case of Queen Victoria, the grief lasts too long and is too deeply felt, then the person's flourishing is negatively affected. Grief should be experienced, but only that required to respect all intrinsically valuable entities affected by the emotional action as they should be, as well as being the best way to respond and feel in the situation, determine what time period range is appropriate. Grief demonstrates to the person that she cared in the right way and should allow her to deal with the loss rather than making her life less than it otherwise would have been.

Each person is obligated to cope with his grief on the grounds of flourishing which includes, among other things, fully facing reality by breaking some of the bonds that tied the survivor to the person who has died and developing ways that will make it possible for the surviving person to find new interests, satisfactions, and creative activities for the remainder of her life (Jackson 1974, 9). As was discussed in the preceding chapter on fomenting and controlling one's emotions and emotional reactions, it is sometimes possible to alter what one feels by behaving or thinking in certain ways. For instance, mimicking how someone with a particular emotional reaction would act can create the same feeling in the person who caused it with her mimicry. Whatever the agent needs to do in order to move her emotional states from grief to coping is required of her so that her ability to lead a good life is not destroyed at the same moment her loved one dies.

As in some cases of guilt, grief can be self-indulgent. Some people grieve more than they should, not because of an actual sense of loss for the deceased but because their actions bring greater attention to themselves. That is, they can focus on themselves without having to consider what is happening to other people around them or what else they should be doing to improve their own lives. This type of inappropriate feeling is selfish rather than justified on any grounds that will lead to thriving for the individual.

Moreover, by being constantly in the throes of grief, self-indulgent mourners draw attention to themselves to the detriment of others. People will wonder at the grief level, and might note such a strong relationship. They might pity the person, which will also give the griever attention. However, the activity is pathetic at best. We, as caring individuals, should support those who have lost others, but we should not enable them to the point that it takes valuable resources from what we should be doing. For example, the over-indulgent mourner might require a great deal of emotional resources that would be better spent on improving our lives or the lives of others who actually need them. While we are attending to the over-indulgent mourner, we could be missing out on helping others with their licit grief or other appropriate emotional reactions.

The obligations to provide emotional support and to not indulge excess grief become stronger as the intimacy of our relationship to the survivor becomes nearer and stronger. Both are based on the basic decency required of people to help in the flourishing of others. If a survivor is going too deeply in mourning, then a friend has a duty to help the person out of it using the methods that will work with the person. After all, we need to care for others with whom we are in caring and nurturing relationships so that they can make their own way in the world. In some cases, that might be a sharp rebuke. In others, it could be taking the person to see professional help. Whichever way it goes, the idea is to nurture people into thriving lives rather than enabling them to devalue themselves.

6.4 Guilt

There are a number of situations in which guilt is generated in survivors and which may be appropriate for them to have. Consider state-sponsored killings such as war and capital punishment. First, in capital punishment cases, those directly involved in the execution or on the periphery of it might feel as if they did something wrong. Jurors could feel guilty for finding that the death penalty should be applied in the case, even though they also believe that the evidence warranted such a judgment on their part. Those who work with the condemned, such as prosecutors, defense lawyers, and prison guards, might feel guilty that they are part of a system that allows another human being to be killed by the state, even if they will continue to be part of the process. The doctor involved in the execution might feel as if she should not have participated, although she ensured that the procedure would function. Even if the doctor's role is labeled as indirect and necessary to minimize the condemned's suffering in a procedure that will happen regardless of what she does, the physician might still believe that such an action violates her Hippocratic Oath. The executioners might feel guilty for carrying out the order of the court. The political officials could have a guilty conscience because they can never truly know beyond a shadow of a doubt that the execution is justified. Members of the public often feel guilt because they are part of a society that still engages in executing other persons. In general, anyone who feels as if something wrong was done, and they are somehow and to some extent responsible for the wrong, will feel guilt as a result of this perception.[8]

Guilt is also associated with more personal takings of life. Abortion, physician-assisted suicide, mercy killing, and passive euthanasia often leave an emotional residue, regardless of whether those involved in the situation feel as if they have performed a permissible or right action. Those who have to make end of life decisions for others, even if they base their choice on clearly articulated instructions left by the person who is no longer competent, often believe as if they helped kill the person when they sign no resuscitation and other life-shortening orders. The guilt can be greater if the proxy has to choose, for example, that hydration and nutrition be removed because those basic necessities to physical life will clearly lead to the other's death. In these cases, reason might inform the proxy that she has done what she should, but her emotional reactions imply strongly that she has acted wrongly or is a bad person for making such a decision. In situations in which there is no clear instruction from the incompetent person, the proxy might feel even more guilt as a result of the worry that the decision is not something the incompetent person, had she been competent, would have wanted.

Finally, there is an issue when someone who is not directly or indirectly involved in bringing about a death still feels guilt when the death happens. "Guilt is the other reason midlife parent loss is ignored – guilt about being released from emotional demands or caretaking burdens…guilt about profiting financially, and - most deeply buried of all – guilt about surviving." (Safer 2008, 51). As populations age, they tend to exhibit diseases which require them to be cared for since those who have the conditions can no longer take care of themselves. In dementia cases, caregivers, generally female, have to give up a great deal of independence to take care of their loved one if the patient remains outside of a professional institution, such as a nursing home. Depending on the length of time and how much the caregiver's interests and needs have to be sacrificed for those of the incompetent loved one, stress and resentment can build. The caregiver's life, or significant components of it, must be put on hold. Even if the dementia sufferer is in a professional institution, the strain of seeing the deterioration can be great enough on

[8]Of course, there will be many who have no guilty feeling at all.

the survivors' emotional health that they might begin to feel as if they are being bad people for feeling this way. They also feel guilty for the relief they feel when the person dies, when they profit financially, and in the case of spouses and other loved ones, that they survived while the other did not. In these cases and others, guilt can occur even when it appears that no ethical principle has been violated. Hence, for a variety of reasons, people can feel guilt even if they did more than was morally demanded of them but less than moral sainthood requires.

That being said, although some feel guilty, that reaction alone does not entail that they should have it. Guilt can be appropriate in two ways: objectively and subjectively. In what follows below, these two justifications will be considered using technical moral dilemmas. The reason for the focus on moral dilemmas is that they fit well with end of life decisions and the guilt generated by them. If technical moral dilemmas exist, then they will have an enormous impact on how we should feel about death and morality. For the latter, proxies would have to accept the fact that at times, such as those where others deaths affect us, there is nothing we can do that is morally permissible. Our hands will be dirty no matter how hard we try. In addition, it might be the case that we need not be connected to the event to be justified in feeling guilt for what happened. Hence, moral dilemmas will be most helpful in sorting out when and why guilt is appropriate in a way that can shed light on feelings associated with others' death and dying.

6.4.1 Objective and Subjective Guilt

When we do not satisfy one of our duties, we are guilty of a moral lapse of some sort that can justify guilt feelings. However, the mere fact we feel guilty in a particular situation does not necessarily mean that we ought to feel that way. As William Neblett says "We can feel guilty in all sorts of ways and for all sorts of things." (Neblett 1974, 653). For example, some people experience guilt feelings for actions that happened before they were born, which, on the grounds of Ought Implies Can, makes the person blameless. Objectively, the person is innocent, yet the person still feels guilty for the action. Thus, guilt might not be a fitting feeling in particular situations, even if the choice made or the action performed was very difficult.

I contend that guilt may indeed be appropriate for the agent to feel, but not on the grounds that the agent has actually failed to do as she should. I will limit my investigation of guilt to the two ways that Earl Conee thinks guilt is appropriate: subjectively and objectively. The former is based on what the agent chooses to feel through some relativistic theory, while the latter is based upon absolute, universal moral principles. The guilt generated by technical moral dilemmas is justified only on subjective grounds. In addition, subjective guilt is sufficient to make the agent attempt to avoid conflict situations, which satisfies Ruth Barcan Marcus' claim that guilt in these situations serves a practical purpose without having to accept the view that there are situations in which the agent cannot act ethically no matter what she does.

First, though, I will explain the difference between two appropriate reasons to feel guilt. According to Conee, the objective appropriateness of guilt (OAG) has three conditions:

> A feeling of guilt about having done A is objectively appropriate for the agent who performed A if and only if the agent performing the act (i) has done something morally wrong by doing A, (ii) believes that A was morally wrong, and (iii) feels guilty about this believed moral lapse. (Conee 1982, 243).

If one or more of these conditions are missing, then the person cannot have objective grounds for guilt feelings.

Although useful in some situations, OAG merely addresses whether any feeling of guilt is appropriate in the situation, but does not mention the fittingness of the feeling's depth.

> Feeling guilt for wrongdoing is normal and natural. But a person may be rational or irrational in feeling the guilt. Rational guilt is fitting, appropriate, and useful. It often leads to apology, reparation, resolution, and character improvement. Irrational guilt, by contrast, is obsessive and unrelenting. It can dominate, incapacitate, and even destroy an individual. (De Grazia, 1998, 78)

Guilt feelings in death and dying situations can lead us astray from morality's demands, so there needs to be a line drawn between fitting and unfitting feelings of guilt. A person might have too much or too little emotional reaction for the situation, which harms, rather than nurtures, her thriving.

To fix the matter, OAG can be revised by adding a fourth condition: (iv) the agent feels guilty to an appropriate degree as determined by what she did, the conditions in which she exists, and how the guilty feeling will foster her flourishing. If it is a minor transgression, then it would not be right for the agent to feel so much guilt that she overdoes it by becoming obsessed with it. By feeling too much, the agent is wasting her resources for improvement and having a thriving life by being over-indulgent or not crafting her life into a narrative that shows good character development. On the other hand, by feeling too little guilt, she is neither erasing the moral debt she incurred by her wrongdoing, nor is she improving her character as she should. For those feelings and emotions that can be controlled, there is an appropriate range on how strongly one should feel in the situation.

In contrast to the objective grounds for guilt, an agent needs not do something morally wrong in order to have subjective grounds for guilt. As Conee says, "Feeling guilty is subjectively appropriate when the belief that one has failed which prompts the feeling fits one's moral principles." (Conee 1982, 243). The subjective appropriateness of feeling guilty (SAG) is:

> The feeling of guilt about having done A is subjectively appropriate if and only if the agent having done A (i) believes A is a moral failure of some sort, and (ii) feels guilty about this believed moral lapse.[9] If Morris is correct, then the appropriateness of the guilt is based

[9]Hebert Morris claims that "if individuals sincerely believe themselves guilty, good reasons exist for accepting the belief." (Morris 1987, 222). Moreover, this guilt is appropriate if "we are

more on convention, and possibly genetically determined structures of the human brain, than anything else.

An agent can do what is morally right, yet still have subjective grounds for guilt. "When someone does what is morally best while neglecting something his morality requires, his feeling guilty is therefore appropriate only because it is called for by morality as he sees it." (Ibid.). To expand Conee's point beyond a consequentialist framework, a person can believe that he has acted rightly, and have acted as he should according to an objective moral code governing the situation, yet still feel as if he has failed to satisfy the morality he has subjectively chosen for himself or is part of his community's conventions on morality (Kavka 1974, 664).

If a person believes that she has not fulfilled all of the conditions of the morality she has adopted, even if they are supererogatory or non-obligatory, she could be justified in feeling guilty according to SAG. Karl Jasper argues, for example, that metaphysical guilt is appropriate when a person has failed to do the supererogatory and believes that she did not do as she should have. Although the person acted ethically, there can be a guilt residue remaining because she thinks she failed in some way, perhaps by not having complete solidarity with a person or group that is central to who she is as a person. Herbert Morris also claims that there is non-moral guilt (Morris 1987, 220). This appropriate guilt can spring from people feeling guilty for having states of mind over which they have no control, surviving when others did not, and for the evil or bad actions of others even though the people feeling the guilt have done nothing wrong themselves (Ibid., 221). For example, many surviving Jews felt metaphysical guilt because they did not perish in World War II with their family and friends, whereas some Germans born after World War II experience guilty feelings because of what their country and its citizens did in the Holocaust. Metaphysical guilt is still with us; it seems to be what many survivors feel when someone they care for dies, especially if the survivor experiences relief that the other has died.

6.4.2 *Guilty Feelings as an Inadequate Indication of Wrongdoing*

Some moral dilemma proponents, such as Ruth Barcan Marcus and Martha Nussbaum, believe that objective guilt is appropriate for an agent when she takes either alternative in a situation of conflicting obligations. These technical types of moral dilemmas need to be distinguished from the everyday variety. By "moral

---

(Footnote 9 continued)

disposed to think the feeling natural, the object cited as occasioning it an acceptable one, and if there is no strong pull to the view the feeling is displaced, there is good reason to regard it as appropriate, [and there is the] existence of widespread respect for attitudes underlying the feeling". (Morris 1987, 224–5).

dilemmas", I do not mean situations in which the agent has to make a hard choice between morally right actions that are equal or where one action is morally better than another, yet both involve evil to a degree such that no one happily performs them. Everyday moral dilemmas are those conflicts in which it is merely difficult to decide what to do or to actually to do it.

The technical moral dilemmas examined below, on the other hand, are those in which every alternative action open to the agent is either morally required or forbidden so that performing any one of them will necessarily entail an objective failure to do one's duty by acting wrongly or not doing what is obligatory. As a result, the agent will not be able to escape some moral residue no matter what action she may perform. The action may even be the best thing she can do with the best mental states possible, but this is not sufficient to prevent her from wrongdoing or being bad in some significant way that justifies guilt.

I will use the two notions of subjective and objective grounds for guilt to show how a person's appropriately feeling guilt does not prove the existence of technical moral dilemmas. It is never the case that it is objectively appropriate for her to feel guilt if she has done the right thing. The guilt an agent will experience from these hard choices is only subjectively appropriate. She knows or should know that the act that she performed was a morally right alternative, yet she could still have violated one of the principles in her own personal moral code about what type of person she wants to be. Nussbaum, Barcan Marcus, and Patricia Greenspan fail to see this distinction and misinterpret the guilt the agent feels to be OAG appropriate. I will show that only SAG is satisfied in the situations Greenspan and Nussbaum explicate, and that SAG can do all the work that we want guilt to do without endangering the ethical enterprise we all support.

6.4.2.1 Nussbaum's Dilemmas

The first type of technical moral dilemma I will examine is one in which the right choice between alternatives is obvious, yet the agent is still (wrongly) thought to have objective grounds for feeling guilt. Nussbaum and Barcan Marcus, among others, make the claim that "it is a better fit with the moral facts that all dilemmas are real, even where the reasons for doing [action] X outweigh, and in whatever degree, the reasons for doing [action] Y." (Barcan Marcus 1980, 193). Basically, it is an irresoluble conflict between moral duties. Barcan Marcus believes that a morally right action X may be the obvious choice, but that does not mean the obligation to do Y is nullified by the mere fact another duty is conflicting with it. An agent can do either X or Y and chooses to do X because the evidence shows it to be a far more important duty, yet she fails in her continuing obligation to do Y. Even if the duty to do Y is overridden in the sense that it is a less weighty obligation than X, the failure to perform act Y leaves what Barcan Marcus calls a "residue" which justifies OAG grounds for guilt.

Nussbaum cites the case of Agamemnon to support her claim that OAG grounds for guilt are justified.[10] In the play *Agamemnon*, the title character is required by Zeus to lead a war party against Troy to punish it for Helen's abduction/absconding with Paris. Agamemnon is a moral man who obeys the god's orders, so he gathers an army and a fleet of ships to sail to Troy. Unfortunately, an angry Artemis has decided to create winds that keep the fleet stuck at harbor in Aulis. Before the fleet can set sail for Troy, Agamemnon must offer a sacrifice to placate Artemis. If he does not, all the expedition's members will perish from starvation, and the war they believe mandatory will flounder before it even begins. Unfortunately, Artemis will allow favorable winds for the expedition if and only if Agamemnon slays his daughter, Iphigenia, as the sacrifice.

Agamemnon is faced with a morally devastating choice. He is required to either slaughter his daughter to assuage Artemis' anger, which gives him the power to fulfill Zeus' command, or to allow his entire war party, including other members of his family, to perish. Moreover, if he does not do what is necessary to go to war with Troy, then another god's divine command is violated. There are only these two courses of action open to him. Either action brings about a great, repulsive evil, even though it is demanded by a god.

Since the reasons for making either choice are complex, and they show that the courses of actions have a number of moral factors, I will form a non-exhaustive list of them and the act to which they are relevant.

Action X: Sacrifice Iphigenia

Reasons for Agamemnon to perform the act:

1. It is the only way to fulfill one god's command.
2. It is the only way to empower himself to fill a different god's command to go to war.
3. It is the only way to save the lives of all members of his fleet.
4. He will not have to desert his necessary expedition.
5. He will save as many members of his family as he can.
6. He will act as a king should act.
7. He will act as a commander should act.

Action Y: Remain stranded in Aulis' harbor by refusing to sacrifice Iphigenia

Reasons for Agamemnon to perform the act:

1. He does not violate the absolute principle that one should not kill a member of one's family.
2. He does not violate his role of father by killing his daughter.

[10] This argument has interesting connections to proxies making end of life decisions for others, especially for their family members.

According to *Agamemnon's* moral code, it is obvious from the above which act is morally right: Action X has the greater moral value because it is the lesser of two very great evils, as well as being required by a god and indirectly fulfilling another deity's command. The Chorus of the play regards the act as necessary for Agamemnon to perform, but they do not make the same claim about Action Y.

For Nussbaum, if Agamemnon feels guilty for performing Action X, then he would have OAG grounds for it. The Chorus and Agamemnon state that both acts open to the latter are morally prohibited even though Action X is morally obligatory. As Agamemnon laments, "A grievous doom is disobedience of the gods and a grievous doom it is if I massacre my daughter…Which of these courses is free from evil?" (Aeschylus 1970, 205–211). Nussbaum believes that Agamemnon, in sacrificing Iphigenia, has done something which is morally forbidden: "Zeus has attached…guilt to [Agamemnon] by placing him, a previously guiltless man, in a situation in which there is open to him no guilt-free course." (Nussbaum 1985, 247). Sacrificing Iphigenia to Artemis is necessary and ought to be done, but it still makes Agamemnon guilty of committing a moral crime (Ibid, 244–247). Agamemnon has slain his daughter; he has violated one of the most fundamental obligations in his life as a father. When Agamemnon sacrificed his daughter, he violated the absolute principle that people should not murder their kin, especially their children. Such a crime will cause the Erinyes, better known as the Furies, to punish the transgressor.

Since Agamemnon has violated one of his primary duties when discharging a different obligation, he must pay some moral price for his action. "[The Chorus] sense, therefore, that their king must and ought to be punished. Even crimes of necessity, if that are of this magnitude cry out for penalty." (Nussbaum 1985, 256). Agamemnon performed the best act he could and followed the god's commands, yet he did something morally wrong in fulfilling his duty. Nussbaum further states "The downfall of Agamemnon…answer[s] to our intuitive demand that even the constrained killer should suffer, in his own person, for his deed." (Ibid.). According to Nussbaum and the Chorus, Agamemnon should be punished even though he did as he should. Since Agamemnon has done something wrong, then one condition of OAG has been satisfied. It would appear to be objectively appropriate for Agamemnon to feel guilty about killing his daughter.

However, there is a problem with Nussbaum's reasoning for the objectivity of Agamemnon's wrongdoing. He, in fact, did commit a wrong act, but not for the reasons stated by the Chorus and Nussbaum. Agamemnon was actually given the choice between at least three alternatives—actions; X is actually two morally different acts, only one of which was morally right. The one he selected was very similar to, but not the same as, the morally right. In the two alternatives that replace X, Agamemnon still kills Iphigenia. What distinguishes the two are the emotions and feelings Agamemnon has when performing the sacrifice. Besides action Y, the other two sections were X′ and X″ as found below.

Action X′: Kill Iphigenia with appropriate emotions and feelings

1. Same reasons as for doing X,
2. The sacrifice is performed with the appropriate feelings of grief, horror, regret, and indignation required by a flourishing life on Agamemnon's part toward the act,
3. Agamemnon develops appropriate mental states in the occasion, such as resolving never to be placed in a similar situation in the future.

Action X″: Kill Iphigenia with inappropriate emotions and feelings

1. Same reasons as for doing X,
2. The sacrifice is performed with inappropriate emotions on Agamemnon's part. For example, he feels the same emotions for his daughter as he would a lower life form to be sacrificed and he is not appropriately distraught, so there is no motivation to become better or avoid such situations in the future.

In the play, Agamemnon has chosen to perform Action X″ instead of action X′. He has, therefore, not done what is morally required of him as a good person who has to do a necessary evil in order to act as he should. This appears to be the main reason that the Chorus should consider him to be blameworthy, that is, if they were not using an antiquated system of inconsistent morality based on non-existent entities with mercurial tempers.

In fact, the Chorus does not blame Agamemnon for the action for which they feel the gods bear a primary responsibility. They call on Zeus here as if to understand the meaning of his violent intrusions into human life (Aeschylus 1970, 160ff). The gods have forced Agamemnon into the evil situation in which he now finds himself. He bears no responsibility for that.

The Chorus holds Agamemnon to be responsible for his attitude toward the act of sacrificing his daughter. What they impute to Agamemnon himself is the change of thought and passion accompanying the killing. When Agamemnon realizes that he is required to sacrifice his daughter, "his mind changed to a temper of utter ruthlessness." (Aeschylus 1970, 220). As Nussbaum writes, "If there is much that [Agamemnon] cannot help, there is much, nonetheless, that he apparently can: his emotional responses to the dilemma, his thoughts about the claims involved." (Nussbaum 1985, 257). The Chorus and Nussbaum are in agreement about this responsibility of Agamemnon's. They all agree that Agamemnon treated his daughter as a mere animal to be sacrificed; Agamemnon has not done what he should have done. First, the best act was X′, which was to sacrifice Iphigenia with the appropriate emotions of grief, regret, and horror. If we believe the common sense principle that in situations in which our choices are limited between actions that will each result in great evil, then we have a duty to do the best we can by taking the lesser of the evils. Agamemnon failed to do what a reasonable person would have done to minimize the intrinsic disvalue; therefore, he is blameworthy. In addition, it is impossible for someone to respect the value of another person when the former is treating her as a mere means. By sacrificing Iphigenia as if she were nothing more than a non-human animal, Agamemnon violates the universal principle to respect intrinsic value as it is. He should have had the appropriate respect and the correct mental states, such as intentions and motive, but he could not have had them since he had made the mistake of downgrading Iphigenia's value to something with far less intrinsic worth.[11] Finally, and perhaps most significantly, it is impossible to flourish as a person who is so callous to the death of another that he could and would treat it with such disrespect. An inability to thrive becomes even greater when we factor in the relationship of father and daughter and what is entailed to be a good father, parent, and person. At the very least, the feelings and emotions that a father should have toward his daughter should create a far different set of emotional responses that would be fitting for the situation. Because he could and did instantiate extremely inappropriate emotional responses, Agamemnon actively damaged his ability to thrive; hence, he acted wrongly by becoming a far worse person than he should have been in this tragic situation. Since Agamemnon did not approach Iphigenia's sacrifice with the appropriate emotions he has committed a morally wrong act on each and all of these three grounds. He is, in fact, blameworthy, and he would be objectively justified in feeling guilt according to OAG if he finally realized that he acted wrongly.

If Nussbaum and the Chorus had merely held the position that Agamemnon was objectively guilty for knowingly performing the wrong action, then there would be little to no controversy with their claim. It might have been, in fact, the case that people's intuitions in the matter would have been generally uniform. After all, it is reasonable to assign objective guilt in such a situation; therefore, most would rightly and reasonably believe that Agamemnon should feel guilt for what he did. However, the Chorus and Nussbaum hold a more controversial position that is much harder to defend, if it can be defended at all.

I will now examine what Nussbaum and the Chorus would say about Agamemnon's grounds for guilt if he had acted as he should have. They claim that there are objective grounds for an agent's guilt even if he has done a morally right and necessary act, provided the morally right act violates an absolute principle or is sufficiently evil. The definition of "sufficient evil" is left vague by Nussbaum, but the examples that she uses involve the taking of another person's life. Perhaps the way to understand it is to say that it is so evil that it is repulsive to the vast majority of reasonable people. There would be some moral residue left over, as Barcan Marcus claims, which justifies more than merely subjective grounds for guilt, even though a duty has been fulfilled. That is, even if the action is mandatory, there are some moral horrors so great that they leave a blemish on the person who performs them.

It is clear that Nussbaum and Aeschylus believe that there can be conflicts between two or more moral principles, which can end in a person acting rightly for one principle and wrongly for the other. The Chorus lives in a world in which morality comes from a

---

[11]Even if the sacrifice had been a goat, indifference to its value as a living animal would also have been wrong.

variety of sources with a few absolute principles, such as the murder of one's relatives being absolutely prohibited, regardless of why the action is necessary. When Agamemnon's son, Orestes, for instance, does as the gods order and slays his mother, while having the appropriate emotions, he still has done something wrong. Even acts which are prescribed for mortals by the gods may not override the absolute ban against murdering relatives. As the Chorus says, "But the black blood that has once poured to the ground in front of a man in death who could call back the incantations?" (Aeschylus 1970, 1018–1021). Nussbaum states that the "Furies of Orestes answer our intuitive demand that even the constrained killer should come to regard himself as a killer and should suffer, in his own person, for his deed." (Nussbaum 1985, 258). Crimes of such a magnitude against absolute principles, even if forced by necessity, demand some sort of punishment, for the agent has performed the forbidden even while doing his duty. Thus, Orestes deserves the fate of being left to the Erinyes (Aeschylus 1970, 1275–1294). Even the constrained killer should feel "an emotion like remorse, closely bound up with the acknowledgment of the wrong that he has as an agent, however reluctantly, done." (Nussbaum 1985, 258). Remorse, of course, is only appropriate if someone has acted in a way that he should not and feels some shame as if his character is somehow defective. Thus, if Agamemnon had chosen the right act and performed it, the Chorus and Nussbaum would still say that Agamemnon would have objective grounds for feeling the guilt he should feel for his act because he is a good moral agent.

I see a number of problems with the Chorus and Nussbaum's position and the bleakness of the moral world they envision. First, intuitions are notoriously unreliable sources of evidence for a position. They could be the result of social convention, personal belief, or some other subjective source, but that does not entail that there is objective, real grounds for a feeling of guilt, especially since Agamemnon would have done what he should have done had he performed action X′. Given intuitions' unreliability, it would not be conducive to practical morality to lay a huge moral burden of guilt on a person who has to make a very difficult moral decision merely or significantly because someone's intuitions demand it. The same conclusion will apply in cases in which proxies have to make end of life decisions for those for whom they care.

In addition, contrary to what Nussbaum and the Chorus believe, not everyone's intuitions demand the punishment that the former think is justified for the constrained killer. Many would agree with my contention (if they can put aside that the world being described and its morality is a work of fantasy) that Agamemnon would have done the right act if he had performed X′. Therefore, there would be intuitions that do not make such a demand because we know that objective guilt is inappropriate for someone who has satisfied the Practical Principle. Moreover, it is rather cruel to tell those facing these moral dilemmas that they are objectively guilty of wrongdoing no matter which choice they make. That insensitivity and morality cannot be conducive to their flourishing.

Guilt is not justified in this situation. In fact, there are far stronger grounds that practical reason demands that the good agent pursuing flourishing, respecting intrinsic value, and doing what reasonable people reasonably believe will produce the best outcome should feel "agent-regret" instead (Baron 1988, 264). Agent-regret is the feeling that an agent has when her action is morally right but produces sufficient evil. "[You] feel regret for what you did, while not believing either that you could have foreseen the situation; or that once in the situation, you could and should have acted differently." (Ibid.). For instance, a constrained killer who feels agent-regret for what she did seems to be a better agent than one who is identical to the former agent except that she does not feel this type of regret for her action. The latter agent seems somewhat callous toward human life, while the former recognizes the value of such life. Moreover, experiencing agent-regret in these situations is better for a person's flourishing than if she felt guilt for doing something wrong. Since the agent did not make a moral error, her belief states about her action would be false, which in turn makes her feel an inappropriate emotion for the situation. Each of these is an intrinsically disvaluable state of affairs, whereas agent-regret is fitting and actually makes the person thrive more. Hence, if Agamemnon had performed X′, then he should feel agent-regret, among other things, for his act. He would realize that his action was justified and sufficiently evil.

Agamemnon may appropriately feel guilty for performing X′, but this guilt will only be subjectively and not objectively appropriate, for Agamemnon has not, in fact, made a moral error. Given Agamemnon's personal moral code, he could believe that he has a moral residue and feel guilty over his believed ethical lapse, but this emotional reaction is not sufficient on objective grounds for Agamemnon to be objectively guilty. Therefore, even if Agamemnon did feel guilt, it would not prove that the moral dilemmas that Nussbaum and the Chorus describe justify objective grounds for feeling guilty.

Subjectively appropriate guilt can do the work of making amends internally and externally that Nussbaum and others want guilt to do without threatening the whole foundations of ethics. Nussbaum and the Chorus believe that a person can fulfill a duty yet still do something morally wrong. "[Such] situations may be repellent to practical logic; they are also familiar from the experience of life." (Nussbaum 1985, 247). Such a position involves a contradiction: Action A is forbidden and obligatory at the same time. Of course, from a contradiction, anything can be derived, such as claiming that a person can be morally obligated to do an action yet still be guilty of wrongdoing for performing that action. Then again, so can the conclusion that if an agent does what is morally required, then she has not done something wrong. A position should be rejected if it entails a contradiction that would make morality pointless because any conclusion is legitimate, or takes away the motivation to be ethical, such as happens with this argument about being obligated and forbidden to act.[12] Therefore, the proposed sufficient condition for objective grounds for guilt based on the Chorus' and Nussbaum's position is false. It is not reasonable to adopt a moral code or morality that states that a person can do as she should yet still be guilty of wrongdoing.

---

[12]Since it poses a problem for both Nussbaum and Greenspan, this problem will be developed in more detail in the section on Greenspan's dilemmas of exhaustive prohibition.

6.4.2.2 Greenspan's Dilemmas

Patricia Greenspan proposes a different type of technical moral dilemma. In this conflict, the two alternatives available to the agent have equal moral status; both are prohibited acts. I will explain this initially stronger-appearing position, and then show why it, too, is inadequate to justify objective guilt on the grounds that there must be at least one morally permissible action in each situation; otherwise, morality is irrational and loses its force to motivate people to do as they should, and therefore, improve their flourishing.

Greenspan believes that objective guilt is justified and appropriate for moral dilemmas in which the agent is at fault for what she does no matter what she does since all her options are ruled out (rather than favored) by the moral code (Greenspan 1974, 118). Objective guilt is fitting because there are "cases in which each alternative is prohibited on serious enough grounds to make it come out as wrong all things considered despite the lack of permissible alternatives to it." (Greenspan 1995, 177). Greenspan calls such cases "dilemmas of exhaustive prohibition." (Greenspan 1974, 118). The main difference between exhaustive prohibition dilemmas and those described by Nussbaum is that Nussbaum's had an alternative that was both morally right and wrong. Greenspan does not introduce this type of contradiction into her dilemmas.

According to Greenspan, exhaustive prohibition dilemmas also differ from dilemmas of exclusive requirements. Exclusive requirements dilemmas are conflicts in which the options open to an agent are two actions, both of which are morally right. For instance, a doctor can save only one of two patients, Patients A and B, because his medical resources are limited. He may either use all of his means, which include time and medical supplies, to save Patient A, or he may save Patient B, but he cannot do both. If the doctor attempts to help both, then both will die from inadequate treatment, so he must decide which one to help.

Not only is saving either Patient A or Patient B permitted in this situation, but doing one of the two is also morally required of the doctor. That is, unlike exhaustive prohibition dilemmas, the doctor is doing something that is his duty to perform. The doctor's "choice is directed towards doing something he should—towards saving A, say—so there is no reason for him to feel guilty about it." (Greenspan 1974, 118). If the doctor did not face the dilemma—suppose that he only has to save Patient A—then that option would be morally required of the physician. On the other hand, if Patient B is the only patient the doctor has, then he should save Patient B. No one would rationally or reasonably question the physician's moral duties in either situation, so if it is a decision between those two being in conflict, then the alternatives' conjunction must not be controversial.

According to Greenspan, objective guilt is inappropriate for exclusive requirements conflicts. There "is not reason for him to feel guilty about … [saving one patient instead of another], though he might feel guilt for not doing more." (Greenspan 1974, 118). It is not the act itself of saving the patient that will give grounds for objective guilt. The guilt for not doing more is also objectively inappropriate if the doctor did all that he could do: "The agent might in fact be left with a residue of guilt; but it would be no more appropriate for failure to fulfill an overridden obligation that it is in cases of simple inability." (Ibid., 121). The doctor does not have objective guilt justification if he performs the act, although he might be permitted to feel guilty on SAG grounds.

On the other hand, Greenspan says that exhaustive prohibition dilemmas justify objective guilt because each alternative is forbidden in its own right. Greenspan uses Sophie's Choice, from the novel of the same name by William Styron, to illustrate the distinction between her two types of moral dilemmas. During WWII, Sophie is required by a Nazi guard to make a choice of which of her two children will be killed. If she does not choose one, both will be executed by the guard. An exhaustive prohibition dilemma would involve the violation of one or more obligations Sophie has. In this particular case, the obligation comes from the general moral principles that a parent should protect her young children from harm, and that a parent should never directly or knowingly cause her child to die or be murdered. Sophie should save her children from being killed, yet she is forced to explicitly choose which child will die, thereby violating the absolute moral principles banning such behavior. No matter which child she chooses, therefore, she will be objectively guilty of wrongdoing because she has performed an action which is forbidden. "There is something, whatever it amounts to or should be called, that gives more than individual psychological backing in such cases." (Greenspan 1995, 178).

However, according to Greenspan, if the guard's question to Sophie were presented positively in the first place, then it would have created an exclusive requirement dilemma. In a version of the same case where the dilemma is one of exclusive requirements, Sophie has to choose which of her two children will live. Either action of the latter dilemma is required according to the same principles which prohibit Sophie's actions in the exhaustive prohibition dilemma. According to Greenspan, if Sophie acts to save either child, then there are no objective grounds for guilt, even though the other child will certainly die as a result of the choice. Guilt is inappropriate because at no moment in the process of reasoning, choosing, and acting did Sophie do anything other than something she should (Greenspan 1974, 121). Sophie, as a parent, would be morally required to save her individual children. If it was not a matter of a choice between the two children, but rather whether to save Child A and only Child A was in danger, then she morally must save Child A. If it was Child B's life that was threatened, then Sophie would be obligated to save Child's B's life. Moreover, as a parent, Sophie is morally required to choose one child to live in order to avoid losing both, and each choice is a moral duty to perform. Given that objective guilt cannot be justified for an agent if she does only what she should, Sophie is not objectively guilty in the exclusive requirement case, for she has merely fulfilled her positive moral duty.

If Greenspan is correct, then in exhaustive prohibition dilemmas, any action that Sophie performs is OAG guilt-worthy. "But as in Marcus' dilemmas, she would have had overriding reason to make the choice [in the exclusive requirements case]; and this is what is missing in the negative case, where the choice is to prescribe death for one of her children, and seems to be forbidden on ground as compelling as those which forbid her letting both be killed." (Greenspan 1974, 122–3). Sophie, in explicitly choosing which child will die, has violated the moral principle requiring her to protect her young children from harm. It does not matter which child she picks; she will still perform a prohibited action. As Barcan Marcus might say, "Sophie is morally damned if she does and damned if she doesn't." There is no alternative open to Sophie which is free of objective guilt because each action is individually prohibited by at least one absolute moral principle. Even if she does the best thing that she can do, treats everyone affected by the action with respect, and does everything she can to promote and nurture flourishing, she will still have satisfied OAG's four conditions.

There are a number of problems with the claim that exhaustive prohibition dilemmas exist. First, the moral distinction between the two types of dilemmas is unwarranted because their morality should not depend on the arbitrary manner in which a question of life or death is asked. Moreover, there seems to be a reliance on the Doctrine of Double Effect. Since the doctrine is false, any position relying on it for credibility will be problematic.

For Greenspan, what separates the two types of dilemmas is what Sophie intends. The effect that Sophie will intend is determined by the way that the guard initially poses his question to her. If the question is expressed in terms of saving a child in the first place, then the dilemma is one of exclusive requirements. In such a case, Sophie can always intend the good effect and merely foresee that the other child will die as a result (Greenspan 1974, 119). If Sophie merely intends the good effect of saving one of her children and foresees that the other child will be murdered when she selects a child to be saved, then she has done something morally obligatory. By keeping her mental states and actions pure, Sophie is able to obey the parental principles prescribing how one's offspring should be treated in thought and deed.

On the other hand, when the question is phrased in terms of selecting a child for death, then the dilemma is one of exhaustive prohibitions (Greenspan 1974, 119). Sophie has to intend that one child will be murdered while merely foreseeing that another will be saved. According to Greenspan, Sophie's end in either case is to save as many of her children as she can. In the negative case, she must do so by impermissible means by violating absolute principles that prohibit either and both of her options. If Sophie intends the evil effect of murder, then she has acted immorally. Of course, by choosing which child will die, then she must intend the murder of her offspring, which thereby violates the principles to protect her children and not to participate in killing them.

Greenspan claims that if the exhaustive prohibition dilemma were to be reworded positively by the guard, Sophie would still have only impermissible alternatives. According to Greenspan, the exhaustive prohibition dilemma cannot be weakened to an exclusive requirement dilemma once Sophie has heard the alternatives presented in terms of murdering one of her children. Even if the exhaustive prohibition dilemma were to be rephrased in terms of which child should be saved, any prescription Sophie makes actually is an indirect way of choosing a child to die (Greenspan 1995). Greenspan's Doctrine of Double Effect says that once Sophie intends the bad effect, she cannot make the act right by merely consciously revising her intention from the bad to the good effect. In part, Greenspan's Doctrine of Double Effect has the criteria that:

> An action, A, is always morally right for an agent, S, performing the action only if (1) S intends merely the good effect of A, and merely foresees any bad effect of A, and (2) S has not consciously revised her intention from A's bad effect to A's good effect.[13]

If correct, this version of the doctrine requires a purity in thinking that other versions do not. The standard interpretations offer agents the opportunity to redeem themselves and make an action ethical if they can alter their thinking to the right or appropriate mental states, whereas this one requires a much higher standard.

The soundness of Greenspan's arguments depends on whether or not this particular version of the Doctrine of Double Effect is true, and there is very good reason to think it is not. The Doctrine of Double Effect sometimes classifies certain acts as morally right when we know they are wrong. G.E.M. Anscombe, for instance, describes several examples of such misclassification. One of them is a servant who holds the ladder for his master who is burglarizing an innocent person's house. According to the doctrine, in order to make his action morally permissible, all the servant must do is direct his intention in a suitable way. Namely, the agent has to intend only the good effect. In the case of the servant, he must merely aim his intention toward the good effect of "avoiding the sack." (Anscombe 1961, 404). The servant may foresee that in order to keep his job, he has to help in the commission of a crime, but that does not make his action of accessory to burglary wrong.

By making an action's rightness or wrongness depend on the intentions of the agent doing the action, a serious problem arises. If the Doctrine of Double Effect is true, then, as Anscombe writes, the correct "direction of intention [will set] everything right no matter what one does", as long as the doctrine's other conditions remain the same, or the question is posed correctly in the first place (Anscombe 1961, 404). Assuming that the action itself is at least morally indifferent, the good effect is not caused by the bad effect, and the good of the good effect outweighs to an adequate degree the bad of the bad effect; if one's intentions are sufficient for morality, then even what we consider to be the most despicable of morally wrong acts can be made morally right by a base, calculated focusing of intention. For example, people may be treated as mere means just as long as the agent's intentions are pure enough, as happens in the case of the innocent homeowner being robbed by the master and servant.

It might seem that the "direction of intention will set everything right no matter what one does" condition is too extreme to bear strict scrutiny. That is, Anscombe's claim might go too far in saying that anything can be set right if one's intentions are

---

[13] I assume that the other conditions of the doctrine remain the same:

1. The action itself must be morally good or indifferent.
2. The good effect cannot be caused by the bad effect, and the good effect must be directly caused by the action.
3. On balance, the good of the good effect must outweigh to an adequate degree the bad of the bad effect.

only directed in the correct way. In fact, Greenspan agrees that in certain cases, the Doctrine of Double Effect misclassifies cases.[14]

In circumstances such as those in Sophie's choice, however, the doctrine is said to be applicable, and correctly classifies moral actions. In both dilemmas there are identical moral consequences, namely, that one child is saved while the other is killed, so that is not the morally relevant dissimilarity. The good consequence does not flow from the bad. Both alternatives in the latter dilemma violate the moral principles prohibiting a parent from causing her child's death. What makes either alternative in the exclusive requirement dilemma morally right is that Sophie always intends only the good effect and merely foresees the bad one, and her actions are good. On the other hand, in both horns of the exhaustive prohibition dilemma, Sophie must intend the evil effect; thus, either alternative involves Sophie in wrongdoing. Hence, the correct intention is not sufficient for morality, but it is clearly necessary.

The problem with the two types of dilemmas is that they might be different in some ways but not sufficiently so to do the work Greenspan thinks they do. In both versions of the choice that Greenspan describes, Sophie knows the same facts about the alternatives in both dilemmas. That is, Sophie fully understands in each case that a child will be killed as a result of her decision. In the exhaustive prohibition version, the wording is different, but the result is the same as that of the exclusive requirement dilemma. One child will live as a result of the choice, while the other dies horribly. Even in the exclusive requirement version, Sophie knows that one of her children will be murdered as a direct result of her selection. This is just the way it is. Although the bad outcomes might not be intended, they are certainly intentional in the sense that all bad foreseen and intended outcomes of an action are intentional (Knobe 2003).[15] Since intentional actions entail moral responsibility for the action's consequences on the part of the agent performing them, then Sophie is equally morally responsible for the death of her child regardless of whether she selected one to live or one to perish. In both, Sophie intends to save as many children as she can, respects intrinsic value as she should, and tries to pursue flourishing in these very tragic circumstances; therefore, she is equally praiseworthy or blameworthy for what she has done. There is no real moral distinction between the two types of dilemmas. Hence, since dilemmas of exclusive requirements involve no OAG, neither can dilemmas of exhaustive prohibition.

Another problem Greenspan's position encounters is that Barcan Marcus' "strong" dilemmas imply dilemmas of exhaustive prohibition, although Greenspan rejects the strong dilemmas and the implications (Greenspan 1974, 117). Since Barcan Marcus' dilemmas imply exhaustive prohibition dilemmas, and Greenspan believes that Barcan Marcus' strong dilemmas yield "no particular recommendations" for moral action, then in order to be consistent, we must also say that Greenspan's dilemmas of exhaustive prohibition yield no particular recommendations (Ibid.). Clearly, for those who believe that they exist, such dilemmas are

[14]Greenspan actually does so in "Moral Dilemmas and Guilt's" fourth note.

[15]This concept will receive far greater development in the next chapter.

supposed to yield particular recommendations. Hence, if it can be shown that the strong dilemmas imply dilemmas of exhaustive prohibition, then there might be little reason to justify treating the latter as if they are morally different from strong or exclusive requirements dilemmas.

Let me begin the argument by assuming a few reasonable, practical premises. First, I will stipulate that there are real world examples of Barcan Marcus' strong dilemmas, even though evidence for them is based more on an intuitive foundation than anything more substantial. There are times in which both options are morally required, although one of them might be the right thing to do all things considered. Next, I will assume that Sophie's choice is such a dilemma if she is presented with a positive question by the guard in the first place. Sophie is told that she must select what child to save, which entails that the other child will be executed. Hence, Sophie can choose between three acts:

Action X: Sophie chooses Child A to be saved from execution.
Action Y: Sophie chooses Child B to be saved from execution.
Action Z: Sophie does something other than choose one and only one child to be saved from execution.

Any other choice or action will result in the deaths of both children, which is understood to be morally forbidden and so morally evil for Sophie to perform that it cannot be a legitimate ethical option for her to take.[16]

Before proceeding further, I need to explain the notion of necessity used in the proof. When Sophie selects one child to live, a consequence of her act is that the other child is executed. According to the way that Greenspan explains the choice, Sophie cannot prevent the one child from being killed after she has chosen to save the other child; an execution necessarily follows either of Sophie's selections. Of course, this is a rather strange necessity since it is not even a causal requirement that the other child dies as a consequence of Sophie's choice, but let us call either death that results from the positive choice a mediated moral certitude. Given what will happen as a result of her decision and that of the Nazi guard mediating the

[16]It might be the case that ordering one child to die is of equal moral value to allowing both to die. Perhaps, working within an evil system, like that of the Nazis, adds negative value to the saving of either child. The additional negative value is sufficient to make it impossible to decide which action is likely to be worse. It might also be the case that collaborating in such a system disrespects the intrinsic value of those involved. The guard is attempting to degrade Sophie by forcing her to make a choice, but Sophie can only be degraded by her choice to collaborate with the guard scheme.

The situation is similar to what happened to those who helped desegregate luncheon counters in the American South in the 1960s. Although having things poured on them and being verbally abused was intended by those doing it to disrespect and degrade the recipients of the treatment, the protestors' human dignity and flourishing as individuals and persons were enhanced rather than harmed by how they responded to the intense provocations. Their refusal to be degraded by allowing themselves to be provoked showed their integrity to their principles of humanity and non-violence, whereas the provocateurs degraded themselves by the inhumanity they exhibited. If she respects herself, Sophie should refuse to choose, according to this line of reasoning. If this is true, then a hard choice remains.

selection's consequences, it is almost a causal necessity for one child to die as a result of Sophie's choice to save the other.

Starting with the definition of exclusive requirements or a strong dilemma, we can readily derive a dilemma of exhaustive prohibition. First, Sophie is obligated to do Action X; Sophie is obligated to do Action Y; and it is not possible for her to do both Actions X and Y, by definition. From here, it follows that if Sophie is obligated to do Action X, and it is necessary that doing Action X entails not doing Action Y, then Sophie is obligated to not do Action Y, according to a plausible and probable deontic principle, which we can label as DP:

> If an agent is morally obligated to do Action A, and doing Action A necessarily entails also doing or not doing Action B, then the agent is obligated to do or not do Action B, respectively.[17]

Since Sophie must not select two children to be saved, then it follows if she selects Child A to survive, then she must not also select Child B to survive, and vice versa. If she chooses both, then she will impermissibly condemn both to death, and this is such an egregious thing that it is not one of Sophie's legitimate options. Once Sophie has made her decision to save Child A, for example, which means she chooses to perform Action X as her duty, then she becomes morally bound not to perform Action Y, which would necessarily undo her ability to perform Action X. That is, Action Y necessarily becomes impermissible to Sophie because of the duty to perform Action X exclusively. Given that impermissible actions are prohibited, it follows that Sophie is prohibited from doing Action Y. Of course, we can also show that Sophie's duty to perform Action Y and save Child B will lead to a prohibition for her to perform Action X. If we conjoined these two prohibitions together, then we have a dilemma of exhaustive prohibitions. That is, Sophie is forbidden to do Action X, she is forbidden to do Action Y, and there are no other legitimate options open to her.

The argument's premises are, for the most part, justified by Greenspan's definitions, the example's formulation, and logic. For instance, according to Greenspan, the selection of one child necessarily causes the other's death. The guard has made it abundantly clear to Sophie that choosing child A will result in child B being killed and vice versa. The other premises, except for the one relying on DP, are likewise verifiable. We can see, therefore, that Barcan Marcus' argument is sound—strong dilemmas and exclusive requirement dilemmas imply dilemmas of exhaustive prohibitions. In order for her position to be consistent, Greenspan must either say that dilemmas of exhaustive prohibitions do not give recommendations for action, or Barcan Marcus' strong dilemmas actually are moral dilemmas in the technical sense. Which alternative is more conducive to Greenspan's position, I will leave for others to determine.

---

[17]Carlo Felice argues that if a person has a duty to help others in dangerous situations or dire need, which requires that the actor keep informed to know when those situations arise, then the actor has a duty to keep informed, *ceteris paribus* (Felice 1990).

6.4.2.3 A General Problem for Technical Moral Dilemmas and Objective Guilt

It is contrary to pragmatic morality that an agent can do what is morally required of her yet still be objectively guilty of doing something morally wrong; such a view weakens the motivation to be ethical and imposes great moral burdens on people who should not have to shoulder them. Suppose that we consider an action that does not involve a fictional world in which moral dilemmas are written into the very fabric of the people's existence but keep the absolute rule that selecting death for one's child is always prohibited no matter what the circumstances. It will surprise no one that some parents have to make a decision about their children's medical treatment which can hasten that child's death. Suppose, similar to the United States District Court for the Eastern District of Virginia's 1994 case of Baby K, an anencephalic infant has enough brain stem development to permit some reflexive actions, such as swallowing and sucking. The parents have to make a decision on whether to sign a Do Not Resuscitate (DNR) order, as happened in the Baby K case. The parents are well aware that their infant will never develop consciousness or feel pain, so all treatment is futile. They also know that the order would likely result in their infant dying sooner than if they did not authorize it. If the parents sign the DNR—even if they do it because they reasonably think it is for the best, they respect intrinsic value as they should, and they have appropriate feelings and emotions for thriving—they still violate the principle of not preserving the life of their child when doing so is simplicity in itself. They would do what is permissible yet still be objectively guilty of wrongdoing by breaking that absolute rule.

A worse case can be found. Infants with severe lung hypoplasia will inevitably die. Life support systems are removed, at times, so that the infant can die while the parents are holding it (Verhagen and Sauer 2005). In these situations, it is a far more immediate death than that of Baby K. Although the parents do not directly kill their child, they still passively and indirectly do so by having life support withdrawn. These grieving parents, too, violate the absolute principle never to hasten one's child's death regardless of circumstances.

If we accept the existence of technical moral dilemmas, then we have two morally insensitive, unintended conclusions we must also accept. First, the parents in these situations should feel guilty because they did something objectively wrong. Since the parents did not do the impossible, although they acted ethically, then in their worst time of need, they have acted impermissibly by withdrawing or withholding their child's life support. When these individuals should be allowed to grieve and feel other emotions and emotional response, and they need to deal with the situation in a psychologically healthy manner and pursue their own thriving, they are justified in thinking that they have done the wrong thing because they *have* done the wrong thing. To make such a claim about these tragic choices seems callous to the suffering of those involved in the situation.

The choice is between two moral codes: one that says that the parents did the right thing and have not violated their moral duties, and one that says that they have done the right thing or what they should and are still guilty of wrongdoing. An adequate moral code should remove burdens on the innocent who are suffering rather than adding to them as moral debts and obligations to make amends for their alleged wrongdoing. It is only the former that is consistent with a practical, pragmatic morality aimed at making the world a better place by pursuing the flourishing of individuals, groups, and communities. The first code eschews intuitive, absolute principles that work only in fictional worlds of two dimensional characters and emphasizes that what is required of us is often very difficult but not heartless to those in tragic circumstances and blind to morality's purpose.

That is not to say that a pragmatic code cannot have absolute moral rules. In fact, many do, such as the principles that (1) it is never morally right to disrespect people affected by an action, and (2) we are always obligated to prevent unnecessary, significant pain and suffering when doing so only sacrifices trivial goods and causes only trivial harm. These two rules, however, are general and flexible enough to apply to all moral situations we encounter rather than being so specific and rigid that they can be used only in very limited situations, such as always prohibiting killing one's family member. We could even keep the specific rule against killing family members if we merely made it more flexible to reflect the complexity of morality in the real world. By adding "in general," the rule becomes "in general, it is morally wrong to kill one's family member, and in the very narrow context, is always morally bad, although it might not be morally bad all things considered." Of course, introducing flexibility of rules to match actual circumstances eliminates technical moral dilemma's existence, but that works out for those who want morality to make the world a better place in an efficient way.

Besides adding unnecessary burdens to the innocent, if technical moral dilemmas did exist, then there would be far less motivation to act ethically and be good people, which would undermine morality's function. In end of life decisions, for example, people would know that despite the terrible situation and their innocence in creating those circumstances, they cannot act ethically. In trivial situations ungoverned by these arbitrary, absolute rules, she can do what she should without residue. In trivial situations in which one of the absolute governs but is easily satisfied, she easily does what she should. However, where morality really matters the most, that is, when non-technical moral dilemmas occur, she is powerless to do more than not being the very worst person, although she will inevitably be a bad one for failing because morality is stacked against her. With technical moral dilemmas, morality works for her in trivial situations, but it is turned against her in too many significant cases. A reasonable person might, therefore, give up trying to respect others, do the best she can, and thrive because regardless of what she does to control her character and morality, there will always be difficult moral situations in which she will fail by necessity. It does not make sense to try to be ethical when one is destined to fail when it matters most.

In addition, instead of guilt being an agent's motivator to make amends and improve her character and life so that, at the very least, she can avoid it again, guilt is a constant condition for those who have to make hard decisions in tragic circumstances that cannot be avoided or escaped. Eventually, with this attitude of defeat, guilt will lose its ability to positively affect behavior and shape character.

Those who have to make such decisions are compelled, so guilt does nothing for them. On the other hand, avoiding guilt would require the agent to run away from any relationship opportunity that might entail she has to make a hard decision about someone else's care. The agent, therefore, would pursue the trivial to avoid the complications that will arise from the complex and meaningful. Of course, this course of action means that the agent never really thrives because she will never encounter any of the hard situations that test her character and morality and make her into a better person from the life lesson learned. The result is that guilt does no good work; it either cannot be avoided or it helps agents avoid dilemmas to the vastly greater detriment to their thriving. Hence, it is far more practical to have ethical principles that do not end in technical moral dilemmas, with their potential undermining of morality's purpose, than it is to adopt intuitive ones that condemn people to wrongdoing no matter what they do or have to have "fixes," such as the Doctrine of Double Effect.[18] This approach becomes even more probable and plausible when we understand that we can achieve the beneficial results technical moral dilemmas are supposed to achieve without damaging morality's ability to do its work.

6.4.3 Avoiding Moral Dilemmas and Evaluating the Appropriateness of an Emotional Response to Death

Objective guilt has been shown to be unfitting when the agent has done all that objective morality requires of her, but there is still room for subjectively appropriate guilt and the benefits it can bring. In these cultural relativism situations, SAG guilt is appropriate when the agent believes that she has failed to fulfill her socially created expectations, although she has done nothing morally forbidden. This is basically a violation of a social rule to which the agent feels strong emotional attachment, or it might be the case that the agent strongly identifies with the community, and anything that affects that entity will cause a positive or negative reaction in her to the extent that her identity is also affected (Armon-Jones 2003, 182). If, for instance, the community's standards are questioned, then the individual citizen might feel that her character is also being questioned.

Existentialism and Individual Relativism are also capable of justifying subjectively appropriate guilt. The agent can subjectively choose or believe, respectively, that she has some moral residue left over because, for instance, she did not act in a supererogatory manner in the situation or feels as if she is not sufficiently good. That is, she has met the standard for being a minimally decent person according to

[18]The Doctrine of Double Effect might strike some as an ad hoc solution to a problem that need not exist. If we eliminate those absolute moral principles that inadequately consider a situation's circumstances, then there is no need for the doctrine. Therefore, the principles will suffice on their own without adding a way to get around them because they are defective by default.

objective morality, but has failed to live up to the self-imposed higher standard. For example, even if the agent believes that she did as she was required, she still may think it appropriate to feel guilty because of her action's negative consequences, especially if they involve the suffering of others or the violation of her personal moral standards (Kavka 1974, 664).

Subjective guilt needs not be a self-indulgent emotion or feeling; it can be ethically justified on a number of additional grounds. Daniel Statman makes a distinction between internal and external justification. A person has external justification for her guilt because it is subjectively appropriate for one or more of the following grounds:

Moral Character: The guilt feelings sensed by moral agents when they justifiably defy a moral rule are testimony to their moral sensitivity. From this point of view, these sentiments are not irrational.[19]

Utilitarianism: Every person needs a moral education to be a decent person. In order to get it, one must have these types of feelings. Of course, we make mistakes about when these feelings are objectively warranted. However, to eliminate all mistakes, we would have to eliminate moral education in its entirety. Mistakes are necessary consequences of an adequate moral education; therefore, it is best to allow them to occur.

Personal Identity: In moral dilemmas, agents must often act against some of the basic values that shape their lives and grant their lives meaning, and that action threatens to ruin the agents' lives. Since these values are essential constituents of the agents' personal identity (values which make them what they are), when agents are forced to act against such values, their personal identity is in serious danger of disintegration and destruction (Statman 1995, 123–9).

For Moral Character, we approve of good people to feel certain ways in particular situations, and it is a credit to their moral character that they feel it as they do. Guilt in death situations, especially when a person has made the final decision to stop life support of a loved one, shows that the person feeling guilt has a good enough character to realize that their justified violation of the moral rule to support and preserve the lives of the vulnerable, especially those with whom we have positive caring relationships and who are dependent on us, is not something that a good person should feel pleasure or inappropriate feelings over. For Personal Identity, to feel regret and other appropriate emotions in the situation keeps those feeling them from having their identity harmed. It is a way of preserving who they are as good people by recognizing their values and the danger to them, as well as creating a mechanism in which they seek to preserve themselves from harm.

Subjective guilt is as capable of motivating us to avoid conflict situations as Barcan Marcus thinks objective guilt is without introducing unnecessary

[19]This seems to be a very large part of Greenspan's argument for practical guilt. (See Greenspan's *Practical Guilt*, 111–5)

contradictions into morality or reducing people's incentive to be ethical. We want guilt to be fitting here because it has an unique ability to achieve desired results: Guilt "provides an element of moral self-threat that is not present in more passive reactions like regret or shame or less in self-oriented feelings like remorse."[20] (Greenspan 1995, 110). This threat produces a strong motivation to make things better and arrange our lives in such a way that we can evade moral dilemmas of both the merely hard and the technical sort. "If we are to avoid dilemmas we must be motivated to do so. In the absence of associated feelings, motivation to stack the deck, to arrange our lives and institutions so as to minimize or avoid dilemma is tempered or blunted." (Barcan Marcus 1980, 197). The agent will try to elude situations in which she must violate one of her subjective moral principles, cause suffering to others, risk community well-being, and therefore, her identity as a community member, because of the subjective guilt involved. The agent will "stack the deck" to avoid such cases of conflict because of the very strong desire not to have these very bad things happen to the agent or those for whom the agent cares. Therefore, subjective guilt can and will do the same work that objective guilt does without making any moral agent think that morality condemns them to forced wrongdoing and evil when they need morality most to find at least one alternative to which they can do what is right, and ease rather than increase their moral burdens.

6.5 Preparing Ourselves for the Death of Others

When discussing a complex issue, there is always a potential for becoming too myopic and losing sight of the forest for the trees. Although objective guilt is appropriate in all circumstances of objective wrongdoing, *ceteris paribus*, and subjective guilt is appropriate at times in others, this result is only based on all things being considered in the situation. That is, feeling a particular emotional response is generally, but not always, appropriate in the relevant circumstances. There are times in which even objective guilt should not be experienced, and a far larger number of instances in which subjective guilt is not only impractical, but also destructive. It, therefore, behooves us to at least sketch out some guidance on how to prepare for an appropriate emotional response for death and dying.

Perhaps one of the clearest and most moving accounts of how the death of those we care for emotionally affects a person can be found in Mark Twain's autobiography. While travelling in Europe, Twain is devastated by the news that his favorite daughter has unexpectedly died.

> It is one of the mysteries of our nature that a man, all unprepared, can receive a thunder-stroke like that and live. There is but one reasonable explanation of it. The intellect is stunned by the shock and gropingly gathers the meaning of the words. The power to

[20] I have claimed that remorse has an element of shame, which would make it much more self-oriented than allowed for in this passage.

> realize their full import is mercifully wanting. The mind has a dumb sense of vast loss – that is all. It will take mind and memory months, and possibly years, to gather together the details and thus learn and know the whole extent of the loss. A man's house burns down. The smoking wreckage represents only a ruined home that was dear through years of use and pleasant associations. By and by, as the days and weeks go on, first he misses this, then that, then the other thing. And when he casts about for it he finds that it was in that house. Always it is an *essential* – there was but one of its kind. It cannot be replaced. It was in that house. It is irrevocably lost. He did not realize that it was an essential when he had it; he only discovers it now when he finds himself balked, hampered, by its absence. It will be years before the tale of lost essentials is complete, and not till then can he truly know the magnitude of his disaster. (Twain 2012, 324)

Death, as has been stated many times, is inevitable for ourselves and others. We do not want others to die, especially those we love but, once again, such an event could happen at any time. Death becomes more likely as people age, become disease ridden, or act in an imprudent manner. Even greater interactions with the external world, such as driving in bad weather conditions or in highly congested traffic areas, can raise the probability of being killed. Although it might appear to be terribly emotionally cold, in order to flourish by protecting our intrinsic value, or possibly enhancing it, each moral agent needs to perform certain tasks that allow her to deal ethically with the loss of those whose demise will affect the agent.

By taking the proper steps to prepare for the death of others for whom we care in some way, then we take positive action to minimize as much as practical the harmful effects of what will happen to our intrinsic moral worth. We need:

1. An education to aid us in understanding what is happening so that we do not feel bewildered and completely powerless: We learn through our own experiences of deprivation and loss;
2. To have an appropriate philosophy of life grounded on flourishing: How best to live one's life should include confronting our own or other's death;
3. Help in building a sound community practice at the time of death: Sometimes they are religious rites that allow one to cope (Jackson 1974, 11).

If it is a loved one who dies after a long illness or at a great age, then grief counseling and establishing support networks for the survivors before the person's end of life could be mandatory for the survivor, provided that the survivor needs them.[21] It is understandable that people facing this situation would fail in this duty, but if they respect their intrinsic value as they should, then they should take whatever reasonable measures are required to ensure their flourishing. The fact that it is a traumatic time can mitigate the duty in some ways—perhaps the experience is too overwhelming for a fragile psyche—but in general, the moral agent should

[21]Sandy Bem spent five years helping her family become accustomed to her death from suicide. Bem had been diagnosed with Alzheimer's, and decided that she did not want a life in which the disease had progressed too far. Bem's daughter, Emily, stated "It was just so obvious that this is about as good as it gets for a human exit…She was surrounded by everyone who loved her, they were telling her how and why they loved her. This is not a bad way to go." (Spiegel 2014)

ensure her flourishing as a moral agent, as long as doing so does not violate the Practical Principle.

In addition, the various types of person who make up the complex whole of the overall individual, including but not limited to the psychological and embedded person, should protect their moral worth as the entities that they are. This preparation demands putting into place the structures that will help them limit the damage done to them. In Oregon's physician-assisted suicide example, survivors felt that their loved one's final wishes were honored, were more prepared, accepted the death more readily, and were less likely to have regrets about how the death happened than those whose loved ones died of cancer or amyotrophic lateral sclerosis (Ganzini et al. 2009). These results could only have happened after careful planning and involvement of all stakeholders, which helps prepare them somewhat for the loss that is coming.

For those individuals faced with a more sudden, unexpected loss of a loved one, there is less of an obligation to create and use complex coping and recovery mechanisms. However, once again, every reasonable person is or should be aware that death is something that cannot be avoided. Depending on the death probability for the person for whom the individual cares as determined by reasonable people making reasonable decisions, there is an obligation to accept the fact that one's loved ones could die in an unexpected way. This duty becomes clearer and stronger based on the loved one's lifestyle or activities. For example, if the loved one likes to take a lot of risk, such as rock climbing in dangerous places or only eating meals with high cholesterol content, then the potential survivor has a greater duty to prepare for the death than someone who cares for a more prudent person. Hence, if the death does occur, the person's shock and intrinsic value decline can be mitigated, at least marginally.

Besides emotional responses, people have obligations to prepare themselves for the death of others, especially if there are dependency relations and vulnerable people involved.[22] Since their loved one's deaths will be a financial strain in many cases because people have not saved enough, then they might have to put money away to prevent financial insecurity. Funerals, for example, are expensive. If they have to be paid for by going into debt, then funeral costs can hinder the survivors' ability to have financial security, which is part of their flourishing. Compare the case of going into debt with one in which the funeral costs are saved for a little bit at a time over a number of years. The savings account, or burial insurance, can cover the funeral and related expenses so that vulnerable survivors do not have to place themselves into financial difficulties up to and including jeopardizing their financial health, which in turn could affect their ability to have positive mental states and avoid negative ones. Life insurance and other forms of resources should also be obtained in order for an individual to protect herself and those for whom she cares. Living and other wills need to be made so that the survivors are not unduly

[22] Islam has long held these financial duties, as well as those that eliminate the person's moral debts whilst improving who she is as a person (Muwahidi 1989).

burdened. Hence, one has a plethora of duties to prepare for the deaths of others for whom the person cares. To act otherwise would be imprudent at best, and she would not respect her true intrinsic worth. There will be other duties, but instead of providing a laundry list of them, these examples give adequate indication of what they are and why they exist.

6.6 Conclusion

It might seem very odd that people have duties to feel a particular way about the death and dying of others. After all, in such chaotic times, any way the person feels should be legitimate merely because the person is feeling it.

However, our primary obligation in life is to thrive and help others do likewise. Thriving, in part, requires us to be certain types of people who behave in a certain type of way. In order to be those people and be motivated to act as we should, we need to have the appropriate emotions and feelings. Since death and dying situations are often chaotic, even with the best preparations we can make, it is incumbent upon all of us to prepare for them so that we not only survive the emotional onslaught caused by them, but also persevere and become even better as persons living flourishing lives.

References

Aeschylus. 1970. *Agamemnon*. Trans. H. Lloyd-Jones. London: Prentice-Hall International.

Anscombe, G.E.M. 1961. War and Murder. In *Moral problems*, ed. J. Rachels, 393–407. New York: Harper Collins.

Aristotle. 1985. *Nichomachean ethics*. Trans. T. Irwin. Indianapolis, Hackett Publishing.

Armon-Jones, C. 2003. The thesis of constructionism. *Philosophy & the emotions,* ed. S. Leighton, 181–203. Peterborough: Broadview Press.

Barcan Marcus, R. 1980. Moral dilemmas and consistency. In *Moral dilemmas*, ed. C. Gowans, 188–204. New York: Oxford University Press.

Baron, M. 1988. *Remorse and agent-regret*, 259–281. XIII: Philosophy.

Conee, E. 1982. Against Moral Dilemmas. In *Moral Dilemmas*, ed. Christopher Gowans, 239–249. New York: Oxford University Press.

Dante Alighieri. 1971. *The divine comedy Vol. I: Inferno*, Trans. M. Musa. New York: Penguin Books.

DeGrazia, D. 1998. Biology, consciousness, and the definition of death, *Report from the Institute for Philosophy and Public Policy*, Winter/Spring.

Felice, C. 1990. On the obligation to keep informed about distant atrocities. *Human Rights Quarterly* 12: 397–414.

Ganzini, L., et al. 2009. Mental health outcomes of family members of Oregonians who request physician aid in dying. *Journal of Pain Symptom Management* 386: 807–815.

Greenspan, P.S. 1974. Moral dilemmas and guilt. *Philosophical Studies* 43: 117–125.

Greenspan, P.S. 1995. *Practical guilt*. New York: Oxford University Press.

Jackson, E.N. 1974. Grief. In *Concerning death: a practical guide for living*, ed. E.A. Grollman, 1–12. Boston: Beacon Press.
Kavka, G.S. 1974. Wrongdoing and guilt. *The Journal of Philosophy* 71(18): 663–664.
Kerrigan, M. 2007. *The history of death: burial customs and funeral rites, from the ancient world to modern times*. Guilford: The Lyons Press.
Knobe, J. 2003. Intentional action and side effects in ordinary language. *Analysis* 63: 190–193.
Konigsberg, R.D. 2011. *The truth about grief*. New York: Simon & Schuster.
Kubler-Ross, E. 1969. *On death & dying*. New York: Simon & Schuster.
Kubler-Ross, E. 1972. *Questions & answers on death & dying*. New York: Simon & Schuster.
Kubler-Ross, E. 1975. *Death*. Englewood Cliffs: Prentice-Hall.
Morris, H. 1987. Nonmoral guilt. In *Responsibility, character, and emotions: new essays in moral psychology*, ed. F. Schoeman, 220–240. Cambridge: Cambridge University Press.
Muwahidi, A.A. 1989. Islamic perspectives on death and dying. In *Perspectives on death and dying*, ed. A. B et al, 38–54. Philadelphia: The Charles Press, Publishers, Inc.
Neblett, W. 1974. The ethics of guilt. *The Journal of Philosophy* 71(18): 652–663.
Nussbaum, M. 1985. Aeschylus and practical conflict. *Ethics* 95: 233–267.
Radzik, L. 2009. *Making amends*. Oxford: Oxford University Press.
Safer, J. 2008. *Death benefits: how losing a parent can change an adult's life—for the better*. New York: Basic Books.
Spiegel, A. 2014. How a woman's plan to kill herself helped her family grieve. NPR, http://www.npr.org/blogs/health/2014/06/23/323330486/how-a-womans-plan-to-kill-herself-helped-her-family-grieve. Accessed 3 July 2014.
Statman, D. 1995. *Moral dilemmas*. Amsterdam: Rodopi.
Twain, M. 2012. *Autobiography of Mark Twain: Volume 1*. Ed. H.E. Smith et al. Berkeley: University of California Press.
Verhagen, A.A.E. and P.J.J. Sauer. 2005. End-of-Life Decisions in Newborns: An Approach From the Netherlands *Pediatrics*, http://pediatrics.aappublications.org/content/116/3/736.full. Accessed 21 June 2014.
Wessel, M.A. 1996. When children mourn a loved one. In *Facing death*, H.M. Spiro, ed. M.G. McCrea Curnen, and L. Palmer Wandel, 77–80. New Haven: Yale University Press.

Chapter 7
Is There a Duty to Die?

Abstract Most discussions about someone terminating her own life tend to focus either on utilitarian considerations or rights. For the latter, a number of people argue over whether we have a right to die, which I take to mean an entitlement to choose to end one's own life. I see at least two problems with the right to die approach. Unfortunately, many arguments and positions are based primarily on the intuitions of those engaged, and therefore often turn into intuition pump battles. Secondly, basing arguments on rights gets us mired in the same problems that any rights talk encounters including, but not limited to, whether they are positive or negative, whether they can be overridden, and what justifies such a right. In this chapter, I am going to argue for a position that makes worries about a right to die moot. If there is an obligation at certain times and in certain circumstances to die, then a right to die is irrelevant. Moreover, if an obligation to die is possible, then it necessarily follows that choosing to die is morally right on some occasions. Doing what is right generally cannot be proscribed by others so it does not matter if there is a right to die.

7.1 Introduction

Besides those duties to oneself that every agent has to prepare for, an accompanying moral obligation goes against numerous moral intuitions and conventions. Many think that people should fight to survive for as long as they can.[1] The value and normative theories developed here support that claim in part by saying that when an intrinsically valuable entity dies, then intrinsic value is lost which, if unnecessary, should not happen. The more intrinsically worthy the entity that ceases to exist, then the greater the negative alteration in the value of the organic wholes of which the entity was a component. Moreover, unnecessary loss of those entities with intrinsic value cannot respect the worth those entities possess.

---

[1]Islam, for instance, prohibits all euthanasia "because it deprives the person of the right to life and the possibility of action". (Muwahidi 1989, 48).

D.R. Cooley, *Death's Values and Obligations: A Pragmatic Framework*,
International Library of Ethics, Law, and the New Medicine 62,
DOI 10.1007/978-94-017-7264-8_7

However, there are instances in which a person *qua* moral agent or one of her organic wholes has a moral duty to die.[2] How this duty is fulfilled depends upon the situation. At times, it might entail that the person takes no further action to remain alive. At other times, more active means, such as suicide, can be prescribed. Since it is the most difficult obligation to establish and entails that there can be a duty to allow oneself to die, I will focus on active suicide duties.[3] Moreover, if there is a duty to suicide or die, then it must be the case that ending one's life earlier than medically necessary is at least permissible in certain circumstances.[4]

Writing about duties to die is fraught with emotional and professional pitfalls. The most obvious of these is being labeled as an unreformed Nazi eugenicist if there is even a hint in one's writing that life need not be extended as far as medical science will permit. Those who make the stronger claim that there is a duty for some to terminate their lives one moment earlier than "nature" requires have reduced their chances to have their ideas heard, much less rationally analyzed. In one example of how unconventional views are attacked on emotional grounds, Felicia Nimue Ackerman (2008) claimed in an op-ed in her local Providence newspaper that certain bio-ethicists want you dead, especially one particular ill, older woman, naming four male bioethicists, including Daniel Callahan, John Hardwig, and me. Lost in Ackerman's reasoning was why anyone, male or female, could rationally believe that moral obligations exist for ending one's life or for an analysis of the actual positions being attacked.

These automatic intuition-dependent responses hinder legitimate discussion of such an important issue, especially in the time of limited medical and social resources, as we have seen from politicians' exploiting non-existent death panels in the American Affordable Care Act health care debate. If we are interested in solving real world problems, then we know that instead of stalking straw man positions which allow one group to be easily demonized as powerful, abusive males versus weak, ill females, healthy youngsters versus the sick elderly, or any of the other illicit dichotomies of "us" versus "them," the philosophical argument each person develops must be scrutinized. For example, in her op-ed, Ackerman does not mention female thinkers who have supported suicide or the people rationally choosing to die, such as Margaret Battin. The view of Dena S. Davis is based on the notion of life narratives: if a person's life has been written by her in such a way that suicide is a fitting outcome of her narrative, and remaining alive in a certain state contradicts that narrative, then taking one's life can be obligatory. That is, this duty would exist for only those with narratives that require suicide for consistency's sake, and it would not apply to those without such a life story.[5]

---

[2]A more standard way to approach death is to argue that it is permissible to die or that there is some sort of right to die. Lebacqz and Engelhardt, for example, argue that there is a *prima facie* right to die (Lebacqz and Englehardt 1986, 85).

[3]I will later argue that allowing oneself to die is generally a form of suicide.

[4]Of course, it is probable and plausible to believe that there are far more instances of shortened life being permissible than rising to the level of a moral obligation.

[5]I might be asserting too strong a position for Davis, but it seems in keeping with her arguments.

Second, Hardwig's duty-to-others argument for a death duty, for instance, has striking care ethics elements (Hardwig 1997). One of the necessary criteria in his reasoning is for the person to be enmeshed in a network of caregivers for whom life has become burdensome. The motive for the suicide duty, therefore, is not merely to maximize utility or beneficence or minimize maleficence but to show love for those who bear the burden of taking care of the individual. By his action, the person who kills himself is saying, "I care for you too much to keep making you shoulder this burden that brings such harm to your life. It is to you I have a duty to die." Although we might disagree with the argument, at least we can understand that it is not implying that, for our selfish purposes, we should send our sickest, most vulnerable community members to some sort of sterile execution room in order to cut costs that prevent us from our trivial pursuits.

In what follows, the arguments are not designed to produce a single, efficient defense against all counter-examples and moral intuitions. Rather, the aim is to provide reasoning that at least one reasonable person would think plausibly and probably entails the conclusion, and that other reasonable people can understand as a reasonable position. In this chapter and the next, I will argue for the existence of duties to die or to suicide based on RPU and QCI combined as the Practical Principle.

7.2 Who Should Die When There Are Limited Resources?

A more subtle pitfall than overreliance on intuitions is to assume that more positive options are always open to the agents performing their actions, even in the face of real world dilemma situations. Christine Overall argues against duty to die obligations based on the existence of better or ethical alternatives than killing oneself. I will first explicate Overall's position and why it works in certain contexts, and then show why, overall, it is too optimistic. Though rare, morally obligatory deaths are sufficient to defeat the conclusion that no suicide can be mandatory.

Overall's work is must reading for anyone who wants to think seriously about death duties, especially given her detailed examination of apologism and prolongetivism. Overall rejects the view of apologism, which accepts death and condemns deliberative attempts to prolong life (Overall 2003, 16). Apologism pessimistically assumes that the "possibility and attainability of high quality living near the end of life" is too small a moral factor to outweigh the potential negative conditions of an unhealthy, low-functioning existence (Ibid., 44). As a result, those who hold this view tend to favor the right to die and obligations to hasten one's own or other's deaths through either active or passive means, as in the cases of suicide and mercy killing.

Overall's view is prolongetivism, which states that measures should be taken to overcome ageism and to work diligently to extend people's healthy, high-quality lives, both of which are highly desirable goals in relation to individual and social thriving. Instead of merely accepting the inevitable and doing nothing to prolong a good life, one should actively pursue the continuance of lives that are clearly worth living. Her argument in favor of prolongetivism is roughly:

1. Many people want to live longer.
2. This life is the only one we ever have.[6]
3. Many old people have been deprived of life's goals.
4. Continuing to live offers the prospect of ongoing opportunities for further experiences and action.
5. *Apologist alternatives to prolongevitism are untenable.*
6. Therefore, prolongetivism is stronger than apologism.
7. Hence, we should rationally adopt prolongetivism over apologism (Overall 2003, 122).

A hint of utilitarianism in other parts of her argument states if people live healthy, high-value lives, then the overall happiness of society will increase markedly (Ibid., 44). Thus, not only is the individual's value as a person respected, society's flourishing is pursued through prolongetivist policies.

Leaving aside the assumption that this life is the only one we will ever live, Overall is correct when she states that many old people have been denied the opportunity to flourish. Perhaps they lived in poverty, which should be interpreted broadly to mean a severe lack of the physical necessities of life as well as poverty in emotional relationships, education, and whatever else is required to make a flourishing existence.[7] On the grounds of respecting human beings and justice, most reasonable people would agree that if there are goods that can be shifted to help the poor achieve flourishing, even though it might mean fewer resources being used for good but not vital outcomes, then there is rational reason to redistribute. Therefore, those who use the status quo to propose the existence of death duties need first to reconsider the possibility of amelioration.

In addition, apologists who argue against any reasonable extension of life seem much like Luddites with their rejection of all technology developed after their childhood. Reasonable people would ask themselves, "Why would prolonging a life that is lived in comfort, health, and well-being be morally wrong to do?" (Overall 2003, 191). Eliminating or reducing such an existence would be a disvalue not only to the person, but also to the community. To incorporate Overall's view into the organic wholes value theory developed earlier, the loss of the intrinsically valuable psychological person, the embedded person, and the devaluing of the organic wholes of which the various people are a part must factor into any ethical decision. Since we care about value and respecting it, reason tells us not to waste it when doing so is driven by an irrational ideology.

[6]If there is an afterlife, then the force of this and other arguments depending on a total cessation of existence will have to be rethought. After all, if the afterlife is likely to be much better, then it would be rational on at least egoistic grounds to pursue self-death.

[7]Poverty is not merely about financial wherewithal. Some people are raised in a financially poor situation but have flourishing lives.

In fact, Overall argues that apologists fail to recognize moral reality and its complexity when evaluating real world situations involving many people and their individual feelings and desires.

> If an eighty-year-old woman wants to continue to remain alive simply in order to spend time with her children and grandchildren, read books, watch television, and enjoy sunshine on a warm spring day, I defy apologists, such as Callahan, to show that such a desire is unreasonable, unjustified, or immoral. (Overall 2003, 51)

Within the scope of the situation, it would be difficult, at best, to prove that such a desire manages to satisfy any one of Overall's three criteria, much less all of them. Of course, the mere fact that the desire is neither unreasonable, unjustified, nor immoral does not entail that one should act upon it, *ceteris paribus*. In crunch situations, as defined by Garrett Hardin, extreme measures that try to bring about the best in situations in which all the alternatives are bleak might be warranted, but the main contention that a person living a good life is morally valuable in the right way holds a lot of reasonable appeal if intrinsic value means as much in a decision procedure as it should. In general, elderly people have positive relationships and have had an enormous number of experiences of the sort listed by Frankena that have improved their intrinsic worth as psychological persons as well as increasing the worth of the organic wholes of which they are members.

How then does one adopt, much less firmly support, a view in which some elderly people have a duty to die either by active or passive means? Overall's arguments are very strong; taking a stance against her can generate internal moral squeamishness, especially since it is literally a life and death issue for some of our most vulnerable community members. The path, I contend, is to agree with her by saying that extenuating circumstances show that deaths, even for those who have fulfilling, highly valuable psychological lives, can at times be morally required. In what follows, I will first show that Overall does not give sufficient consideration to pertinent circumstances that would affect the moral calculation on death obligations, and that Hardwig's position on a death duty can be expanded to avoid the charge Overall makes against him.[8]

First, suppose that in a particular situation, there are only sufficient goods to allow five people to live, but there are 10 people who need the resources. If all 10 people attempt to use the goods, then every one of the 10 will die. Even if four commit suicide so that the others have more time to live, the remaining six are doomed to failure from too few resources. Also assume that each of the 10 is morally identical in regard to desert, merit, future ability, and any other morally relevant feature or combination of features that might otherwise give us an opportunity to say that one is better than another. That is, the values of their psychological and embedded persons are the same. By removing the factors that allow for unlike beings to be treated differently, this will need to be a judgment of

---

[8]Hardwig could agree with Overall, and then state that his criteria already include a condition that the person will not have the flourishing life Overall describes.

equal among equals. Although simplistic, this is the actual situation faced in many cases in regard to health resources in both the developing and developed world.

The definition of a life worth living is assumed but often unspoken.[9] A person having a physical life worth living has her basic physical needs of food, hydration, and shelter met, but no more than that. Let us call this the ***subsistence life level***, which does not include flourishing, for that would require many more resources to have the agent's psychological and other needs of life satisfied as well. Perhaps the subsistence life can be compared best to that of Peter Singer's notion of a marginal-utility life, in which a loss of one intrinsically valuable thing will make the life not worth living.

Overall's argument works for each individual in the case above when considered as a discrete individual person, but it does not succeed when evaluating the situation in a larger context or as a whole. If we consider each individual as an individual, then we would say that each should be kept alive using the resources she needs, but we are not incorporating what happens in the overall circumstances if each person's needs are met for a short time. It is true that likes are being treated alike. However, when we consider the overall situation and the limitations set in place, matters become much more difficult to decide as happens with too few medical resources, such as transplant organs, that need to be distributed among many people. Not everyone can have all that is needed, so some rationing or prioritization, as occurs in private and public insurance programs, is justified; otherwise, the system becomes overburdened and fails. For this case, even though each discrete individual as an individual should receive the resources, it is unclear what we should do ethically for the group as a whole. We do know that difficult decisions must be made in this case that will not only fail to prolong life, but may hasten death as a consequence.

In a case showing a different perspective, suppose that we have the same resources, but only five people who need them. The options here are between subsistence living for all and a more flourishing existence for fewer people. That is, four people could be capable of flourishing if they took the fifth's portion and used it for themselves, which, of course, would be fatal to the fifth person. The four's lives would thrive in the manner Overall stipulates when she states:

> In considering the possibility of extending life, I am not referring to mere biological survival [subsistence level of life]; I am speaking of the continuation of life with the human capacities for emotion, perception, thought, and action in tact to at least some minimal degree. (Overall 2003, 15)

Although this definition is vague, it is sufficient to put in context what kind of existence qualifies as a flourishing life. It entails that the life is "human life lived in at least a minimum of health, comfort, and well-being," which would require a number of the resources that are mandated by a subsistence living (Ibid., 191).

9This ambiguity is a mistake: To leave a central term of this importance too vague or ambiguous will make disagreements rife as people talk past each other instead of discussing the same issue.

Overall's type of flourishing life requires more intrinsic worth than one in which only the most basic needs are met.

In the case of the five people in the subsistence level of life versus the four flourishing at a higher level, the extra resources are sufficient to give the four the energy to do more activities that allow their psychological and other needs, such as security related to health, comfort, and well-being, to be fulfilled. In other words, they are not staying barely alive; they have many more intrinsically valuable states of affairs than they otherwise would have. The additional states will, in turn, contribute greatly to the psychological and embedded persons' intrinsic worth and hence allow them to thrive. Unfortunately, the fifth member of this group will die as a result of the four's depriving her of sustenance, but the survivors' life values will be greatly improved in some significant ways. The ethical question is whether it is morally permissible for the four to take the resources from the fifth if they know that doing so will kill her.[10] The answer is clearly no.

These two cases show a disconnect between Overall's view and the positions of those she criticizes. Her focus is on situations dealing solely with the second type of case. The only difference is that all the individuals occupying her arguments are capable of having flourishing lives, but someone wants to have more than is necessary for such an existence.

> But longer lives, freed of the fetters of ageism and supported by medical research to sustain the quality of life, would open up the potential for activity and accomplishment by those whom we now regard as old. (Overall 2003, 36)[11]

This is a fine, caring idea with which no reasonable person in her right mind would disagree. Overall is correct for these cases.

However, the first type of case is more problematic for prolongevitism. What happens in crunch situations in which resources are limited for medical research and care, or when others cannot have the basic goods they need for subsistence living, much less lives that flourish? First, we must acknowledge that decisions have to be made for groups divided by age, global position, social membership, resource availability, health care need, and a significant number of other variables which will impact the ability and instantiation of the intrinsically valuable states of affairs laid out earlier. The very young and very old require a great deal of medical resources, while those in the middle age groups, in general, do not need as much due to their relative good health. Furthermore, the developed world uses an enormous amount of its resources on medical care, whereas the developing world does not, probably because the poor there lack ready access to it. If we consider each group in isolation, then we can see the force of the argument for medical resources being used for them as individual groups in isolation to receive as many goods as they need.

---

[10]Some will argue for the permissibility of the four taking resources from the fifth, possibly on the grounds of maximizing utility, but the quantitative versus qualitative issue is not fundamental to this discussion.

[11]Even in cases in which there are insufficient resources, Overall has an optimistic outlook for society's ability to overcome this difficulty.

The elderly, as elderly, should get them. The very young, as very young, should get them. However, there are competing groups in the real world, which is a far larger organic whole that needs to be considered when we think about scarce resource allocation. Given that resources are insufficient to meet the overall needs, much less the demands of each set of people, then it becomes apparent that difficult, tragic decisions have to be made; some people will not receive the care they need to prolong their life as much as possible.

Overall contends that money being moved from military spending, tax incentives for the wealthy, and corporations will be adequate to pay for the additional health care needed to make the 10 people cases into five people cases (Overall 2003, 88–9). She also states that improving health care delivery would also generate revenues. To be fair to Overall, she acknowledges that:

> Within the limits of this book, it is not possible to evaluate all the alternative priorities that society might adopt, and social needs and goals could justifiably vary from nation to nation. But it is also not obvious that we must assume, a priori, that various priorities must necessarily be incompatible with each other. Moreover, if priorities can be changed, part of the case for doing so must be that the old priorities are no longer (as) important. (Overall 2003, 36–7)

Overall is correct that incompatibility cannot be established using mere a priori reasoning because many times multiple goals can be satisfied simultaneously while working for one end, as in the case of poverty and crime reduction. Evidence of the mutually beneficial result of lowering poverty and crime rates can only be established through empirical means. Moreover, due to the inter-connectedness and inter-relatedness of social functioning and actions, unforeseen benefits and costs can accrue from one policy change as it affects relationships, markets, and other social forces. As with the environment, we should never forget that we cannot do just one thing. Whatever action we take will have ripple consequences that will affect, either simultaneously or over time, other nodes and relationships in the overall organic wholes of a particular society and the world, thereby altering the whole. Scientific research is necessary to determine how things actually work out in real societies and other organic wholes. We have seen that well-meaning efforts based on a priori reasoning about central, universal components of a moral issue, such as human nature, have produced plans that failed miserably, while carefully crafted, empirically justified schemes have led to positive results.

It is acknowledged that not enough consideration is given to redistribution and broader thinking in decision processes. Asserting that these revenue streams can address the costs of giving everyone the health care they need for a flourishing life, even in wealthier developed countries, lacks evidence that such resources would actually cover costs. An economic analysis would have to be completed to show, first, that there is sufficient revenue, and second, that the diversion of it from other areas would not be so detrimental to the society as a whole that it would be unable to produce the health care resources in the future. It is easy to say that corporations should not receive incentives because that fits nicely with certain ideologies against the rich and powerful being able to exploit the weak and vulnerable,[12] but it needs more than a priori evidence that acting in this matter will be able to achieve the goals we seek rather than destroying the economy and harming everyone much more than they otherwise would have been.

In addition, we have to be pragmatic when talking about changes in social policy. It might be the case that in a more perfect world, these policies can be altered with few negative consequences, but in our less than ideal reality, that might not be the case, as indications from the United States' delayed attempt to cover uninsured citizens through the Affordable Care Act give evidence. Not only is there a drive afoot to repeal many of the health care law's beneficial features, but also portions of it seem to be less efficient than alternatives, such as single payer system, would have been. Although it is likely to protect citizens in the long run when they no longer have private insurance, many are opposed to single payer healthcare or mandated health insurance. This is not a unique situation. In all societies, different political factions have power to different degrees, which they use to pursue ends that they believe to be in their self-interests. If it is practically impossible to alter society to move resources from, say, the military, then it is more pragmatic to discuss what actually can be done in the real world than to ponder how the world should be in less likely circumstances.

Although it might be called ageist, discussing the relative value of lives is necessary for making social policy decisions when using the government to create and support policies. Overall argues that comparing deaths between the young and old, as many do, is illegitimate. Instead of the badness of a death being dependent upon a person's age and the unfilled potential the person has:

> Present prospects and the potential for the future are the most significant [factors of death.] On the one hand, the present and future prospects of enjoying a high quality of life are a good reason to want to prolong one's life and are a justification for taking steps to promote the extension of that life. On the other hand, having already enjoyed a high quality in the past does not make one less entitled to a longer life. (Overall 2003, 49)

As stated previously, if considering the organic wholes of an elderly person as a psychological person and lower-level embedded person to the corresponding organic wholes of a young person, then the elderly person will have far greater intrinsic value than the young person. She possesses mental states and other intrinsically valuable states of affairs, such as relationships, that, from sheer number alone, outweigh the young person's. Quality of lives value would also favor her in this regard.

However, social resource distribution requires examination of much larger organic wholes, such as those of societies and worlds. The problem is that some

[12]It should be noted that if we are going to have discussions about redistribution, then there cannot be a privileging of developed countries over developing countries. The mere fact the former has great wealth and power does not mean that individuals lucky enough to be born in them deserve the resources they take for granted as a right. I take it this is a point that Battin makes in her argument for some people to no longer take resources if they have already had their fair share.

elderly people have already used a great deal more resources than children have just because the former have been around longer. If everyone has an equal right to a flourishing life, not merely a right to life, and it is impossible to have flourishing lives for all, then someone will have to lose in the competition for scarce resources. In these contexts, having had a high quality life ***does*** alter the moral desert of the person who had it, especially when comparing that life to one who has not been afforded the same opportunity. That is, the comparison would be made between someone who has always lived in poverty compared to one who has lived in abundance. A somewhat similar analogy can be drawn to someone who has had an excellent supper, and wants another one tomorrow. If that person reserves the resources required for a meal, which are the last resources that can be used for another person's sustenance today, then when another person is waiting to fulfill her basic dietary and aesthetic needs today, we would be right to condemn the former person's thoughtlessness toward and disrespect of the other, as well as what it says about someone's character to take excess when others are in need. Now, if there were plenty for all, no reasonable person would be upset rationally over the excess food intake of the first person, but when there is insufficient for everyone, then taking more than one needs shows a callous disregard for the needy.

Of course, in the case of scarce resource allocation, if the elderly person has not had a flourishing life while the younger person has lived life to an excess, then reasonable people would be more likely to say that the elderly person should not be sacrificed to save the younger. The reason for this conclusion is clear. By excessively frontloading them, the younger person has wasted the resources she needed for a flourishing life, while the elderly person has not been able to have her basic needs met in order to thrive. Both the desert and flourishing life factors should be considered in any moral principle utilized to make decisions about medical resource allotment, but even then, it will pertain to relatively few cases in comparison to the overall social reality. If the majority of society's elderly have had flourishing lives, or at least the opportunity for such which was wasted by them, then in general, resources should be allocated to those groups who more need them to procure their own thriving lives.

Another limitation on resource allocation is based on the group's size. Since there are insufficient goods to be distributed according to need, we must maximize the impact of what little there is. For example, some illnesses have very few victims and considerable costs to alleviate, if possible. Given that breast cancer affects more people than, say, Unbalanced Translocation Chromosome 15, resources might be better spent on the former than on the latter. The tradeoffs will be based on how many people are helped by the distribution, to what degree they are benefited, and how they should be benefited. There is no one right way to determine which is the best outcome because there are three variables to be maximized in some way, and not everyone will agree on a particular decision no matter what it is. However, if the choice made is something a reasonable person would reasonably make or understand, then it is morally defensible.

7.3 Is Choosing to Die Morally Identical to Choosing Suicide?

Michael Cholbi recognizes that suicide's standard definition relies too heavily on a mistaken notion of intention and how it works in suicides. Steven Luper, for example, defines suicide as "an act performed by someone who expects, correctly, that it will be fatal, and who does it in order to die." (Luper 2009, 174). Others have included the intention clause as necessary for an act of taking one's life to count as a suicide (Beauchamp 1986, 326; Fairbairn 1995, 84; Margolis 1975, 26). Although it would make discussions about suicide far easier if we could require *intention to suicide* or something sufficiently similar to help sort takings of one's own life into suicides versus not suicides, it is not pragmatic to force reality to fit the theory rather than the other way around. As Cholbi shows in various examples, a person does not primarily have to intend her own death for an action of self-killing to be a suicide, although the person has to do more than merely foresee the end of her life. The result is a fuller understanding of what suicide is so that we can better evaluate its morality for far more cases of choosing to die, devise improved treatments for those who are suicidal, and possibly change attitudes about suicide's permissibility. At the very least, we can eliminate unnecessarily narrow conceptions about the taking of one's own life that stymie progress in our moral and social interactions.

Although Cholbi goes a long way in improving how we think about suicide and intentions, I contend that his definition should be broadened.[13] Suicide is the deliberate taking of one's own life, but there is no requirement to make the deliberation fit too high of a standard that will limit the work we want to do with the term. The non-rational acquiescence definition developed below is plausible, and challenges Western traditional views of what counts as suicide.

7.3.1 *Is It a Suicide Only if It Is Intended as a Suicide?*

Before the question about whether a suicide must be intended can be answered, it is necessary to understand what it means to intend something. In too many instances, what an intention is tends to be deemed common or intuitive knowledge. As a result, no formal attempt is made to help understand exactly what the person means when she is using the word. This, in turn, leads to people speaking past each other instead of understanding the actual positions and arguments.

Contrary to common practice, I will investigate suicide's definition in this section (since it does so much work when trying to determine if an action is truly a suicide), whether suicides can be rational, and how people think about suicides when it comes to this action. The proposed definition of intention found below is

[13]The approach I am taking here is similar to Peter Windt's (1980) Wittgensteinian analysis.

one which I think can be readily substituted, without change in meaning, wherever suicide is formally discussed.

Intentions are very complex mental states of affairs and are made even more difficult to understand as a result of their intimate connections with or interdependence on other mental states of affairs, such as motivations. However, it is safe to say that one necessary part of any intention is some sort of desire. When someone intends to do something, then there is a commitment on the person's part to bring about the state of affairs. That is, if I intend a state to come about, I have to want in a certain way for it to come about, and that desire has to be a strong enough motivation to cause me to carry out an action to achieve my desired result, all things considered. However, intentional desires[14] are not merely desires or wants, such as sexual wanting or a longing to go abroad, which have no particular intended object. They are also not purely motivational desires, which are the undirected desire reasons that cause individuals to form intentions, such as a hatred causing someone to plot revenge against the hated party.

In addition, Alan Donagan argues that in order to intend a state of affairs, the agent must have some sort of plan to achieve the state. Intentions must be directed toward a sufficiently specific end rather than be like motivations, which are directionless. This does not mean that the agent's plan needs to be completely formulated or "full-blooded" as Donagan calls it (Donagan 1987, 94). An agent, for example, may not have time to develop a detailed plan; or some areas of the scheme may have to be left vague by design, since not all of the situation's circumstances are known; or there may be a lack of time to adequately prepare. Sometimes, plans call for improvisation as the agent goes along.

Donagan describes a chess match between Karpov and Kasparov, which serves as an illustration of why improvisation is necessary and why intentions need not have a complete plan of action in order for them to be intentions.[15] Suppose that Karpov intends to win a chess game, and there are various ways that he can carry out his plan. Since Karpov does not know how Kasparov will respond to Karpov's moves, he must be ready to improvise on the spot in order to win the game. The real world, in general, is like that. Unless the situation involves very few variables, then much like playing chess games, plans of action in the world are too complex to be fully worked out in advance or must be contingent because they depend on what others will do, but there is still some planning done in order for the mental state to be an intention (Donagan 1987, 99). This planning, no matter how minimal, is what allows us to rationally attempt to satisfy the motivating desires rather than merely committing random physical events in a haphazard manner.

[14]These are not desires formed intentionally, but rather desires that commit the agent to carrying out the intention.

[15]This is also relevant to suicides and other takings of life.

Taking into account how people use the word and the need for a pragmatic definition, I propose that:

S intends X = df.

1. S desires or wants X to come about,
2. For all S knows, X is a state of affairs that she can bring about either alone, or with the assistance of others or a group,
3. As a result of her motivating desires and reason, S is committed by her intentional desire to bring X about either on her own, with assistance from others, or as part of a group,
4. S has some sort of plan for achieving X.

The second condition recognizes that to intend something, the person has to believe it is possible for that state to be brought about by her in some way. If any of the four conditions are not met, then S does not intend X to come about.

Before moving on to the definition's advantages, it is worthwhile to note an element that will be of greater focus later in this chapter. Because S has chosen to bring X about and planned for it, S has ***approved*** of X in some way. Again, we have Hume's notion of approval and disapproval, or more precisely, emotions being the motivator to action. I do not mean that the agent approves of every part of the means or end itself. Instead, I maintain that the agent approves of the conjunction of the overall means and ends in some meaningful way when she intends to bring about a specific end that is part of the conjunction.

One of the definition's advantages is that it recognizes the fact that an agent can intend X to come about, even when the intended state is never obtained. Perhaps an agent intends to take a trip to Europe, but circumstances are such that she never is able to go because her life is cut short. Though she never fulfills her intention, she still has it up until she ceases to exist. This feature captures a reality about how our minds work. We can formulate intentions, and they can lie dormant until a time in which they can be carried out. At times, they will or can never be carried out, but they still exist as intentions.

If an agent intends to suicide—that is, she fulfills the definition's conditions—then no one would disagree that the person has suicided. That is, the person intends to suicide if and only if she wants for the state of affairs in which she commits suicide to come about, she believes that she can bring about this state of affairs primarily on her own, she is committed to bringing about this state of affairs primarily on her own, and she has a plan for bringing about this state of affairs. If the person does this, then it is a very clear taking of one's own life as a case of suicide.

Do suicides have to be intended in this particular way or in any other way that we may define "intention"? My definition of intention has four rather stringent requirements for what counts as a suicidal or other intention. However, taking one's own life might be a suicide but it might not involve the agent explicitly planning on taking her own life or being overtly committed to taking her own life as an intended suicide would. For example, the person might desire an end to her pain, and that is the state of affairs she seeks to bring about rather than intending the state of affairs in which she is dead by her own hand. Committing suicide is not intended as such by her, one might argue, because it is merely part of the plan but not the intended outcome. In addition, Kant and others talk about individuals sacrificing themselves for others, which appear to be suicides for all intents and purposes without the stigma attached by being labeled as taking one's own life. Below, I will also argue that those who choose to die by refusing treatment or committing some other act of omission are suiciding, although it might not be as explicitly obvious as those who intentionally suicide through an act of commission. Therefore, the suicides class includes more self-killings than those committed with an intention to suicide. These also need to be part of any adequate definition of suicide.[16]

7.3.2 *Suicide as Intentional but not Intended*

Our suicide definition should not be too exclusive or too inclusive. According to Cholbi, suicide is self-killing, but merely killing oneself does not entail that a suicide has occurred. Those who smoke, eat the wrong types of food, or do not exercise can be thought of as killing themselves, but few would think that they are committing suicide, even though they are likely shortening their lives. To commit suicide, there has to be a greater commitment to the person's death than that. More precisely:

> (S) Suicide is intentional self-killing: a person's act is suicidal if and only if the person believes that the act, or some causal consequence of that act, would make her death likely and she engaged in the behavior to intentionally bring about her death (Cholbi 2011, 21).

Although Cholbi spends much time defining all the relevant terms in this formulation, the most interesting part is what it means to intentionally bring about one's own death without intending one's demise.

Cholbi has a way of understanding intentionality that links the notions of intention and intentional in a significant way. An outcome, such as death, may be intentional, although it is not itself intended as required by many psychologists and others working in the field who define what suicide is. That is, a person can intentionally kill herself yet not intend that she dies. "A person's self-killing is intentional just in case her death has her rational endorsement in the circumstances in which she acts so as to bring about her death" (Cholbi 2011, 28). This definition fits Cholbi's intuition about a case he labels "Foxhole Jumper" (Ibid., 16). The jumper is an unfortunate soldier who throws himself on a live grenade in order to save the lives of other members of his unit. Before he performs his leap, Foxhole Jumper is aware of the consequences of his action, but does it anyway because he

[16] I am not going to argue that there is one definition of suicide.

also understands that if he does not act in such a manner, then his comrades will be killed in the blast.

According to Cholbi, Foxhole Jumper commits suicide without intending it. In fact, although the soldier did not intend the state of affairs in which he dies as a result of his heroism, his death was intentional because of his rational endorsement of that probable outcome.

> He knows that dying is the near certain result of his jumping on the grenade, and absorbing the grenade blast is his means to saving his comrades. And in that split second in which Foxhole Jumper jumps on the grenade, the foreseeability of his death almost certainly informs his deliberation, even if that deliberation takes place so rapidly as to nearly escape his notice. And unless he is also gravely depressed, for instance, his death counts in his mind as a reason *against* jumping on the grenade. He jumped on the grenade despite this danger. In essence, Foxhole Jumper endorses his own death - neither as a means nor as an end, but as a foreseeable result of the act he chooses as a means: a death that has his rational approval. (Cholbi 2011, 27)

Some unpacking of this rather long quotation is useful in seeing what needs to be done in order to have a rational endorsement of one's own death. Firstly, the person must foresee that his action will be a significant cause of his death. If Foxhole Jumper thought that he would merely receive a serious but non-fatal injury, then there would be no rational approval of his death. It is also here that we see that the death need not be intended by the person. Secondly, the person must put the situation into some form of context he understands. In this case, it is saying that if he throws his body on the grenade, then his death will rescue his unit's members. Finally, there is a weighing of the evidence, although very brief, for the two alternative actions. If the Foxhole Jumper does not make his leap, then his unit's members will die and he might or might not die or be injured. If he leaps, then he will expire, but the members of his unit are spared. From these two alternates, he chooses the latter, which implies that he accepts that particular action with its foreseeable and intended consequences over its competitor.

Cholbi seems to need the rational endorsement condition in order to account for the intentionality part of his definition of suicide. Rejecting the position that a death has to be intended for it to be a suicide requires Cholbi to find a similar enough internal mechanism to link the agent to her self-killing so that the action is a suicide. That is where intentionality comes into play. "Everything a person intends, or that is part of her intention, whether her end or means, is thereby intentional. But not everything intentional is thereby a component of a person's intention" (Cholbi 2011, 27). Foxhole Jumper does not intend his death according to intention's definition, so it cannot be a suicide on those grounds. Since Cholbi still wants to call what Foxhole Jumper does a suicide, then the death has to be intentional but unintended. Hence, the claim that Foxhole Jumper rationally endorses his own death appears to be necessary to do the work Cholbi wants it to do.

7.3.2.1 Is Foxhole Jumper's Death Actually an Intended, Rather than an Intentional, Suicide?

Roderick Chisholm and Henry Sidgwick argue that the intended action is indeed intended, but the conjunction of the intended action and certain foreseen consequences are also intended. Therefore, Foxhole Jumper, as well as others, cannot direct his intention only to the good effect of saving his comrades from being blown to bits by the grenade; he must also intend all the foreseeable results of his action, which includes his death. Since Cholbi claims that the Foxhole Jumper gives his rational endorsement to his death, he must foresee it in a way that would make it part of the conjunction of the state of affairs the Jumper intends and every other effect he foresees. Of course, according to Sidgwick and Chisholm, if the agent intends the conjunction of the intended and foreseeable, then he can be held morally responsible for both the intended and foreseeable consequences of his action. If they are correct, then Cholbi's Foxhole Jumper intended the overall state created by his action of jumping on the grenade. We would, therefore, need only intention's definition and a conjunction rather than an additional criterion of commitment to classify each and every self-killing as a suicide.

Given that it might provide a simpler and more useful way to classify the Foxhole Jumper's action as a suicide and provide some insight on how to classify most choices to die, let us consider Sidgwick and Chisholm's claim about overall intention and the argument for it. Sidgwick contends that an agent intends all foreseen effects that he may not desire as well as those that he does. "It is most convenient to regard 'intention' as including not only such results of volition as the agent *desired* to realize, but also any that, without desiring he foresaw as certain or probable." (Sidgwick 1907, 60). According to this view, it is not merely convenient to regard intention in this way; for as Sidgwick later says, if an agent intends one effect and other consequences are foreseeable, probable outcomes of his actions, then he intends the conjunction of the outcome desired and the foreseeable. (Ibid.). Therefore, the probable foreseeable effects are not merely ***intentional***; they are ***intended*** as part of the overall state of affairs that will be brought about by the agent's action.

Chisholm states much the same position with his Diffusiveness of Intention Principle, although he does not require the person to believe that the foreseen consequences are certain or probable. The principle is:

> If a rational man acts with the intention of bringing about a certain state of affairs p and if he believes that by bringing about p, he will bring about the conjunctive state affairs, p and q, then he *does* act with the intention of bringing about the conjunctive state of affairs, p and q. (Chisholm 1990, 636).

According to the principle, intention is not distributive over the conjunction. That is, the rational agent may intend the conjunction of a state of affairs he intended to bring about and those states that are merely foreseeable results of his action, but that does not entail that the agent intended each of the foreseeable states of affairs on their own or in their own rights (Ibid.). The agent may consent to the foreseeable consequences, but consenting is far different from intending (Ibid., 637). For instance, intending appears to require a stronger commitment lacking in mere consent.

To use one of the examples Chisholm describes, an agent may intend to drive off in a car parked at the corner. The agent also foresees that in driving the car, which does not belong to him, he will be driving another man's vehicle. According to Chisholm, the potential driver intends to bring about the state of affairs of driving the car, and he intends to bring about the overall state of driving the car parked at the corner and driving the vehicle which belongs to another man. However, the driver does not intend to bring about the state in which he drives another man's vehicle because that is merely a foreseeable effect of the action to drive the car.

Although initially appealing in its simplicity, a significant problem with this claim is that it does not adequately capture the agent's intentions in these matters. As George Pritcher writes, "It is a strange logical principle that makes so important a distinction between two conjuncts on the sole basis, apparently, of the order in which they happen to be written." (Pritcher 1970, 667). The fact that an agent intends one state to occur and foresees another state resulting from his action does not entail that he intends the overall state of affairs or the conjunction of effects. Saying otherwise is a misuse of the word "intends", especially if "intention" is defined with my four conditions, or given any of its standard meanings. In the actual world, there is neither commitment nor desire to see the conjunction come about in these situations.

In the Foxhole Jumper case, the agent merely intended the action of jumping on the grenade but did not intend the consequence of his death in conjunction with saving his comrades' lives. Therefore, this is an example of taking one's own life which is a suicide but cannot be considered an intended suicide because the mental states involved in regard to the self-killing do not rise to the level of satisfying intention's definition for suicidal actions.

7.3.3 Intentional Suicide and a Broader Standard

Although Cholbi builds a strong argument, we should be a bit concerned about the level of rationality and commitment Foxhole Jumper has to make Cholbi's definition work.[17] We can stipulate the certain mental states that run through Foxhole Jumper's head before he makes the decision to act, but in the real world, can such a thing actually happen? Given that most grenades detonate in five seconds, and from that meager handful of seconds, we must deduct the time the grenade spent getting

[17]Part of this argument appears in my review of Cholbi's *Suicide* at *Notre Dame Philosophical Reviews*.

to the foxhole in the first place, there is not a great deal of time to do the necessary foreseeing and evaluating required for a rational endorsement.[18] Let us stipulate that Cholbi's definition can be done in some cases, but there are others in which there is insufficient temporal resource to meet his standard. Although we know that five per cent of attempted suicides report that they spent less than one second forming the intention to kill themselves, they are in significantly different situations (Simon et al. 2001). The Jumper is not suicidal, whereas the others have already thought about it even though they have not formed the intention that will lead to action.

Even with a lack of time, we still want to say that Foxhole Jumper acted as a moral or rational agent in the situation, especially if we think that he acted in a way that was supererogatory. If he did not act, but merely exhibited behavior the way that a trained dog would, then we cannot claim that his action was right, wrong, permissible, obligatory, or supererogatory. His action definitely would not be heroic if it was merely a reflex born from an uncontrollable impulse on his part.

How can the Jumper's behavior rise to the level of an action while still accounting for the very limited time and other constraints which could very well make it impossible for there to be a rational endorsement? My suggestion is that we broaden what counts as "intentionally" to include not only rational endorsement but non-rational acquiescence as well. Non-rational acquiescence can do the same heavy lifting as rational endorsement does in regard to classifying actions as suicides while not running into the same difficulties.

Rational endorsement is too stringent a standard in the Foxhole Jumper case on the grounds that, in many situations, there is no real time and insufficient information to make a rational endorsement. By a rational endorsement, I am assuming that the evidence on the whole sufficiently supports the endorsement over its competitors. An endorsement is irrational if the evidence on the whole sufficiently shows that the endorsement is defeated by a competing endorsement with far superior evidence. Perhaps there is an endorsement that is better supported given the actual situation. A non-rational endorsement is the result of insufficient evidence for or against an approval. In this particular case, the agent's desires provide primary motivation to make the endorsement or not, rather than the approval being based solely on the evidence. Non-rational endorsements happen in cases in which the evidence is too evenly matched or where there is insufficient evidence to conclude one's endorsement is supported over the others. What drives these approval types is primarily desire or emotion based, such as being based on care for oneself or others.

A much stronger case can be made that Foxhole Jumper's endorsement is non-rational, but is being made by a rational or reasonable person. Given that Foxhole Jumper is a rational person who uses appropriate means to achieve his goals, it follows that his behavior is an action. We can say that the action is rational in that manner. However, the lack of time and information for reason to do its job of weighing evidence makes it unlikely that Foxhole Jumper is influenced by reason

[18]This lack of time objection can also be used against Chisholm and Sidgwick's position. It is difficult to see that there is an intention formed in these cases.

sufficiently to make a rational endorsement. At best, his approval is non-rational because it is based primarily on his desire to save his unit members' lives. He cares for them in some motivating way; therefore, he wants to save them even though it will cost him his life. Foxhole Jumper is a rational agent and his reasoning makes his action rational, but the reason he acts is primarily because of care or some similar incentive.

More importantly, it is hard to believe that Foxhole Jumper endorses his own death. An endorsement shows some form of positive feeling toward whatever is being endorsed as well as approval, as we can see in political and product testimonials. The person doing the endorsing likes whatever is being endorsed and is trying to persuade others to feel the same way. Endorsing, therefore, is a very active and positive thing to do. If Foxhole Jumper endorses his death, then he is not merely accepting the inevitable.

Although it is possible for the Jumper to feel this way, it makes little sense on the grounds of how morality works for us. Our moral platform and full blown morality incline us toward preserving rather than ending our lives. Based on life's value to us, we surrender it reluctantly and with regret only in the most pressing circumstances, although we still consent in some way to the ending of our lives as a permissible action or necessity for our morality. If Foxhole Jumper had a chance to tell us what he feels about his death, then it is unlikely that he would give such positive authorization of it. Although he would approve his expiration in some way because he has chosen to do it, it does not seem as if he would endorse his own death.

We need, therefore, something that commits Foxhole Jumper to his sacrifice so that it will qualify as a suicide, but not so strong that it is an endorsement of his death. Acquiescence is the solution to our problem. Acquiescence merely shows that the person accepts what happens to him but does not think positively of it. It is an agreement/commitment without the positive power of an endorsement.

S acquiesces to Y, a conjunction of X and Z, = df.

1. S intends X,
2. S foresees that Z is a likely effect or means of bringing about X,
3. S is committed to bringing about X even though S knows or rationally believes that bringing about X will likely result in Z or Z is a means to X,
4. S passively accepts Z for the sake of bringing about X.

Acquiescence is a compliance which implies that the agent has chosen to do something. In the case of acquiescence, the agent chooses to passively accept the foreseeable means and effects and the overall state of affairs, thereby making himself morally responsible for the whole, although he does not intend the whole. Acquiescence is also distributive over the conjunction. When we acquiesce to an overall state of affairs, then we acquiesce to each foreseen component of the overall state regardless of whether each component is intended. When we get vaccinated against the flu, for example, we do not rationally endorse the pain from the inoculation, but we do acquiesce to it so that we can get the benefit of not being laid up for a week with a nasty bout of the illness. We accept the pain for the benefit we reap.

Acquiescence better captures the morality and reality of Foxhole Jumper's situation than does rational endorsement. The Jumper would never endorse his own death because that would not recognize his true value as a person and individual. If he felt positively toward his death the way an endorsement requires, then some of the worth of the actual sacrifice Foxhole Jumper has made is lost. Autonomously endorsing what is about to happen to him makes it somehow better than what it should be; he is undervaluing himself or is trying to dismiss some of his own value through the use of his autonomy. Of course, at times, autonomy allows us to sacrifice our self-interests for others or for other ethical pursuits and reasons, but autonomy does not have the power to forgive the agent for not recognizing his worth and treating it in the proper manner. That is, we can waive our rights, and by doing so, we forgive the negative value associated with violations of those rights, but we cannot illicitly devalue ourselves while waiving our rights because that is an unforgivable offense. We cannot disrespect the very thing that makes us persons in the first place.

The Jumper foresees that his death is a means to the end of saving his friends, and he acquiesces to it as a likely result of his intended action. By recognizing that he will die if he continues, and accepting through acquiescence that state of affairs with the gravity it deserves, then what is actually happening when one soldier performs such heroism is more accurately captured. This is a true sacrifice that recognizes what will be lost and why it is being given up. Foxhole Jumper is merely resigning himself to his chosen fate for his bravery above and beyond the call of duty. He is not waiving any of his value or devaluing his true worth. Hence, if we adopt non-rational acquiescence, then Foxhole Jumper's action is still a suicide, but one that fully captures what happens in the sacrifice.

I will not provide a definitive definition of suicide here for a very good reason. Given the various disagreements by some of the very best thinkers on the subject, there is excellent justification to believe that there is not one definition that will work, but this result should not be surprising. Pragmatically, we need to acknowledge that there are many definitions of suicide that do work in the situations human persons need them to work in. At times, the focus might be on mental health professionals who need to function efficiently at their jobs. The popular belief is that all suicides are insane or mentally disturbed to such a degree that intervention is justified (Battin 1996, 4). It is the case that changes in personality are at least strong predictors of depression, which can influence rationality (Chow and Roberts 2014). It is also true that many of those contemplating suicide lack autonomy and rationality as a result of immaturity or intellectual impairment or psychological disturbance, such as depression or psychosis; therefore, their choices and wishes stemming from these states need not be honored (Fairbairn 1995, 179–80). If the person is not sufficiently rational, then mental health professionals clearly have a role to play.

However, one definition does not always work in all situations we encounter because human behavior is so very messy in its artificial complexity. The definition appropriate for the medical community working with people seeking their help or committed to them might not work as well when considering larger contexts, such as those affecting societies, that will require a broader definition. Peter Windt argues that the intention to die is not always a wish to die, and a wish to die is not necessarily a significant contributor to a death that occurs. Therefore, there is a real danger of labeling all suicides and potential suicides as a behavioral or medical condition, which would force us to intervene to save a "victim" rather than appreciating that the person is rational and making a decision that should not be interfered with (Windt 1980, 46). At times, the mental health professional definition does not work, and we need to find one better able to function in the circumstances.

Given the above, a general understanding of what suicide is will be helpful here. In order for an action to be considered a suicide, the person significantly causing the action to happen acquiesces to her death being a means to or a reasonably foreseeable consequence of achieving an end she intends her action to achieve, and the action actually does cause the person to die. Of course, this understanding will be attacked for being far too inclusive, but it does have the benefit of including all actions that any reasonable person might think of as being a suicide.

A controversial result follows from the broadened definition of suicide: those individuals who refuse treatment while knowing the results of such an action, or who choose hospice that does not extend their lives to their "natural" end, commit intentional suicide. Suppose we have a man who has decided that he will no longer fight his cancer. He has undergone two rounds of radiation and chemotherapy, but he is told by his physician that another is required, and this one might not produce a remission. On the grounds of believing he has led a good, flourishing life (and that is not likely to be the case in the future) and a strong desire to avoid putting himself and his family through any more of the process, he refuses further treatment. It is a suicide because he intends the refusal of treatment, foresees that refusing treatment will cause his death sooner rather than later, is committed to ensuring no further treatment is received by him even though he knows refusing will likely hasten his death, and passively accepts his death as a result of his refusal. So, although his death is not intended, it is still intentional.

Palliative care is often preferred over euthanasia but there is good reason to believe that palliative care is a form of passive euthanasia, at least insofar as the former term is often used.[19] For palliative care, further treatment is generally ruled as futile, and the patient's pain and other needs, such as thinking that their remaining lives are worth living and they are not a burden, are managed to the patient's death (Omipidam 2013). There is no attempt to prolong life even though

[19]Sometimes, there seems to be confusion about what palliative care is. If it is merely a treatment or management of various kinds of pain from various sources, then it is not a form of euthanasia. Palliative care, in this sense, will also apply to a number of non-terminal cases, and is generally preferred by the Practical Principle if everything that is intrinsically valuable is respected as such and it is likely to result in the best outcome. However, if palliative care means an alternative to prolonging life, and that seems to be what it is in many proponents' arguments (Pereira 2011; Omipidam 2013), then it is a form of passive euthanasia even if its primary intentions are directed toward managing pain and making the person feel positive about herself (Steffen and Cooley 2014). The main difference between it and active euthanasia is the difference between passive and active euthanasia.

that is something that could be done and life could be extended. Passive euthanasia, on the other hand, sometimes includes faster dying processes such as no longer hydrating the patient. The only true dissimilarity between palliative care and passive euthanasia in these cases is the method used or the practitioner's primary intentions, but the end results are the same and known by those involved. In addition, the outcome of death is intentional because there is at least a non-rational acquiescence to it as the highly probable result.[20]

Although many suicides perform an act of commission to end their lives, acts of commission are not essential to the notion of what a suicide is or its permissibility. Both omission and commission can be the intentional self-taking of life. There is no moral difference with regard to the rightness or wrongness of the actions between an act of commission in which a person ends her life through an overdose or asphyxiation, and those in which she allows herself to die (Brock 1998; Prado 2008; Tooley 1980).[21]

How, then, should we account for the many reasonable people who maintain that there truly is a moral difference between acts of commission and omission in the takings of one's own life, such as Holland? (Holland 1969, 80).[22] I think that the mistake stems from a confusion between determining the rightness or wrongness of an action and the act's goodness or badness. In James Rachels' famous example of Smith and Jones killing their respective six-year-old cousins in order to inherit the cousins' estate, both acts are clearly wrong, but there actually is a moral distinction between them (Rachels 1975, 78–80). The respective actions of drowning and allowing a child to drown in both cases are unethical and forbidden, but the act of commission is significantly worse because of the mental states that have to exist in each. A person who can hold a child underwater must be constantly aware of what he is doing as he is doing it. He has to overcome the constant, forceful perceptual data coming into his mind that should be generating empathy. Unless he is a master of self-distraction or disassociation, he has to be exquisitely and immediately aware at every moment that he is murdering a child as he presses the child's head underwater and the boy thrashes about to save himself, yet he does not feel what the vast majority of people would feel in the same situation.

---

[20] A question arises as to what "true" suicides from acts of commission actually intend. My suspicion is that they intend the cessation of pain and suffering or some other end rather than their deaths. If this is true, then it becomes even clearer that these cases are morally identical and therefore should be labeled as suicides.

[21] Gavin Fairbairn argues that refusal of treatment is a case of suicide, whereas jumping on a grenade is not (Fairbairn 1995, 114–7). What makes the two morally different is that in the latter, the agent really has no choice in the method of saving his friends; therefore, he was compelled in a significant way. Of course, the problem with Fairbairn's 'only one method' argument is that the grenade jumper had the option of preserving his own life, even though that would have resulted in his friends' deaths. Therefore, the grenade jumper committed suicide because there was a different, legitimate option open to him at the time.

[22] Some writers argue that there is a moral distinction between the two. J.P. Bishop, for instance, claims that since voluntary active euthanasia's purpose is to die, then there is no meaningfulness in the dying process as there is in passive euthanasia (Bishop 2006, 223).

In the case of allowing a drowning, the person could distract himself from what is happening, although, as Rachels says, the person is monitoring the situation to see if he needs to step into finish the job, so to speak. In this case, the focus could be on other things, such as what he is going to do with the money, how the bathroom's tile is patterned, or some other immediate item his mind can center his thoughts upon, while the monitoring is more indirect than in the case in which the agent directly confronts the victim by holding the child's head under water.

Returning to the simulation theory of mind reading, a person who is actively drowning a child has to block the empathy and intentions that allow him to be a moral person in the first place. This individual is much more of a moral monster in a way that someone who is thinking about other things is not. In the latter's case, there is not a direct rejection of what makes us capable of being moral agents, although it is far from a laudable situation.[23] Both murders are wrong and evil, but the one who performs the act of commission is much worse than the other.

7.3.4 *Different Types of Suicide*

Cholbi's Foxhole Jumper suggests an interesting avenue of inquiry that might eventually result in conclusions that are too worrisome to adopt, but the pursuit of the issue could help our understanding of what suicide is and its morality. Foxhole Jumper is a sacrificial suicide that is set apart from those in which the agent takes her own life primarily for her own sake or needs and desires or a lack of ability to have her needs met: egotistical and anomic suicides, respectively (Durkheim 1951). Given the existence of sacrificial and anomic suicides, do other types of suicide not fit the common understanding of the term—killing oneself when overwhelmed by some mental illness or distress—but qualify as taking one's life in the relevant way?

First, there is a distinction between irrational and rational suicides. Irrational suicides are morally wrong actions or neutral events. For the former, an agent has done something wrong by choosing a course of action that is prohibited by the evidence available to the agent and by morality at the time in which the agent makes the decision. For example, she is diagnosed with a fatal, painful disease by a doctor she knows does not specialize in that area, but based on such flimsy information, she decides to take her life to avoid what she believes will be an unbearable existence. What is interesting about these self-killings is that we can hold the agent morally responsible in a manner of speaking even though she acted irrationally. As a rational and reasonable agent, she should do what is right, especially in situations of such importance. By failing in her duty of due diligence in gathering information and making the best decision to her and her narrative, she

[23] I think that this difference is what motivates some to use the Doctrine of Double Effect to distinguish cases of permissible and impermissible actions when they are really considering better or worse situations and people.

has disrespected her true worth as well as not doing her best when she could have easily done so. Hence, her action is morally wrong.

A different form of irrational suicide, of course, is that driven by mental illness or some other disabling condition that renders the person unable to make rational decisions as a moral agent in regard to her medical health.[24] In these situations, the entity can still be a moral agent overall or in other areas in which the person's judgment can be rational, but is merely a moral subject when it comes to those issues involving her mental incapacity. The entity is a moral subject because her decision-making ability is too limited or defective to produce rational or reasonable outcomes. Hence, what the moral subject does as a moral subject is not an action but a physical event. This fact entails that these suicides are neither morally right nor wrong, and may be permissibly interfered with by outside agencies, such as the state.[25]

What a rational suicide is seems to be a matter of debate among those who believe that suicides can be rational, although everyone who argues that there are such things implies that they are morally permissible.[26] The universal rightness of rational suicides might stem from how they are defined. For Prado, rational suicide is:

> A. when suicide is a genuine option for the agent, chosen after deliberation consistent with accepted standards of reasoning and unimpaired by error, false beliefs, or lack of relevant information;
> B. *and* the agent's motivating values are cogent to others, not unduly contravening the agent's interests;
> C. *and* suicide is in the agent's interests, not causing more harm than continuing to live. (Prado 2008, 40)

Thus, a rational suicide is an ethical suicide. Basically, the decision to take one's own life is rational because the decision procedure is perfect and it is understandable to others, so the decision is reasonable and it passes the ethical egoist test of being in the agent's interests.

Although Prado's definition does apply to a number of suicides, the standard might be far too high to capture all rational suicides. Two immediate problems for Prado's view are that many of our major life choices involve false beliefs and sometimes lack relevant information because we are imperfect reasoning beings

---

[24]The mere fact that a person is mentally ill in regard to an unrelated condition or at risk of suicide does not automatically render the person incompetent when it comes to making decisions about medical treatment or suicide. Moreover, it does not automatically permit outside agents, such as physicians or the state, to interfere in legitimate decision making (Callaghan et al. 2013, 384).

[25]Neill Allen argues that "suicidal citizens have the right to require public authorities to take reasonable precautions to avoid their death if they ought to know of a real and immediate risk to life." (Allen 2013, 356). Apparently being suicidal is sufficient to eliminate autonomy, but not the positive rights associated with remaining alive the individual is uninterested in having fulfilled.

[26]Jeanette Hewitt notes that the rationality of suicide seems to be linked to whether it is performed to alleviate extreme physical or mental suffering. Suicide to stop extreme physical disorders are more likely to be justified than are those from mental suffering, in which it is automatically assumed that psychological conditions render the person incompetent. As a result, intervention is more often thought to be obligatory in the latter when not justified in the former (Hewitt 2013).

living in a less than perfect world. [27] For all of that, we still act rightly or wrongly and rationally, even though we have failed to satisfy a higher standard. As we saw in the Foxhole Jumper case, the Jumper is a rational person making a rational decision overall. Clearly, he has insufficient time and information for his decision, as well as a choice that is probably not in his best interests.

Perhaps it would be best to make the conditions for rational actions and being rational much more general. For Battin, rational suicide is examined in light of any other rational action performed by a rational being. To be a rational action, then the agent of the act must:

1. have the ability
   a. to reason;
   b. a realistic world view;
   c. an adequacy of information;
2. avoid harm;
3. act in accordance with his fundamental interests. (Battin 1996, 115)

Suicides can be rational if all five conditions are fulfilled but, as Battin recognizes, how these criteria are understood is often a matter of individual ideologies. For instance, those who believe that all suicides are irrational will argue that a person contemplating suicide is unable to fulfill conditions 1a and 1b, perhaps due to mental illnesses or disturbances such as depression.

Although there are objective elements to each of Battin's criteria, we should not forget that subjective evaluations about the person's life by the person are not only morally relevant but should receive considerable weight and respect by others in the decision procedure. "[P]ersons are the best judges of the proper balance of values in their lives. An element of respecting others is respecting their right to take their lives in the absence of contravening duties." (Lebacqz and Englehardt 1986, 85). How people live their lives is generally up to them unless there is an overriding interest to prevent them from doing so, as in the case of enforced seat belt use being a social good of sufficient interest that the state may intervene in people's behavior in their own cars. This relative permissiveness based on how a person views the value of his own life is not a new approach to the matter of dying and suicide. In Athens, all a citizen would have to do is plead his case before the Senate that his

---

[27] Brandt's approach has the same problem as Prado's: "The basic question a person must answer, in order to determine which world-course is best or rational for him to choose, is which he *would* choose under conditions of optimal use of information, when *all* of his desires are taken into account." (Brandt 1975, 69).

Although an ideal way to make a decision, Brandt's standard is far too high for humanity. It is asking us to act as we would as perfectly rational people in a perfect society, but we are clearly not perfectly rational creatures with perfect access to all relevant information, yet we are still moral agents.

life was intolerable, and he would receive permission to take it if the Senate found him competent:

> If your existence is hateful to you, die; if you are overwhelmed by fate, drink the hemlock. If you are bowed with grief, abandon life. Let the unhappy man recount his misfortune. Let the magistrate supply him with the remedy, and his wretchedness will come to an end.
> (Libanius as cited in Alvarez 1998, 65)

We should note that the Greek process seems to fulfill Battin's five conditions. In pleading his case, the individual shows to like citizens his reasoning. They must judge his ability to reason, whether he has a realistic world view as determined by him, an adequacy of information, that his death will avoid harm through the elimination of hateful existence, and whether he is acting in accordance with his fundamental interests of no longer suffering. What is missing is a requirement to prove that what he is doing is something that others would do, an action that the majority would approve of, or a requirement to establish that by an objective standard, his life is not worth living. That is, for the latter, it does not matter what most others think about the value of his life or its potential using some objective standard. Instead, it is what the rational and emotionally unhappy individual thinks about it and whether he can get the Senate to agree.

It is clear that the mental states of the agent performing the action will alter the reality of any situation, as was argued previously. Suppose, for instance, that there are two worlds with two different agents. The only difference between the worlds is that one agent has one mental state that is not the same with the other agent.[28] Let us further suppose that each of the mental states is relevant to the situation in which the agent decides to act and then acts. In our example, each agent is obligated to fulfill a promise to a loved one. In the first agent's case, the agent resents having to do what she swore she would. In the second, the agent is pleased to be discharging her promise. The actions appear to be identical, but we would not be wrong in saying that they actually are different. First, they have different values based on the state of affairs. The agent with the bad attitude makes her world less valuable as a result of her negative view, whereas the other agent improves her world's worth, *ceteris paribus*. Second, when we describe the actions, we have to do so differently if we attempt to capture the whole of the action. The first action is performed with ill-grace, but the second is one that is more laudable. In fact, they fall under different classes of actions. Perhaps the difference is not as great as would be the case had the intentions been different, but what the act truly is cannot be the same if the moral values of the two are not the same because of the agent's mental states. Hence, all suicides share a sufficient number or set of characteristics to fall under the class heading of suicide, but the various particular actions will be unique beyond the simple variables of being done in different times and spaces. Each mental state involved in creating and performing the action must be identical which, given the unique characteristics of human persons, is possible, but highly unlikely.

---

[28] I think this example might be too simplistic. In order to have the different mental states, other states would have to be altered between the worlds so that the former mental states can exist.

Given the variety of mental states that can be associated with a suicide, it is clear that each one is unique in a significant way because there is at least one important variable it instantiates that is not shared in common with all other takings of one's own life.

Although each suicide is unique, it is possible and useful to create broad categories of different types of suicide based on the mental states involved. For instance, motives matter. An agent might be motivated by her honor, patriotism, courage, a desire for self-affirmation of liberty or morality, a distaste for life, pain, humiliation, or loneliness (Minois 1999, 48–9). Agents might take their lives from "a shattering loss of someone or something highly valued." (Phipps 1987, 73). Some kill themselves as an act of revenge (Ibid., 76). There are prudential and altruistic suicides, the latter being exemplified by the Foxhole Jumper case (Luper 2009, 176). Care for others seems to play a large role in a number of takings of one's own life. Derek Humphry writes about some elderly who "Quietly, unceremoniously, with tremendous care not to inconvenience or shock others … self-destruct." (Humphry 1998, 313). As we will see in the next chapter, those who have strong care relationships with others often do not want to be a burden on them, and commit suicide primarily out of care.

We should also recognize that people's motivations to die or suicide could be pluralistic; there could be a variety of motivations without any one of them being primary or sufficient on its own. Hence, the types of self-killings increase as each relevant variation is included.

Other mental states relevant to the action, such as intentions, will have an impact on what the suicide is in particular and how it can be classified in general. Some might simply intend to die. The vast majority can have a far more complicated set of intentions, such as intending to take one's own life in order to punish those who the suicide believes have been insufficiently supportive or harmful to them and to prevent the suicide's life from continuing to be too great a burden to him. Again, we have to recognize the complexity of these types of human actions.

Suicides can also vary based upon how they are performed. Both assisted and unassisted suicides are self-takings of one's own life, but in the former case, the taking has to be aided and abetted by another person. Perhaps the suicidal agent needs to gain access to medications that only the second person can obtain. There are also preemptive, surcease, assisted surcease, and requested or voluntary euthanasia suicides (Prado 2008, 29). Preemptive suicides avoid a medical condition the patient does not want to instantiate. Surcease ends the life of someone with a chronic condition he does not want to live with by the person's own hand, whereas in assisted surcease, the person needs someone else to help the agent take his life at his own bidding. Finally, requested or voluntary euthanasia is performed on the person asking for death, but the person is so disabled that he cannot assist in the method used to kill himself (Ibid.)

Most controversially, there are different types of lives that can be lost; therefore, there might be different types of suicide or losses of lives based on what is being killed. Let us begin by thinking about the possibility that each person has different types of lives. We have already seen those in which the person's physical life is taken, but other lives would also fall under the realm of suicide: mental and moral suicides. First, a mental suicide would be a situation in which the agent ends her mental life as a person. That is, she destroys her mind intentionally or in some internal way identical to Foxhole Jumper's state of mind that allows us to call his action a suicide. A moral suicide, on the other hand, is the taking of an agent's moral life using the same sort of mental states that make a taking of an agent's physical life a suicide. The difference between a moral and mental suicide is that the person retains her mind but has acted in such a manner that her moral life is dead beyond redemption. We can imagine instances in which a person does a wrong and horrifically evil action or set of actions, as a result of which she incurs such a moral debt that she can never have a flourishing life as a moral agent. I will consider each of these in greater detail below.

Mental suicides pertain to instances in which an agent chooses to end his mental life. Although these might seem odd at first, because we focus on physical suicides that take both mental and physical lives by ending the life of the body, it does make sense to discuss the loss of mental life. Often, when someone has entered into a permanent vegetative state (PVS), survivors will remark that the person died the moment she entered that state. What is being said gives comfort to survivors so that they feel less guilty about what has happened or can begin their mourning process in order to get a start on the grieving that is justified when the PVC entity in a permanent vegetative state's body ceases to function as an organism.

Instead, to be a person requires certain abilities, which include but are not limited to being rational, making decisions, and guiding the person's life and actions to attempt to achieve goals the person has. Given that the entity cannot do any of these necessary things required to be a person in the first place, then the entity is no longer a person. That is, the person has ceased to exist, but ceasing to exist is not the entire story. A life has been lost. That is, if life means that an entity is able to interact with its environment, to alter itself to respond to stimulus from that environment, and to have functions as a living entity, then it is alive. Given that persons fulfill these conditions, as well as create their life narratives, they have mental lives, as we can see when people legitimately discuss someone having a rich mental life rather than one that is bereft of intellectual stimulation often thought as necessary to flourishing.

We can also intelligibly talk about someone having a flourishing physical life without mentally thriving and mentally flourishing when the physical life is not capable of doing likewise. In the first type of case, a person might focus too much effort on her body and making it as powerful an engine as possible, which prevents her from spending much time improving her mind. Although her body might be an exemplar of physical fitness, we would reasonably claim that she is not flourishing overall. Since the body is thriving, then it must be something else that is preventing her from flourishing as a whole person. Of course, it is her mental life that is languishing. In the second type of case, a person's mind might be doing very well at the same time her body is failing her. Those who suffer chronic physical diseases, such as **amyotrophic lateral sclerosis,** can have a very active mental existence or life even though their body is incapable of following suit.

How a person commits mental suicide will be very similar to that of physical suicide. In both cases, the suicide might be intended or intentional. In an intentional, self-focused mental suicide, the person intentionally seeks the end of that mental life. For example, some individuals with mental illness choose to forego their medical therapy because of the treatment's ill effects on them. Some describe the state as being in a constant fog without interest in life. Without the drug, the person is fully aware that she will cease to be a rational creature, but the effects of the psychiatric treatment are too great for her to bear. She would rather surrender her rationality and her psychological life so that she can cease to exist as a rational being but continue as a moral subject or a different person who is happy, although clinically insane. The more likely case would be for individuals not to intend this sort of suicide, but merely acquiesce to it when they decide not to continue to exist in their perpetual mental fog induced by their drug regimen.

Moral suicides occur when an agent makes it impossible for herself to exist as a good person.[29] Granted, such suicide cases are rare because the individual has to be irredeemable as a result of the action, but they could occur as long as what the person has done is sufficiently evil that she cannot regain a classification of being a good person. Given that the moral suicide cannot pay her moral debt or do whatever it takes to be good, she is morally dead as a good person. However, she can become even more evil or less evil, but the option of being a good person is no longer open to her.

Although it would be easier for the psychological community and others if all suicides fit into a far more restrictive definition, the truth of the matter is that there are many more self-killing variations than can be fit into a single, simple theory. This is not necessarily a bad result of broadening suicide's definition. It gives us the opportunity to think about our morality in different ways that might be far more pragmatic. For example, for sacrificial suicides, it allows us to value the action more accurately than we otherwise could. These types of taking of one's own life are far more audible than those performed out of a desire to revenge or harm others. Different types of suicide allow us to better evaluate and formulate plans on how to prevent, treat, or deal with them.

7.3.5 A Potential Problem and a Conclusion

Given suicide's negative cache, many would prefer to limit suicides only to those actions in which the agent clearly intended to suicide. By broadening what counts as a suicide, Cholbi and others like him might be thought to misclassify actions that are unfortunate self-takings of life that do not rise to intentionally killing oneself as a suicide. Those who perform such acts are not being wrongly destructive of their

[29] I have argued in "Crimina Carnis and Morally Obligatory Suicide" that those who would engage in murder, terrorism, rape, pedophilia, genocide, and other morally horrific actions can lose their moral lives.

lives or even doing anything that gives rise to concerns that the person is doing something bad. Since they did not intend their deaths, they cannot have committed suicide or anything else as disgraceful. They are merely sacrificing themselves. Moreover, the survivors can focus on the hero's actions rather than suffering from the thought that she committed suicide to save their or other people's lives or flourishing. If the person simply sacrificed her life, then she has not endangered her soul or committed a wrong in order to do what is right or to bring about a good. Instead, the sacrifice is right and good in itself, unlike suicides which are inherently evil because they are self-destructive.

Although suicides are always bad in the sense that there is a significant loss of intrinsic value and tend to be wrong on the grounds of not respecting that worth, we should not allow our conventional beliefs and emotional reactions about them to serve as a legitimate objection to broadening the class of suicides. One of the reasons many people eschew suicide is not because of reasoned positions but rather from habit and indoctrination into societies and cultures that took St. Augustine's and their religions' arguments against self-killings as fundamental social mores. To challenge that type of thinking causes a great deal of discomfort because it strikes close to the heart of who we are as individuals. These beliefs are central components in our ideologies, so they are part of our individual essence or very important to who we are. To have fundamental beliefs questioned is often to be confronted with questions about our identity and whether some of our most sacred or fundamental ideas about ourselves are true. This disturbs our self-image, but at times such a disturbance is necessary so that we grow to become better people who understand the world more clearly. We have to throw off Hedgehog thinking for Fox cognition. Broadening the class of suicides will cause us to label more actions as suicide, which, in turn, might make us reconsider suicide's rationality and moral permissibility. No matter what we do, however, we should never lose sight of the intrinsic value lost when any decent person suicides.

References

Ackerman, F. 2008. Bioethicists push for your death. *The Providence Journal*. http://www.projo.com/opinion/contributors/content/CT_ackerman30_09-30-08_KTBN7NQ_v10.1a566ce.html#. Accessed 24 Aug 2014.

Allen, N. 2013. The right to life in a suicidal state. *International Journal of Law and Psychiatry* 36: 350–357.

Alvarez, A. 1998. The history of suicide. In *Last rights: Assisted suicide and euthanasia debated*, ed. M.M. Uhlmann, 57–74. Grand Rapids: William B. Eerdmans Publishing Company.

Baier, A.C. 1970. Act and intent. *Journal of Philosophy* 67: 648–658.

Battin, M.P. 1996. *The death debate: ethical issues in suicide*. Upper Saddle River: Prentice Hall Inc.

Battin, M.P. 2000. Global life expectancies and the duty to die. In *Is there a duty to die?*, ed. J.M. Humber, and R.F. Almeder, 1–22. Totowa: Humana Press.

Beauchamp, T.L. 1986. What is suicide? In *Ethical issues in death and dying*, ed. R.W. Weir, 323–329. New York: Columbia University Press.

Bennett, M. 2000. Do we have a duty to die? In *Is there a duty to die?*, ed. J.M. Humber, and R.F. Almeder, 41–60. Totowa: Humana Press.

Bishop, J.O. 2006. Euthanasia, efficiency, and the historical distinction between killing a patient and allowing a person to die. *Journal of Medical Ethics* 32: 220–224.

Bogen, J. 1998. Suicide and Virtue in *Virtuous Persons, Vicious Deeds*, ed. A.E. Hooke, 645–50. Mountain View: Mayfield Publishing Company.

Boyle, J.M., Jr. and T.D. Sullivan, 1977. The diffusiveness of intention principle: A counter example, *Philosophical Studies*: 357–60.

Brandt, R. 1975. The morality and rationality of suicide. In *A handbook for the study of suicide*, ed. S. Perlin, 61–76. New York: Oxford University Press.

Brandt, R. 1992. *Morality, utilitarianism and rights*. Cambridge: Cambridge University Press.

Brock, D. 1998. *Life and death: Philosophical essays in biomedical ethics*, 144–183. Cambridge: Cambridge University Press.

Callaghan, S., et al. 2013. Risk of suicide is insufficient warrant for coercive treatment for mental illness. *International Journal of Law and Psychiatry* 36: 374–385.

Callahan, D. 1975. Vital distinctions, mortal questions. In *Arguing euthanasia*, ed. J.D. Moreno, 173–195. New York: Simon & Schuster.

Chisholm, R. 1990. The structure of intention. *Journal of Philosophy* 67: 633–647.

Cholbi, M. 2011. *Suicide: The philosophical dimensions*. Ontario: Broadview Press.

Chow, P.I., and B.W. Roberts. 2014. Examining the relationship between changes in personality and changes in depression. *International Journal of Psychiatry* 51: 38–46.

Davis, L.H. 1979. *Theory of action*. Englewood Cliffs, NJ: Prentice-Hall Inc.

Davis, D. 1998. Why suicide is like contraception. In *Physician assisted suicide*, ed. M.P. Battin, R. Rhodes, and A. Silver, 113–122. New York: Routledge.

Davis, D. 1999. Rational suicide and predictive genetic testing. *The Journal of Clinical Ethics*, Winter: 316–323.

Donagan, A. 1987. *Choice: The essential element of human action*. New York: Routledge & Kegan Paul.

Durkheim, E. 1951 *Suicide: A study of sociology*. Ed. George Simpson, trans. J.A Spaulding and G. Simpson, Glencoe: Free Press.

Fairbairn, J.G. 1995. *Contemplating suicide: The language and ethics of self harm*. New Fetter Lane: Routledge.

Hardwig, J. 1997. Is there a duty to die? *Hastings Center Report* 21, no. 2: 32–42.

Hewitt, J. 2013. Why are people with mental illness excluded from the rational suicide debate? *International Journal of Law and Psychiatry* 36: 358–365.

Holland, R.F. 1969. Suicide in *Talk of God*. Royal Institute of Philosophy Lectures, Vol. 2, 1967–8, ed. G.N.A. Vessey, 72–85. London: Macmillian.

Hume, D. 1740. *Treatise of Human Nature*. London.

Humphry, D. 1998. The case for rational suicide. In *Last rights: Assisted suicide and euthanasia debated*, ed. M.M. Uhlmann, 307–316. Grand Rapids: William B. Eerdmans Publishing Company.

Kant, I. 1964. *Groundwork of the Metaphysic of Morals*. Trans. H. J. Paton. New York: Harper & Row Publishers.

Kant, I. 1996. *The Metaphysics of Morals*. Trans. Mary Gregory. Cambridge: Cambridge University Press.

Kissell, J.L. 2000. Grandma, the GNP, and the Duty to Die? In *Is there a duty to die?*, ed. J.M. Humber, and R.F. Almeder, 191–204. Totowa: Humana Press.

Lebacqz, K., and H.T. Englehardt Jr. 1986. Suicide and covenant. In *Ethical issues in death and dying*, ed. R.W. Weir, 345–353. New York: Columbia University Press.

Libet, B., et al. 1983. Time of conscious intention to act in relation to onset of cerebral activity readiness-potential: The unconscious initiation of a freely voluntary act. *Brain* 106(3): 623–642.

Luper, S. 2009. *The philosophy of death*. Cambridge: Cambridge University Press.

Margolis, J. 1975. *Negativities: The limits of life*. Columbus: Charles E. Merrill Publishing Company.
Minois, G. 1999. *History of suicide: voluntary death in western culture*. Baltimore: John Hopkins University Press.
Muwahidi, A.A. 1989. Islamic perspectives on death and dying. In *Perspectives on death and dying*, ed. A. B et al., 38–54. Philadelphia: The Charles Press, Publishers Inc.
Omipdam, B.A. 2013. Palliative care: An alternative to euthanasia. *BMJ Support Palliat Care* 3: 229.
Overall, C. 2003. *Aging, death, and human longevity*. Berkeley: University of California Press.
Pereira, J. 2011. Legalizing euthanasia or assisted suicide: the illusion of safeguards and controls. *Current Oncology* 18(2): e38–e45.
Phipps, W.E. 1987. *Death*. Atlanta: John Knox Press.
Prado, G.C. 2008. *Choosing to die: elective death and multiculturalism*. Cambridge: Cambridge University Press.
Pritcher, G. 1970. 'Intending' and side effects. *Journal of Philosophy* 67: 659–668.
Rachels, J. 1975. Active and passive euthanasia. *The New England Journal of Medicine* 292: 78–80.
Searle, J. 1983. *Intentionality*. Cambridge: Cambridge University Press.
Sidgwick, H. 1907. *The method of ethics*, 7th ed. London: Macmillian & Co. Ltd.
Simon, O.R., et al. 2001. Characteristics of impulsive suicide attempts and attempters. *Suicide and Life Threatening Behaviors* 32(1 Suppl): 49–59.
Steffen, L., and D.R. Cooley. 2014. *The ethics of death: Religious and philosophical perspectives in dialogue*. Minneapolis: Fortress Press.
Tong, R. 2000. Duty to Die. In *Is there a duty to die?*, ed. J.M. Humber, and R.F. Almeder, 133–158. Totowa: Humana Press.
Tooley, M. 1980. An irrelevant consideration: Killing versus letting die. In *Killing and letting die*, ed. Bonnie Steinbock, 56–62. Englewood Cliffs: Prentice-Hall Inc.
Windt, P.Y. 1980. The concept of suicide. In *suicide*, ed. M.P. Battin, and D.J. Mayo, 39–47. New York: St. Martin's Press.

Chapter 8
A Duty to Suicide

Abstract Even if a person accepts that it is permissible to end one's life earlier than if all efforts were made to preserve it, there is no guarantee that the person would also believe in an obligation to commit suicide. After all, an obligation to die would seem to remove the autonomy we want people to have when making end of life decisions. We would also have to be concerned that such a duty, if it existed, would be misused in order to eliminate the more vulnerable members of our society including the terminally or chronically ill and the elderly. In this chapter, I develop a Kantian argument for a suicide duty. In order to respect oneself and to preserve one's moral life, at times it becomes mandatory to end one's physical existence. Although a discordant position to take, I will show that such an obligation can be part of a very good life narrative, and can be done in a way that nurtures communities and caring relationships.

8.1 Introduction[1]

In August 2014, Gillian Bennett committed suicide to avoid the onslaught of her dementia. Her suicide letter, which has been widely distributed by the Internet, is an example of a person who has autonomously and caringly examined the relevant moral factors, such as her narrative of who she is and her relationships, and then come to the decision that she wanted her life to end on her own terms. Portions of the justification for her decision follow:

> I want out before the day when I can no longer assess my situation, or take action to bring my life to an end. There could also come a time when I simply must make a decision based on my deteriorating physical health. I do not like hospitals—they are dirty places. Any doctor will tell you to stay out of them if you possibly can. I would not want a fall, a stroke, or some unforeseen complication to mess up my decision to cost Canada as little as possible in my declining years.

[1]This chapter is a further development of arguments first seen in Cooley (2006, 2007, 2013).

D.R. Cooley, *Death's Values and Obligations: A Pragmatic Framework*,
International Library of Ethics, Law, and the New Medicine 62,
DOI 10.1007/978-94-017-7264-8_8

> Understand that I am giving up nothing that I want by committing suicide. Each of us is born uniquely and dies uniquely. I think of dying as a final adventure with a predictably abrupt end. I know when it's time to leave and I do not find it scary.
> There are many ethical issues here: life extension radically alters people's ideas of what it is to be human—and not for the better. As we, the elderly, undergo manifold operations and become gaga while taking up a hospital bed, our grandchildren's schooling, their educational, athletic, and cultural opportunities, will be squeezed dry.
> The heart of the problem is arithmetic: The post-World War II Social Welfare State, created at a moment when the baby boom was still gestating, is built on a generational Ponzi scheme. As life expectancy increases and birth rates decline, the population pyramid is being inverted—and in some countries that is causing the entire economy to topple.
> All members of my immediate family are in Vancouver: daughter, son, two granddaughters and four grandsons. All know that it matters to me not to become a burden to them, or to Canada. I have discussed my situation with them all. In our family it is recognized that any adult has the right to make her own decision. (G. Bennett 2014)

Although Bennett did not argue for a duty to die or suicide,[2] the moral factors she considers in her letter capture the essence of the discussion of rights and obligations to end one's life. What makes her letter and reasoning more powerful is the fact that they are not mere theoretical discourses by someone wholly untouched by the issue, but rather they are based on her personal experience and narrative which give them an authenticity that should carry some weight in the conversation.[3]

Based on the elements from Bennett's suicide letter, I want to develop and defend the view that there are instances in which a moral obligation to take one's own life exists even though it goes against the grain of the survival instinct and social learning.[4] Although relatively rare when considering that everyone dies, these suicide duty cases provide evidence for morality being something above and beyond the naturalistic workings of the human brain because they cause us to reject physical existence in favor of moral values, such as preserving our dignity as persons.[5] That is, if our minds were merely the result of evolution that protects the physical organism no matter what the cost, then there would be no question about a moral duty to take one's own life—such an obligation would be impossible. However, the fact that such a duty can be reasonably defended on rational grounds entails that morality is something above and beyond mere evolutionary adaptation.

In contrast to the standard justice and utilitarian arguments,[6] rights/autonomy positions rarely require an individual to commit suicide. Instead, they focus on the

---

[2]She did argue that living wills and similar devices are mandatory.

[3]See Mary Rose Barrington's (1980) "Apologia for Suicide" for another reasoned argument for suicide in these circumstances.

[4]Battin provides an excellent comprehensive development of most arguments for and against suicide in *The Death Debate: Ethical Issues in Suicide*.

[5]Scott Shershow argues that terms such as "dignity" and "sanctity" are imprecise words (Shershow 2014). I agree that the terms are inherently vague, but reject his deconstructivist approach in favor of a pragmatic one based on objective standards, as far as that is possible, and what works in the situation to achieve the goals of those involved in the conversation.

[6]The utilitarian argument states that if the best thing a person can do is to suicide, then that is her duty. On the other hand, the justice arguments might be based on various distributive justice

moral permissibility or right to do so, and then leave it to the individual to choose her own path. Whichever decision the agent makes, generally, is morally permissible for her, as long as she decides autonomously or exercises her right in a proper manner.

Kantian arguments for morally obligatory suicide are even more unusual.[7] Many believe that Kant thought suicide was absolutely prohibited conduct, mostly on the grounds that no agent could consistently will the generalized form of any suicide maxim based on self-love as a law of nature.[8] In the *Groundwork of the Metaphysic of Morals* (*Groundwork*), for instance, Kant states that self-love makes it impossible for people to consistently will the generalized form of the maxim: "If my life looks to be on the balance more evil than good, then I will end my life out of self-love" as a law of nature (Kant 1964, 89).

Nonetheless, Kant sometimes speaks as if some suicides are morally obligatory.[9] In *Crimina Carnis*, he seems to condone some suicides. "Suicide is the most dreadful [conduct,] but it is not as dishonorable and base as the *criminal carnis contra naturam*." (Kant 1979b, 170). A plausible interpretation of Kant's views states that, under certain conditions, not only is the person permitted to kill herself, but also that she is required to do so as a duty to herself *qua* moral agent.[10]

Although it is important for a true understanding of Kant's overall ethical theory to know how he thought it functioned, for the purpose of this discussion, it does not matter what Kant literally intended when he wrote the passages concerning suicide. I am more interested in formulating a plausible interpretation of his position in relation to Bennett's that can be used to evaluate real life situations than in exhaustively defending the interpretation as Kant's actual stance on suicide. Hence, my version of the principle does not attempt to become a definitive reading of Kant's ethics, although it does seem to capture a small portion of it.[11] In addition,

(Footnote 6 continued)

principles. In one, if a person has used more than her fair share of healthcare resources, then she has an obligation to refrain from using any more, which will passively end her life.

[7]See Brassington (2006).Thomas De Quincy, for instance, argues that men have such a duty if the alternative is having their nature dishonored through corporal punishment (De Quincy 1856, 231).

[8]According to Christine Overall, the taking of one's own life does not respect life's intrinsic value in the proper way (Overall 2003, 74–5).

[9]Richard Brandt, among most others, claims "Kant was opposed to suicide in all circumstances." (Brandt 1979, 466). Alan Soble, on the other hand, adopts the view that Kant endorsed suicide but was heartless and misogynistic as well. (Soble 2003, 1–2).

[10]Some may wonder why PP is not being used in this argument. In fact it is; I will show that such suicides respect the intrinsic value of those things affected by the action and that it is something that at least one reasonable person would reasonably believe will likely maximize utility. Satisfying these two conditions makes the action morally right.

What makes some suicides mandatory is that they are the only way an action can satisfy the Kantian QCI; therefore, I will spend more time developing this claim than showing RPU is satisfied.

[11]I will not address Emile Durkheim's claim that society for social purposes can impose a duty of suicide (2003). Instead, I will take it as a given that Kant's claim that no one can bind a person to commit suicide is true (Kant 1979c, 150).

my final position on mandatory suicide will incorporate elements of narrative theory and care ethics to humanize Kant's rather cold morality. I will not argue that every person has duty to suicide, but that there can be such duties for certain people if the increasingly more common situation justifies it.

8.2 A Kantian Argument for Obligatory Suicide

Even though it is more than obvious from his work that Kant is opposed to the vast majority of suicides and on several occasions explicitly rejects this type of action, it is not clear that suicide is ***always*** impermissible under Kant's theory and principles or that Kant actually thought that all suicides were morally wrong. According to Kant, by rejecting their human rationality and acting as animals would not, human beings degrade themselves below the level of animals (Kant 1979b, 170). In fact, he believes some actions are so terrible that:

> [They] make man unworthy of his humanity. He no longer deserves to be a person. From the point of view of duties towards himself such conduct is the most disgraceful and the most degrading of which man is capable. Suicide is the most dreadful, but it is not as dishonorable and base as the *criminal carnis contra naturam*. It is the most abominable conduct of which man can be guilty. So abominable are these *criminal carnis contra naturam* that they are unmentionable, for the very mention of them is nauseating, as is not the case with suicide. (Kant 1979b, 170)

If some actions are so morally horrific that the mere mention of them make moral agents physically ill, then it seems plausible to start thinking that Kant is implicitly advocating suicide for individuals who are not able to control their desires for sex with themselves, members of their own sex, or animals. At the very least, Kant seems to be saying that although suicide is generally or always wrong, the taking of one's own life is not as bad an alternative as any *criminal carnis contra naturam*.

In "Care for One's Life," Kant adds support to the case for morally obligatory suicide by explicitly stating, "There are duties which are far greater than life and which can often be fulfilled only by sacrificing life…The moral life is at an end if it is no longer in keeping with the dignity of humanity" (Kant 1979a, 154–6). That is, sometimes suicide is necessary to preserve the agent's moral agency.[12] In order to treat everyone affected by an action as an end in itself—in other words, to do the agent's duty—it might be necessary to commit suicide. If an action is abominable and degrading to the extent that it is actually worse than suicide, then it seems as if it would be obligatory for an agent to kill herself in order to avoid the greater dishonor that makes all moral agents physically ill, at least according to Kant.

The seeming inconsistency between Kant's apparent ban on all suicides and his prescription of others can be explained by examining the distinction Kant makes

[12]Kant's position can be compared to that of Buddhists'. "Buddhism sees death as not the end of life, but simply a transition; suicide is therefore no escape from anything." (Becker 2000, 499).

between physical and moral life. A person's physical life can be sacrificed in order to protect her much more important moral life, which gives her human dignity in the first place. In *The Metaphysics of Morals (Metaphysics)*, Kant contends that the physical life is guided by the same self-love to preserve life that any animal has. In fact, Kant says that "The *first*, though not principal, duty of a human being to himself as an animal being is *to preserve himself* in his animal nature. The contrary of this is willful *physical death* or killing oneself" (Kant 1996, 176). As can be seen, Kant is making a clear distinction between the duties human beings have to themselves qua animal beings and those human beings have qua human persons. Furthermore, although the self-love principle is first, it is obvious that Kant thinks that there are more important principles that can override it. Given Kant's distinction, the likely candidates for this role are principles that apply to the preservation of human persons qua human persons.

The moral life is the life of a moral agent qua moral agent (or human being qua human being.). In "On Servility", Kant claims that human beings in the system of nature have much the same ordinary value as any other animal, "[b]ut a human being regarded as a person, that is as a subject of a morally practical reason, is exalted above any price…he possesses a *dignity* (an absolute inner worth) by which he exacts *respect* for himself from all other rational beings in the world." (Kant 1996, 186). The feature of dignity, also known as humanity, is a feature that human persons do not share with animals. Moral agents have the potential or actual rationality to act as moral agents, while animals lack this capacity. The intrinsic value of being a person is what makes it morally wrong to treat humans affected by our actions as mere means, i.e., to degrade their value; at the same time, it allows us to use animals as we see fit as long as doing so does not endanger our humanity.

A plausible working interpretation of Kant's position on suicide can now be formulated. Generally, it is morally wrong for an agent to take her physical life because it ends the agent's ability to make decisions in this sphere of existence and violates the principle of self-love. However, the principles governing human persons qua human persons always trump the principles of human persons qua animal beings. If it is a choice between ending the agent's physical life and the agent's moral life by degrading her humanity, then it is at least permissible, if not obligatory, to end one's physical life. Moral life suicide is never permissible, but the agent can always take his physical life as long as it is performed to preserve his moral life, much as one can have a gangrenous limb amputated to save one's physical life.

The examples Kant describes in the sections on suicide show that very stringent conditions must be met in order for any agent to have a duty to commit physical life suicide. In "Care for One's Life," there are two men: an innocent honorable man who is accused of treachery even though he has done nothing wrong and a man who has actually committed degrading crimes. Both men are found guilty and forced to choose either to become a galley slave or to be executed. According to Kant, "A man of inner worth does not shrink from death; he would die rather than live as a scoundrel of contempt…but the worthless man prefers the galleys, almost as if they were his proper place." (Kant 1979a, 155). In other words, the worthless man has already committed moral suicide. He has rejected his humanity, which requires freedom, and has made himself a slave to be used as one would an animal. Well, perhaps he is not being used as an animal, for one should not treat animals with the contempt that ought to be used on an entity that once had human dignity as a person but acted in such a way as to destroy that dignity. The honorable man, on the other hand, still has his moral life, which he will lose if he becomes a slave. In order to protect his human dignity, he must sacrifice his physical life.[13]

Similar examples are described in the Metaphysics and "Suicide". (Kant 1979c, 149, 1996, 178). The case most relevant to bioethics is Kant's example of the man with hydrophobia. Kant argues that a man with rabies who will inevitably be reduced to the stature of a mere animal is morally permitted to kill himself before he enters that state. Of course, the man must perform the action with good intentions and motives; in this case he takes "his life lest he harm others as well in his madness", but once again, it appears that a person ought to sacrifice his physical life in pursuit of preserving his moral life.

The examples described above provide insight into four conditions that must be met in order to require a moral agent to commit physical suicide. First, the person has to be honorable and innocent of any acts that degrade his humanity or moral life. That is, he still has his moral life. Second, the agent's intentions and other mental states regarding the situation must be virtuous; i.e., he is sacrificing himself for noble reasons. He takes his own physical life out of duty to preserve his moral life with all the required mental states that such an obligation entails.[14] Third, the agent must be externally coerced to either degrade himself or die. In the cases in which suicide is morally required, the degradation of the agent's moral life is almost completely out of the control of the moral agent. The innocent and honorable man who is faced with becoming a galley slave or dying has no other choice than death because, according to Kant, becoming a galley slave will necessarily degrade his human dignity and moral life. That is, he must either choose death or be ruined in his moral life. The innocent, honorable man must know that a rational agent believes the proposition that "To live is not a necessity; but to live honorably while life lasts is a necessity" and would act accordingly (Kant 1979c, 152).[15]

Fourth, as far as possible, moral agents should not directly cause their deaths, but should try to have others kill them while the agents are defending their moral and physical lives. Lucretia's suicide, Kant states, was morally wrong, even though it

---

[13]Not everyone will agree with Kant's claim that an innocent person commits moral suicide when he chooses to become a slave. Most people would probably agree that a slave is still a person, although owned by another.

[14]For the purposes of space, I will not try to list and defend the required mental states. Rather, I will stipulate that there must be certain mental states, given Kant's theory, but what they are is a question for another time.

[15]Kant's psychological theories are often primitive at best. He assumes a vast control over behavior that does not in reality exist, such as in cases of internal compulsion. The issue of internal compulsion and Kant's inadequate psychology will be addressed in the section on psychotic sexual predators.

was committed on the grounds of modesty and honor preservation, because she had the option of defending herself to the death (Kant 1979c, 149–50). This interpretation of Kant's condemnation of Lucretia's act makes it consistent with the seemingly contradictory claim in "Care for One's Life."

> The moment I can no longer live in honor but become unworthy of life by such an action, I can no longer live at all…If, for instance, a woman cannot preserve her life any longer except by surrendering her person to the will of another, she is bound to give up her life rather than dishonor humanity in her own person, which is what she would be doing in giving herself as a thing to the will of another. (Kant 1979a, 156)

If there is an opportunity to avoid death by the agent's own hand, such as dying in battle to preserve her moral life, then that option should be the first one taken. However, if the only way to save an agent's moral life is for her to end her physical life, then such a duty is required of the agent.[16]

8.3 Morally Obligatory Suicide for the Soon-to-Be-Demented

Dementia-causing illnesses, such as Alzheimer's disease, can alter sufficient individual characteristics not only to change personal identity, but also to eliminate moral agency's necessary features, including rationality. Other essential characteristics can be found in Callahan's distinction between full and limited/impaired selfhood. Full selfhood includes rationality in addition to "the capacity to have feelings and to be aware of them…and to enter into relationships with others." (Callahan 2001a, 96). Limited and impaired selfhood lacks one or more of the indispensable elements. (Ibid.) People with full selfhood are clearly moral agents, while those with limited and impaired selfhood might or might not be depending on how far their disease has progressed. Although it is sometimes difficult for us to correctly classify entities with full, limited, and no selfhood, it is clear that some people with full selfhood eventually progress into a state in which they lose their rationality, and hence, their moral agency. In other words, the demented no longer possess the selfhood that makes them moral agents, which is one reason why they are prevented from entering into contracts or making crucial decisions about their self-interests. Of additional moral concern is that their slow death process takes 8.6 years for the average elderly dementia patient, and 10–20 years for the average Alzheimer's patient (Pence 2007, 245). Therefore, not only are they moral subjects who are former persons, but also the new entities have to remain in that state for a number of years.

---

[16]The fourth condition does not work well. In end of life decisions, I will argue later, it is wrong to burden others by requesting and using their assistance unless the agent cannot kill herself by herself. On these grounds, I contend that the first three conditions are sufficient when all three are met.

The overall dementia problem will grow worse over time. "Increasing life expectancy will mean that more individuals will close their lives severely senile and requiring expensive custodial care, and many will find it unreasonable to set aside resources for care under such circumstances." (Engelhardt 1995, 45). Ironically, improved medical treatments have raised rather than lowered dementia patient numbers. Instead of dying from other ailments prior to the mind-wasting disease setting in or becoming severe enough to be noticeable, those with dementia-causing illnesses live until they lose much of their personal identity, be it as a psychological or embedded person. As medical treatments for other life-terminating diseases increase in efficacy, greater numbers of moral entities will become former persons. Moreover, given current medical science, the demented, with their "greatly limited human possibility," have no hope of regaining their moral agency or reestablishing the psychological or embedded people who once existed (Kass 1995, 226).[17] As I have argued previously, the loss of moral agency is not only a cessation of existence, but also death.

In some cases of those who will become demented, it is a rational duty to die physically before dying psychologically, and hence, morally.[18] The moral agent should select self-inflected death before she becomes incompetent because she owes it to herself as a moral agent to respect her dignity as a person and as a moral agent. Furthermore, since it is perfectly rational to select a better world in which a moral agent can remain a moral agent than one in which her body is physically alive but her moral status is severely degraded, she will select the former with equanimity (Brandt 1979, 478–9). By choosing to take her physical life in the right way, and then completing the action at the proper time and in the proper manner, the agent respects who she is as a moral agent, while she is still a moral agent.

Finally, by choosing to end her physical life herself and not imposing undue emotional, financial, or other burdens upon them, she respects those who would have had to kill her or watch her body die after she, as a person, is dead, which was a worry of Bennett's. According to Gregory Pence, the ancient Greeks believed it was a duty to act in this manner. They thought it base and ignoble to place a heavy burden on society or family if doing so is only to give a short extension of life to one citizen (Pence 2007). Therefore, suicide or euthanasia could be construed as a duty every citizen qua citizen has to her fellow citizens.

Unlike the Greeks, John Hardwig bases his obligation to die on duties to loved ones; a person is more likely to be required to die if living would impose a significant burden on his family and loved ones. This duty becomes stronger or weaker depending on eight factors, including the individual's age, fullness and richness of the life she has led, the resources to care for herself, and her ability to adjust to her illness (Hardwig 1997). Furthermore, a person has a greater obligation

---

[17]Kass argues vehemently against suicide on the grounds that people are unable to make free decisions. The illness and its consequences are too coercive to allow for free choice (Kass 1995, 228).

[18]Feldman states that actively pursuing death can be a rational action (Feldman 1992, 222). Marilyn Bennett claims it is rational to die before a loss accrues (Bennett 2000, 49–50).

to end her life if there is an increased likelihood that the part of her that is loved is likely to be lost or seriously compromised, her family has already made significant contributions to her, and her loved one's lives are already difficult and impoverished. Committing suicide in these extenuating circumstances can be considered a beneficence given to those for whom she cares.[19]

Although a duty to die is owed to the self, a person's suicide does not necessarily treat another affected by it as a mere means, even though it can harm family members and others in an agent's caring community. Kant argues that each rational agent has an imperfect duty of beneficence to others (Kant 1996, 151). "To be beneficent, that is, to promote according to one's means the happiness of others in need, without hoping for something in return, is everyone's duty." (Kant 1996, 202). In other words, we have an obligation to nurture members of our caring community consisting of entities emotionally caring for the well being of each other (Holmes 2007).[20] To fulfill this imperfect obligation, we as agents must first "realize, in ourselves and others nurturing *goods* such as caring, considerateness, compassion, sympathy, and love." (Ibid.) Given the requirements of care for our community and what they entail, oddly enough, certain suicides can satisfy these obligations. Killing oneself in these circumstances both promotes the individual's interests and shows respect for and honors the person's family (Beauchamp 1986). If the suicide is performed in the correct manner, then no person is treated as a mere means, although those for whom the person cares and those who care for her are emotionally injured. On the other hand, not killing herself would show selfish callousness. She causes unnecessary harm to others by imposing significant burdens upon them rather than autonomously solving the problem.

Given the reciprocal nature of care, duties exist for members of the community other than the future demented agent. As members of a caring community, survivors are obligated to understand why an agent acts to preserve her own moral life and to support her in her difficult decision. Doing otherwise is unethical. To expect another to live physically without human dignity is to generally treat the ill individual as a mere means by burdening her at a very difficult time with others' less vital goals and desires. Although it might be comforting to have the loved one's living body to remind the survivors of their relationship with her, it does not allow the moral person a dignified exit from life as a moral person. Hence, if community members are truly caring, they should support death duties for these agents. It, therefore, becomes each member's duty as a fellow community member to understand why the person did what she did, to be supportive and, hopefully, to find some solace in knowing that the act was performed out of respect and caring for her, rather than the agent merely pursuing selfish ends.

[19]Menzel expands Hardwig's scope to include others burdened by the continued life. If there is a high burden placed upon others and a low prospective given to the individual for staying alive, then according to Menzel, there is an obligation to die (Menzel 2000, 97).

[20]The combination of reason and emotion recognizes that reasonable decisions, which are part rational and part non-rational, are how people actually make ethical choices. An overly rational approach is impossible; no one would ever be able to choose one alternative over another. On the other hand, an overemotional response ignores necessary evidence.

8.4 Why Suicide Is Morally Preferable to Mercy Killing or Being Allowed to Die as a Moral Subject

During the time period between when she knows she will become demented and she is no longer a moral agent, three alternative actions are open to the agent; two of these lead to moral demise before physical death. The impermissible two both require the agent to remain physically alive and succumb to dementia over time, eventually as a moral subject, but the methods by which the individual dies are different. In one, she can merely die of natural causes without direct moral agent interaction, while the other involves mercy killing in which the agent arranges with someone else to end the former's physical life after her moral agency has terminated as a result of her illness.[21] Dworkin, for example, believes the latter is permissible because he maintains that advanced directives made by dementia patients before they fully succumb to the illness should be obeyed, even though the original person has ceased to exist. The support for his position is that we have an obligation to respect the former person as an end in herself.[22] That is, our duties to people do not cease the moment that the people perish. As Pence states, if we do not honor an advanced directive, "We *disvalue* the former person and her life [for example,] by continuing to spend money on her in ways she forbade." (Pence 2007, 258–9). Somehow, our failures can reach back in time to affect the former person, or the former person can in some way be disvalued when the duty is left unfulfilled.

However, Dworkin, as are Priorists, is mistaken about a failure to respect the former moral agent by not killing the moral subject. If moral agency is lost, then, by definition, the remaining entity is no longer numerically identical to the moral agent who was this particular psychological or embedded person or any moral agent.[23] She qua moral agent or whichever person she was cannot be demeaned or treated as a mere means because she no longer exists to be affected in any way.[24] At best, when we do not perform mercy killing, it is her memory or reified distributed persona that is disrespected, but memories and other mental states about another are not the same things as the person herself, nor is the reified distributed persona an actual person per se. It does have a moral standing as a moral entity, but disrespecting it does not rise to

---

[21]One arguable legal method of killing oneself is the intentional starvation of the individual. The greatest benefit to this course of action is the medical assistance and advice the individual would have during the process, which not only makes it less violent and more effective, but also creates a caring, supportive atmosphere.

[22]Dworkin's (1995) position contradicts Callahan's claim that no one can place death's power in another's hands (Callahan 2001b, 224).

[23]This issue has been convincingly addressed by Rebecca Dresser (1989, 1995) and John A. Roberston (1991).

[24]Even if there is life after death and the entity that lives on is identical to the person, not killing the person's body shows no disrespect to the person. The independent departed is no longer a part of the dyad; hence, it has no right to decide the disposition of the body. What is done to the body, therefore, is not done to the person.

the level of demeaning an actual person. The intrinsic value of the latter is far greater than the former, primarily because it is a person.

On Kantian grounds, it is clear that neither of the moral-death-first options is ethical. Since she is morally required to be able to will what is right and good, the agent must still be a person in possession of her moral life to be able to take it; hence, she must not be demented. After her disease progresses too far, she can no longer act as a moral agent because she is a moral subject. By not making and carrying out the decision to end her physical life while she is still a moral agent, the individual is rejecting the superior value of being a moral agent and the particular psychological and embedded person she is over being a former person. In other words, although being a former person might still have intrinsic worth, and being a moral subject certainly does, the moral agent treats herself as a mere means by not respecting her true worth as a moral agent with full selfhood. She also does not respect who she is as a narrated psychological person and embedded person because her moral agency is an essential part of those various entities. The only way she can treat herself as Kant's theory requires is by making the decision to end her life for the right reasons and at the right time, and then carrying it out in a fitting way.

Finally, under Kantianism, no individual can ethically use an advanced directive to prolong her moral life. Although it is intuitively appealing to believe that the longer length of time she lives as a person makes her actions permissible, Kant's example of the innocent man forced to choose between slavery and execution shows the fundamental error. Recall that the man is unjustly condemned for crimes he did not commit. The options available to him are state-imposed death or becoming a galley slave. Kant asserts that the man should select execution because if he chooses to be a slave, then he is unworthy of his humanity. On the other hand, a man who has already sacrificed his humanity should choose slavery because it is appropriate for him. Given that it is the individual's ***choice***, not the choice's actual consequences, that counts, those who use advanced directives lose the very humanity that they seek to preserve when they make the choice. By imposing a moral obligation on others and trying to maximize their agent-utility to eke out as much time as they can, they are fully aware that they will lose their humanity in the end and they act as the galley slave rather than as the innocent man. Those who select suicide would rather die as a full person with moral life intact than extend their physical lives by becoming slaves.[25]

---

[25]Claiming that the person's continued intrinsic value as a person in her various forms overcomes the loss caused by the disrespect she shows to herself by not committing suicide does little to defeat a duty to kill oneself. Considering only the person as psychological person's and moral agent's intrinsic values, then a bare state of affairs in which she exists has more worth than a bare state of affairs in which she does not. To respect intrinsic value and to try to bring about the outcome that reasonably will produce the best outcome, the bare state of affairs in which the person exists is preferable to the alternative. This is a simplified view that does not reflect the much more complex reality of which the individual is part.

When considering the overall psychological and embedded person, the Practical Principle actually establishes that suicide is morally right in the circumstances. Moreover, since it is the one alternative likely to satisfy the Practical Principle in many situations, it becomes a duty for them.

8.5 Some Objections, Responses, and Further Development of the Obligation Argument

Of course, such a controversial interpretation of Kant's position is bound to draw criticism on many different grounds, especially due to the soon-to-be-demented component. I will focus on some of the more difficult, and therefore interesting, objections to the overall position, which leads to refining the argument to include care ethics and narrative provisions.

8.5.1 Women and the Suicide Duty

Rosemarie Tong's potent criticism of Hardwig and those who have a similar position[26] is that the duty to commit suicide will unfairly fall upon the shoulders of women as a whole.

> Assuming that Gilligan's analysis is with merit, and that society as a whole as well as women in particular view women as more caring, giving, nurturing, and self-sacrificing than men, it is reasonable to hypothesize that the duty to die might rapidly be "feminized." Women might perceive themselves as imposing unreasonable burdens on others simply because they typically live longer than men, suffer from more chronic conditions than men, and require more long-term care than men. (Tong 2000, 147)

I argue that the duty to die is held in common by all moral agents facing non-reversible dementia, yet Tong does make several legitimate points. Women have a greater probability of becoming demented just by living long enough for the disease to begin its onset. Social expectations tell us that women will sacrifice themselves for others for whom they care instead of being a perceived burden.

If we accept that there is a duty to commit suicide, we should be prepared either to accept the consequences of such a duty to individuals and society or, if the consequences are too morally repulsive, to abandon such a duty as theoretically legitimate but impractical to implement. Battin, for instance, is concerned with circumstantial and ideological manipulation that can make suicide a rational, but not an ethical, action (1994, 196–8, 1996, 146–7). Both forms of manipulation alter a particular set of conditions or core social beliefs and conventions so that suicide becomes a rational choice for an individual in those circumstances or certain groups of people within the society. Circumstantial manipulation, for example, can occur when an individual's particular living arrangements become so extreme, e.g., unremitting torture or living conditions that make the life far from being worth living, that suicide is the rational choice for the individual to make to escape her coercive circumstances. On the other hand, ideological manipulation occurs when social values are altered so that suicide becomes the social norm for certain

---

[26]Davis (1999, 118).

circumstances, such as being too great a burden on the family or avoiding a painful end to one's existence. (1994, 200). For both manipulation types, the "choice remains crucially and essentially voluntary, and the decision between alternatives free". (Ibid.) If this is correct, then we, individually and as a society, would be unable to classify such suicides as irrational because they are reasonable and rational given the morally repulsive circumstances. The result would be that many who would not choose to die in other circumstances would do so merely because social pressure and their unquestioned ideologies incline them to act in such a manner or to avoid an unnecessarily ugly living situation created by others. Hence, the manipulated suicides would be rational for the person killing herself but unethical overall because of what others have done to create such coercive environments.[27]

Because suicide causes the termination of a valuable human person, manipulation should be a worry for us. Taking one's own life when it is not the result of meaningful choices, such as those that result from coercion, cannot be a moral duty. No one will deny that.

However, we can expand the Kantian suicide duty criteria from the four in the Kantian formulation above to an additional 14 to eliminate as much of the cause for concern about manipulation as we can. First, Kant's criteria are objective enough to avoid the problem of moral relativity created by social conventions changing to require suicide for such non-justifying reasons as being the victim of rape (Battin 1994, 197). The person's moral life has to be worth saving; the only way that life can be saved is through the loss of the physical life; and the agent can treat herself and all others affected by her actions with the respect that they deserve. Hence, the social conventions about suicide do not matter.

Second, we will add Hardwig's nine criteria to Kant's four to create a set with far more care ethics, justice, consequentialism, and nuance than dreamed of in the far more reason-oriented approach. Hardwig's nine criteria are:

1. If continuing to live will impose significant burdens on your family and loved ones,
2. As you grow older,
3. If you have lived a full and rich life,
4. If your loved ones' lives have already been difficult or impoverished,
5. If your loved ones have already made great contributions to make your life a good one,

[27]There is a debate over whether euthanasia has led to a slippery slope. On one side, it is claimed that since euthanasia was legalized in the Netherlands, there does not appear to be a slippery slope developing, although there is a contention about what euthanasia is and what it is not (Battin et al. 2007; Norwood et al. 2009) Rietjens et al. 2009). On the other, it is claimed that there are a large number of euthanasia cases left unreported (900 people annually are euthanized without asking for it), and that euthanasia is being used impermissibly for persons and moral subjects who are not terminally ill (Pereira 2011). I contend that euthanasia would occur regardless of its legal standing, and it is far better to regulate it than to allow it to remain hidden, and therefore open the door to far more abuse than already may exist.

6. If you cannot make a good adjustment to your illness,
7. If you cannot make significant contributions to the lives of others, especially your family,
8. If the part of you that is loved will soon be gone or seriously compromised
9. If you have lived a relatively lavish lifestyle instead of saving for your illness or old age (1997, 38–9).

According to Hardwig, as each criterion gains additional moral weight by becoming more likely or being fulfilled to a greater degree, then it becomes more likely that there is a duty to die. As his or her presence is reduced, then it is less likely that someone has an obligation to die.

Third, although the care approach in Hardwig's criteria is much more subjective and would put greater pressure on women in caregiving relationships to commit suicide to help those for whom she cares, Sheila Mullet's criteria ensure that women and other caregivers should not be given short-shift by themselves or others. To do what is right, we must make sure that the caregiver is fulfilled by the activity, her unique and particular individuality is used in the activity, her care is not the result of exploitative gender roles, her care is reciprocated with caring, and the activity happens and is sustained as a result of conversation and consciousness-raising (Mullett 1988, 119–120). Hence, manipulating and exploiting the weak are not morally permitted, but empowering them is. If these moral boundaries and requirements are transgressed, then we can rightly declare that those who committed such actions are objectively guilty of morally reprehensible actions, just as it was morally reprehensible to expect women to commit suicide because they were raped or because their husband died.

Although women have historically been forced by social pressures and their own illegitimate buy-ins of repressive social mores to act against their self-interests, it is not necessarily the case that a suicide duty will continue such oppression or make women perform coerced behavior. That is not to say that social pressure on women to kill themselves, if it is done out of sexist or other morally wrong beliefs, should not be condemned and fought. Denying every woman the duty to die based on what might happen to her, however, takes away part of what makes her a person in the first place, viz. her being a moral agent with power over her flourishing and life. Granted that protecting those needing it is morally right in general, and a moral duty in many cases, that safeguard must avoid becoming illicit paternalism. To do what is right, we should respect women as moral agents and for the psychological and embedded persons they are by not wrongfully interfering with the enjoyment of their rights or the performance of their duties.

Having a duty to die can be a positive good for women as moral agents. As Davis writes, "to seize control of one's destiny, perhaps especially under circumstances of great tragedy and limits, is to be active and dominant" (Davis 1999, 115). That is, women can overcome the historical association of women as passive beings at the mercy of their own bodies. (Ibid.) By taking charge of their lives as full persons with moral obligations to fulfill, women are able to decide for themselves how "the final chapter of [their lives are] written," rather than passively accepting what their bodies do to them (Ibid., 114). In other words, by satisfying the duties that any moral agent has, women participate as moral agents rather than as almost moral agents or moral subjects.

8.5.2 Harm to Others

It is difficult to prove that reasonable people would confirm the existence of a suicide duty for every person who will become demented in her lifetime. After all, it is hard for people to agree on many trivial matters, much less one of the most important actions some people must take. However, I do not have to demonstrate that all reasonable people will agree that there is a duty, only that some will, and that the duty exists for at least some people, viz. individuals who do not have any extenuating circumstances that justify them staying alive without their moral life. Many individuals who are embedded in positive relationships with others have or should have accepted what will happen and understand the decision, although if they are caring, they will regret that it needs to be done. These are the individuals who have the suicide duty.

The issue of harm caused to the loved ones of a suicide has been raised to defeat any duty to kill oneself. Rabins states that "the family and friends of those who commit suicide are often permanently and grievously damaged when a loved one or close acquaintance commits suicide, whether or not there was a 'good reason'". (Rabins 2007, 48). If we respect others for the ends that they truly are, this line of reasoning suggests that we should not cause them unnecessary, excessive injury by killing ourselves. Moreover, provided that the harm suffered is likely to be lessened or avoided altogether, the individual would have an obligation to stay alive for as long as she can.

Three plausible ways, when taken together, provide an adequate response to this criticism. First, the suicide duty is not absolute because a number of individuals may permissibly stay physically alive for the interests of others. Perhaps their families need more time to accept the inevitable, and this inability to understand and accept is not caused by some moral defect they have, such as being self-centered or willfully uninterested in the well-being of others. In these situations, the person can sacrifice herself by losing her identity as a moral agent or psychological person after the dementia sets in, but she retains her human dignity while she lives as a moral agent due to the fact that her sacrifice is noble rather than venal. She does not destroy her dignity by trying to stay alive as long as she can and leaving the burden of taking care of her on her loved ones and society which causes her to deserve a lower moral status. Rather, she stays physically alive for others because she cares for them. The agent is in pursuit of their flourishing, the thriving of the organic wholes of which she is part, and the flourishing of the organic wholes of which her moral subject will be a component. This exception fits in with Tong's notion of death as a gift given to loved ones (Tong 2000, 152). Although Tong would disagree that there is an obligation to die, in these rare conflict situations, two competing alternatives (if they did not exist simultaneously for the agent) would both be duties. However, in these dilemma cases, the agent may select either alternative and still do the right thing.

One can respond to the unethical harm through grief argument in two other ways. First, if there is a duty to oneself to die, then the harm caused to others is permissible. No one ever said that doing a right thing, especially in difficult situations such as these, leaves everyone with a positive or neutral outcome. Suppose two people, Person A and Person B, are stranded in a blizzard with only enough food to keep one of them alive. Furthermore, assume that the food is owned by and is in the possession of Person A. If Person A eats it, then she indirectly harms Person B by killing/allowing him to die from starvation. However, Person A has done nothing wrong in preserving her life, even though she caused B pain, suffering, and death. It might not have been A's duty to stay alive, since she could have acted altruistically by sacrificing herself and giving the food to Person B, but her action was morally permissible. Even in some cases in which Person A intends to kill Person B, there is no moral blame to be laid at her door. Suppose that B is eating the food that A brought, which would be understandable in the circumstances. If the only way that A can keep the food for herself is to murder B, then the choice is tragic but ethically justifiable on the grounds of self-defense and private property rights to the food. Person A has the responsibility to take care of her own flourishing, which, in this case, requires that she keep the food for herself. Therefore, the mere existence of severe harm to others does not entail that anyone was treated illicitly.

Second, and more importantly, the existence of a great deal of permanent and grievous damage to people as the result of some suicides does not establish that they ought or have a right to feel this way. It is the naturalistic fallacy to derive an "ought" from an "is," as has been pointed out by many over the years, including Hume and G.E. Moore. Therefore, to prove that the harm is illicit requires us to investigate the reasons why these individuals suffer these feelings.

I contend that any permanent and grievous damage felt by them is the partial result of social learning which introduces a form of relativity into thinking about the taking of one's own life. Suicide has a "culturally variable nature." (Margolis 1975, 25). Whether to call an act a suicide, or even how to evaluate such an activity, is subject to our cultural memberships. "Elective-death deliberations and their assessment…are as much affected by coincidental-culture values as by foundational-culture values" where coincidental-culture is one defined by its group membership based on shared interests and objective and foundational-cultures as a culture that has a deep history and values (Prado 2008, 189). We are raised and reinforced by a foundational-culture to believe that death and suicide are always very bad things or are wrong so that the word itself cannot be normatively neutral (Fairbairn 1995, 31). To show the power of foundational-culture, even after the local populace ceased degrading the corpses of suspected witches and vampires, local communities still buried suicides in crossroads with a stake through its heart, such as in England's last recorded case in 1823 (Alvarez 1998, 58). Coincidental-cultures have even greater power on our beliefs and actions. Our coincidental-cultures reinforce, and perhaps magnify, our already held beliefs as we interact with people who think like us and with whom we closely identify, which might not always happen in the far larger and less binding foundational-culture. As a result of these social forces, and our evolutionary adaptive desire to survive, suicide has such a negative stigma[28] attached to it that it is difficult to discuss or analyze it without normative bias.

Beliefs about elective-death, especially suicide, were not always this way. Prior to the rise of Christianity and St. Augustine's work, suicide was not only permitted, but it was also encouraged in certain circumstances and by some groups such as the Stoics (Dowbiggin 2005, 8–9; Rosen 1975). Suicide could be honorable in order to avoid capture, protect one's reputation, to honor another, or as a form of suttee, and it was permitted in order to avoid a life not worth living due to some ailment (Rosen 1975, 5–14). Since the loss of faith and increase in doubt about Christianity and other religions' teachings in the 20th century, there has been far greater acceptance of euthanasia and taking one's own life (Dowbiggin 2005, 67). That is, as foundational and coincidental cultures change, so too do attitudes toward suicide's morality and worth.

There also seems to be an underlying belief that those diagnosed with a dementia-causing illness are fragile victims requiring full paternalistic protection from themselves and others. Something similar is seen in Slavoj Zizek's analysis of NATO actions in Kosovo:

> What we encounter here is again the paradox of victimization: the Other to be protected is good *in so far as it remains a victim*…the moment it no longer behaves like a victim but wants to strike back on its own, it magically turns all of a sudden into a terrorist/fundamentalist/drug-trafficking Other. (Zizek 2000, 60)

A victim who takes control of her life and throws off paternalism by others causes the above-mentioned discomfort to those wanting to help her. As a result, the moral agent who wants control over the most vital central aspects of her life is labeled as a danger. This is one of the reasons why claiming that those who will be demented have a duty to kill themselves is so controversial. By rejecting the helpless victimhood paradigm, the suicide duty challenges fundamental social beliefs about the status of those diagnosed with dementia-causing diseases. It preserves their moral agency and allows them to be in charge of their life story rather than being its victim.

Of course, what is socially learned can be altered. Given that social attitudes have changed, it follows that unless we are full of hubris, we cannot merely assume that our contemporary views about suicide and death are superior measures of morality. In fact, they might be unjustified if a better social belief is available to us. That is, if a belief works better at promoting and sustaining flourishing, perhaps as

[28]The negative stigma might affect available statistics on suicide because of the desire for survivors to hide what happened from outside judgment (Gorowitz 1975, 213). Even today, people tend to be reluctant to put the cause of death as suicide in newspaper obituaries or discuss how the person died.

understood through RPU and QCI, then there is a great incentive to favor it over one that does not achieve the same beneficial results.

If we changed our society in such a way that people accepted certain types of suicide as a reasonable, rational, honorable action, then far less damage would occur. I am not arguing that there will never be negative emotions, such as regret, remorse, guilt, and grief—some legitimately felt while others are not—but that understanding of why the person did what she did as a right action can alleviate much of it in the same way that those emotions are reduced when families and friends are part of the process when patients refuse further treatment for their illnesses.[29] Instead of allowing oneself to succumb to the disease, the person ends her life at the time and in the circumstances she controls, which, if the family respects and cares for her, might be tragic and laudatory to them rather than causing them permanent damage. The suggested alteration to social beliefs could create a community that is much better than the one we now have. Instead of people being wracked by guilt, overburdened with emotional and financial costs, or incurring other avoidable expenses, there can be a loving society in which caring support is given and the person finishes her life as a moral agent acting as a moral agent.

8.5.3 Moral Duties, Comfort, and Power

A suicide duty can be comforting and empowering. As Dena Davis writes, "people have a strong interest in ending their lives with a final chapter that is consonant with the narrative as a whole". (Davis 1998, 318). That is, some individuals desire to die in a way that reflects how they lived. They do not wish to be someone slowly but surely progressing to the state of former personhood. They do not want to be in a living body that must rely on others for massive support, regardless of what might be in the latter's best interests. Rather, these agents desire to end their physical life as a person and take an active rather than a passive role, which allows them to remain "active and dominant" (Davis 1999, 115). This is what happened in Bennett's case.

Many individuals want to be remembered as someone who not only lived her life as she saw fit, but also ended it in the same way. If the death does not fit with the life as a whole's narrative, then it might be the case that the individual lived as a hypocrite, which is one of the reasons I think Aristotle claimed that a person's life can only be evaluated by another person after the former person has died. A hypocritical end shows that her intrinsic value as the psychological and embedded person she was might not have been what others thought. In order to select an incongruent death, she must have had the mental states and qualities that enabled her to make the choice. Those states and qualities show a lack of authenticity, which quite possibly negatively affected her relationships with others,

[29] A more detailed discussion of appropriate emotional states appears in an earlier chapter.

thereby deceiving the others into believing the relationship was something it was not. Hence, those with whom the person has caring relationships feel disrespected by her deception and hypocrisy as well as her intrinsic value as a moral agent, psychological person, and embedded person devalued.

The requirement to end one's own life while still a person or moral agent leads to an alleged problem that has been raised by some critics. No one would disagree that there is no clear moment when a patient progresses from full selfhood to being a former person. The line between the two is indeterminate at best, and it might even be the case, as argued earlier, that both states, i.e., personhood and moral subjectiveness, exist simultaneously. Given this lack of clear definition and the value of being a person, it would be wrong, according to critics, to end one's life early. Since the dignity of a person's humanity must be preserved no matter what the cost to her physical life and to maximize her time as a person, she should passively allow dementia to render her a different person or a moral subject/former person.

A similar criticism first given by Rhodes (2007), and then taken up by others, is based upon the same problem about how the argument works. The objection is basically that when a person commits suicide, she destroys that which should be preserved, i.e., her moral agency. Therefore, there cannot be an obligation to suicide because the justification for it is inconsistent. Suicide to preserve moral agency actually destroys it.[30]

The first response to this objection ignores Kant's ontological view of souls. In order for morality to make sense in the first place, Kant states that there must be an afterlife in which people who deserve reward or punishment receive what is coming to them.[31] After all, some people perform actions that are not or cannot be punished sufficiently to eliminate the wrongdoer's debt, whereas other people are not or cannot be rewarded for the good they have done because there is a lack of resources or time, such as in the case of sacrificing their lives in order for others to live. In these situations, there has to be a surviving soul that is sufficiently like the person, the moral agent who bears desert, so that it can be affected appropriately by the entity passing out reward and punishment (Kant 1993). If Kant is correct, then the moral agent does not end with her suicide, and there is no inconsistency with killing the body to preserve the moral agent.

Secondly, even if we do not accept Kant's life after death view, the inconsistency claim arises from a basic misunderstanding of my timeline and the internal nature of the ethical. There is not an exact moment in time in which the death must occur. Further diagnoses have to be sought to provide sufficient evidence of what will happen to the person so that she can make an informed decision about when to end her life. Since this matter is of the greatest importance, her belief in her condition and what is entailed by it should be a rational and reasonable belief in which the evidence is overwhelmingly in favor of the belief's proposition being true. After the person has

---

[30]See Altman's (2011) and Perry's (2007) articles.

[31]This is actually Kant's (1993) argument for why it is rational to believe God exists (See the *Critique of Practical Reason*. Part I, Bk. II, Chap. V).

that assurance, she has to settle her affairs, including bringing those who care for her into the process of her end of life. Finally, but not least importantly, since the person does not become demented overnight, she has the opportunity to live fully as a person for as long as she can justify it, provided that she knows there is a limitation to her window of opportunity. When her disease has progressed far enough so that recovery is impossible, but she is still assured of being a full person, she has the duty to kill herself. Either the suicide duty exists or it does not for a person, but her decision and action time is limited to an acceptable range.

Let us not forget that morality is internal. It is the ***choice*** of what the person is and its connection to respecting the person for who she is, as well as respecting those affected by the person's action, that matters more than the actual consequences of the action in which she takes her life. That is, the choice is essential to the person's and action's morality. When an individual chooses to take her own life in the right way at the right time and for the right reasons, then she treats herself as she should be treated. It is only when the person does not select correctly that she disrespects herself. Whether or not she dies makes the overall action of the choice better or worse, but it does not make it right or wrong. Hence, the objection that suicide destroys the thing that was of value fails to recognize that the entity with a moral life is already gone the moment the individual fails to make the correct choice.

8.5.4 Certainty and Consequences

Although my position is not primarily a consequentialist one, the issue of certainty and probabilities has been raised. If there is a death duty, according to some, then the evidence that a negative outcome will occur should be so great that the person knows it with absolute certainty. However, as Delia Outomuro states:

> Medicine is based on probabilities and we do not know with certainty whether a person with prognosis of dementia…will develop the disease or not. Neither is it possibly (sic) to determinate (sic) when the disease will start or when the patient will lose personhood. (Outomuro 2007, 65)

Everything Outomouro states is true, but why should certainty be the standard applied to this or any other actual world situation? No one would deny that being absolutely guaranteed about a belief would be a wonderful state to obtain, but no belief about the physical world that depends on empirical evidence rises to that lofty level. For example, I might be dreaming that I am at work and believe it with all my heart because images with great force appear veridical, but I am not actually there. Given this doubtfulness, it is still the case that everyone can permissibly make decisions significantly affecting hers and others' lives. In medicine, physicians treat patients in emergency situations without being certain what is actually wrong with them. Medical personnel make their decisions based upon the best evidence available to them at the time, but while their treatments might work for the vast majority of patients, through no fault of the medical personnel's own, there is no guarantee that they will not harm a significant number of others. People try to do the best they can, but producing the best outcome overall is not necessarily the same thing as doing what is right in the circumstances. It all depends upon how a reasonable person would act or evaluate the situation.

The duty to die does not require absolute certainty, but since the consequences are dire, it demands sufficient evidence for a reasonable person to believe reasonably that the outcome will be dementia. In its earliest stages, what will happen is uncertain, unless it is one of the rare genetic disorders for which there are excellent tests available (Davis 1999, 316). As the disease progresses, though, the outcome becomes more and more apparent. "Those in the final stages...lose their ability to communicate, fail to recognize loved ones and become bed-bound. Alzheimer's disease is ultimately fatal" (Alzheimer's Association 2013, 3). Since it would be irrational to try to cling to every second a person can be a rational being when there is a great chance that her mental condition will deteriorate too much for her to do her duty, she should take her own life while cognitively whole. Given the relatively slow progression of the disease, she will have more than enough opportunity to put her affairs in order and prepare her caring community for what she will do while she still has her moral agency and knows that she will not recover from her affliction.

8.6 Conclusion

To argue that those who are among the most in need of protection by others have a moral obligation to end their lives can be thought to be too cold and uncaring. We intuitively believe the weakest members of our society should be nurtured and cared for rather than required to kill themselves. Those who will be demented can be among the weakest members. Hence, we should not force them to act in ways not in their best interests.

Two factors mitigating the coldness and lack of care charges should be considered. First, although one has suicide obligations, concern should be limited about some people forcing others to take their own lives. The duty to die is not an enforceable or maximal sense of duty obligation (Narveson 2000, 38). That is, unlike duties to fulfill legitimate contracts, suicide obligations do not entitle anyone to require an agent kill herself (Menzel 2000, 97–8). Hence, although choosing not to kill oneself might be unethical, the most that can be done about it is community disapprobation for a duty unperformed.[32]

More importantly, as members of caring communities, we should nurture and care for others, including when they are at the end of their moral lives. Suicide is many times a hostile act toward another, but in the right circumstances, it can be a pragmatic act of narrative dignity to the self and all others affected.

---

[32]Even if the Kantian argument is insufficient to prove a duty to die, at the very least it shows suicide in these situations is morally permissible and should be legally permitted.

As the number of ill and elderly people increase without a sufficient increase in resources devoted to fulfilling their needs, questions of healthcare rationing and death duties will also become more urgent. However, I take Mill seriously when he states that:

> [T]he only way in which a human being can make some approach to knowing the whole of a subject, is by hearing what can be said about it by persons of every variety of opinion, and studying all modes in which it can be looked at by every character of mind. (Mill 1972, 88)

If reasonable people at these relatively early stages can create an honest dialogue devoid of excess emotion and flawed reasoning, then making policy decisions now will allow people to have sufficient time to consider and adopt reasonable pathways to promote and sustain their flourishing.

References

Altman, M.C. 2011. *Kant and applied ethics: The uses and limits of Kant's practical philosophy*. Malden, MA: Wiley-Blackwell.

Alvarez, A. 1998. The history of suicide. In *Last rights: Assisted suicide and euthanasia debated*, ed. M.M. Uhlmann, 57–74. Grand Rapids: William B. Eerdmans Publishing Company.

Alzheimer's Association. 2013. *Alzheimer's disease facts and figures*. http://www.alz.org/downloads/facts_figures_2013.pdf. Accessed 30 Aug 2014.

Battin, M. 1994. *The least worst death*. New York: Oxford University Press.

Battin, M.P. 1996. *The death debate: Ethical issues in suicide*. Upper Saddle River: Prentice Hall Inc.

Battin, M., et al. 2007. Legal physician-assisted dying in Oregon and the Netherlands: Evidence concerning the impact on patients in 'vulnerable' groups. *Journal of Medical Ethics* 33: 591–597.

Barrington, M.R. 1980. Apologia for suicide. In *suicide*, ed. M.P. Battin, and D.J. Mayo, 90–103. New York: St. Martin's Press.

Beauchamp, T.L. 1986. Suicide. In *Matters of life and death*, ed. Tom Regan, 77–124. New York: Random House.

Becker, C.B. 2000. Buddhist views of suicide and euthanasia. In *Living well*, ed. Steven Luper, 498–500. Fort Worth: Harcourt Brace and Company.

Bennett, G. 2014. Goodbye and good luck! http://deadatnoon.com/index.html. Accessed 15 Feb 2015.

Bennett, M. 2000. Do we have a duty to die? In *Is there a duty to die?*, ed. J.M. Humber, and R.F. Almeder, 41–60. Totowa: Humana Press.

Brandt, R. 1979. The morality and rationality of suicide. In *Moral problems: A collection of philosophical essays, 3rd edition*, ed. J. Rachels, 460–89. New York: Harper & Row, Publishers.

Brassington, I. 2006. Killing people: What Kant could have said about suicide and euthanasia but did not. *Journal of Medical Ethics* 32: 571–574.

Callahan, D. 2001a. Terminating life-sustaining treatment of the demented. In *Bioethics*, ed. J. Harris, 93–108. Oxford: Oxford University Press.

Callahan, D. 2001b. When self-determination runs amok. In *Ethics at the end of life*, ed. R. Borgen, 223–228. Belmont: Wadsworth.

Cooley, D. 2006. *Crimina carnis* and morally obligatory suicide. *Ethical Theory and Moral Practice* 97: 327–356.

Cooley, D. 2007. A Kantian moral duty for the soon to be demented to commit suicide. *American Journal of Bioethics* 76: 37–44.
Cooley, D. 2013. A Kantian care ethics suicide duty. *International Journal of Law and Psychiatry* 36: 366–373.
Davis, D. 1998. Why suicide is like contraception. In *Physician assisted suicide*, ed. M.P. Battin, R. Rhodes, and A. Silver, 113–122. New York: Routledge.
Davis, D. 1999. Rational suicide and predictive genetic testing. *The Journal of Clinical Ethics*, Winter: 316–323.
De Quincy, T. 1856. *Essays in philosophy*. Boston: Houghton Mifflin.
Dowbiggin, I. 2005. *A concise history of euthanasia*. Lanham: Rowman & Littlefeld Publishers Inc.
Dresser, R. 1995. Dworkin on dementia—elegant theory, questionable policy. *Hastings Center Report* 25(6): 32–38.
Dresser, R. 1989. Advance directives, self-determination, and personal identity. In *Advance directives in medicine*, ed. C. Hackler, R. Moseley, and D.E. Vawter, 155–170. New York: Praeger.
Durkheim, E. 2003. *Altruistic suicide*. http://www.suicidereferencelibrary.com/28.html. Accessed 30 Aug 2014.
Dworkin, R. 1995. Life is sacred. In *Arguing Euthanasia*, ed. J.D. Moreno, 246–251. New York: Simon & Schuster.
Engelhardt Jr, H.T. 1995. Fashioning an ethic for life and death in a post-modern society. In *Arguing Euthanasia*, ed. J.D. Moreno, 42–50. New York: Simon & Schuster.
Fairbairn, J.G. 1995. *Contemplating suicide: The language and ethics of self harm*. New Fetter Lane: Routledge.
Feldman, F. 1992. *Confrontations with the reaper*. New York: Oxford University Press.
Gorowitz, K. 1975. Case registers. In *A handbook for the study of suicide*, ed. S. Perlin, 213–226. New York: Oxford University Press.
Hardwig, J. 1997. Is there a duty to die? *Hastings Center Report* 21(2): 32–42.
Kant, I. 1964. *Groundwork of the Metaphysic of Morals*. Trans. H.J. Paton. New York: Harper & Row, Publishers.
Kant, I.1979a. Care for one's life. In *Lectures on ethics*, trans. Louis Infield, 154–157. Indianapolis: Hackett Publishing Company.
Kant, I.1979b. Crimina carnis. In *Lectures on ethics*, trans. Louis Infield, 169–171. Indianapolis: Hackett Publishing Company.
Kant, I. 1979c. Suicide. In *Lectures on ethics*, trans. Louis Infield, 148–54. Indianapolis: Hackett Publishing Company.
Kant, I. 1993. *Critique of Practical Reason*. Trans. Lewis White Beck. New York: Macmillan Publishing Company.
Kant, I. 1996. *The metaphysics of morals*. Trans. M. Gregory. Cambridge: Cambridge University Press.
Kass, L.R. 1995. Death with dignity and the sanctity of life. In *Arguing euthanasia*, ed. J.D. Moreno, 211–236. New York: Simon & Schuster.
Margolis, J. 1975. *Negativities: The limits of life*. Columbus: Charles E. Merrill Publishing Company.
Menzel, P.T. 2000. The nature, scope, and implications of a personal moral duty to die. In *Is there a duty to die?*, ed. J.M. Humber, and R.F. Almeder, 95–114. Totowa: Humana Press.
Mill, J.S. 1972. *Utilitarianism, on liberty and considerations on representative government*. Ed. H. B. Acton. London: J. M. Dent & Sons, Ltd.
Mullett, S. 1988. Shifting perspectives: A new approach to ethics. In *Feminist perspectives*, ed. L. Code, S. Mullett, and C. Overall, 109–126. Toronto: University of Toronto Press.
Narveson, J. 2000. Is there a duty to die? In *Is there a duty to die?*, ed. J.M. Humber, and R.F. Almeder, 23–40. Totowa: Humana Press.

Norwood, F., G. Kimsma, and M.P. Battin. 2009. Vulnerability and the 'slippery slope' at the end-of-life: A qualitative study of euthanasia, general practice and home death in The Netherlands. *Family Practice* 26(6): 472–480.
Overall, C. 2003. *Aging, death, and human longevity*. Berkeley: University of California Press.
Pereira, J. 2011. Legalizing euthanasia or assisted suicide: The illusion of safeguards and controls. *Current Oncology* 18(2): e38–e45.
Pence, G. 2007. *The elements of bioethics*. Boston: McGraw Hill.
Perry, C. 2007. Suicide fails to pass the categorical imperative. *American Journal of Bioethics* 76: 51–53.
Prado, G.C. 2008. *Choosing to die: Elective death and multiculturalism*. Cambridge: Cambridge University Press.
Rabins, P.V. 2007. Can suicide be a rational and ethical act in persons with early or pre-dementia? *American Journal of Bioethics* 76: 47–49.
Rietjens, J.A.C., P.J. van der Maas, B.D. Onwuteaka-Philipsen, et al. 2009. Two decades of research on euthanasia from the Netherlands. What have we learnt and what questions remain? *Journal of Bioethical Inquiry* 6(3): 271–283.
Rhodes, R. 2007. A Kantian duty to commit suicide and its implications for bioethics. *American Journal of Bioethics* 76: 45–47.
Roberston, J.A. 1991. Second thoughts on living wills. *Hastings Center Report* 21(6): 6–9.
Rosen, G. 1975. History. In *A handbook for the study of suicide*, ed. S. Perlin, 3–29. New York: Oxford University Press.
Outomuro, D. 2007. Morality of suicide in dementia. *American Journal of Bioethics* 7(6): 64–65.
Shershow, S.C. 2014. *Deconstructing dignity*. Chicago: University of Chicago Press.
Soble, A. 2003. Kant and sexual perversion. *The Monist* 86(1): 55–89.
Tong, R. 2000. Duty to die. In *Is there a duty to die?*, ed. J.M. Humber, and R.F. Almeder, 133–158. Totowa: Humana Press.
Zizek, S. 2000. *The fragile absolute*. New York: Verso.

Index

D.R. Cooley, *Death's Values and Obligations: A Pragmatic Framework*,
International Library of Ethics, Law, and the New Medicine 62,
DOI 10.1007/978-94-017-7264-8

The manufacturer's authorised representative in the EU is Springer Nature Customer Service Centre GmbH, Europaplatz 3, 69115 Heidelberg, Germany. If you have any concerns regarding our products, please contact ProductSafety@springernature.com

Printed and bound by CPI Group (UK) Ltd, Croydon, CR0 4YY
22/09/2026
02221058-0015